Planar Antennas for Wireless Communications

WILEY SERIES IN MICROWAVE AND OPTICAL ENGINEERING

KAI CHANG, Editor
Texas A&M University

A complete list of the titles in this series appears at the end of this volume.

Planar Antennas for Wireless Communications

Kin-Lu Wong

WILEY-INTERSCIENCE

A JOHN WILEY & SONS, INC., PUBLICATION

Published by John Wiley & Sons, Inc., Hoboken, New Jersey.
Published simultaneously in Canada.

For general information on our other products and services please contact our Customer Care Department within the U.S. at 877-762-2974, outside the U.S. at 317-572-3993 or fax 317-572-4002.

Wiley also publishes its books in a variety of electronic formats. Some content that appears in print, however, may not be available in electronic format.

Library of Congress Cataloging-in-Publication Data:

Wong, Kin-Lu.
Planar antennas for wireless communications / Kin-Lu Wong.
p. cm. -- (Wiley series in microwave and optical engineering)
Includes bibliographical references and index.
ISBN 0-471-26611-6 (cloth)
1. Microstrip antennas. I. Title. II. Series.

TK7871.67.M5 W6524 2003
621.382'4--dc21

Printed in the United States of America

10 9 8 7 6 5 4 3 2 1

Contents

Preface

Planar antennas, including microstrip and printed antennas and other types of antennas that are flat in appearance and have a low profile, have recently received much attention for application in cellular communication systems, such as the global system for mobile communication (GSM; 890–960 MHz), the digital communication system (DCS; 1710–1880 MHz), the personal communication system (PCS; 1850–1990 MHz), the universal mobile telecommunication system (UMTS; 1920–2170 MHz), and wireless local area networks (WLANs) in the 2.4 GHz (2400–2484 MHz) and 5.2 GHz (5150–5350 MHz) bands. Many novel designs of planar antennas for related applications such as internal mobile phone antennas, base-station antennas, WLAN, or Bluetooth antennas, and so on, have been reported very recently, especially since the year 2000. The author and his graduate students at National Sun Yat-Sen University, Kaohsiung, Taiwan, have contributed to many of the novel antenna designs for wireless communications. These designs along with many other advanced antennas reported recently by antenna researchers are scattered throughout many journal and conference papers and are also available in some patents. It is the intention of this book to organize these advanced designs in the area of planar antennas for wireless communications.

This book is organized into seven chapters. Chapter 1 presents an introduction and overview of recent advances in planar antenna designs for wireless communications. Chapters 2 and 3 describe the advanced designs in dual-band and multi-band internal mobile phone antennas using planar inverted-F antennas (PIFAs) and very-low-profile planar monopoles, respectively. Novel designs of base station antennas for cellular communication systems are presented in Chapter 4. The antenna designs for WLAN or Bluetooth applications are discussed in Chapter 5. Among other types of planar antennas, dielectric resonator (DR) antennas have attractive features of low loss, compact size, and good performance for operating at higher frequencies, for example, in the operating bands above 5 GHz, and their promising designs for wireless communications are addressed in Chapter 6. Chapter 7 is devoted to the integration of antennas for different operating bands.

In Chapter 2, an introduction of recent advances in PIFA for internal mobile phone antennas is first given. Novel PIFA designs using a slotted patch, a meandered patch, a folded patch, a spiral strip, two or more shorted patches, parasitic shorted patch loading, chip inductor and chip capacitor loading, and so on, to achieve dual-frequency operations in the 900 and 1800 MHz bands or triple-frequency operations in the 900, 1800, and 2450 MHz bands are described, and detailed antenna designs and experimental results are presented and discussed. The physical concepts and design considerations of these antennas to achieve dual-band/multi-band operations with a compact antenna volume (as small as $5 \times 10 \times 24\,\text{mm}^3$ or less for operating in the 900 and 1800 MHz bands with an air-layer substrate) are also addressed. For most cases, the simulation results obtained with the commercially available simulation softwares are shown for comparison. The design of using an L-shaped ground plane to reduce the antenna's backward radiation is also discussed.

Chapter 3 discusses very-low-profile monopole designs for dual-band or multi-band internal mobile phone antennas. The monopole designs presented have an antenna height less than 15 mm (about 4% of the wavelength at 900 MHz), much less than that (one-quarter wavelength) of a conventional straight monopole antenna. Some designs can even have an antenna height as low as 7 mm from the ground plane (about 2% of the wavelength at 900 MHz). These designs are very promising for placement within the mobile phone housing; that is, a concealed antenna for the mobile phone can be obtained. In addition, the monopole design with reduced backward radiation is also described, which can enhance antenna performance and reduce the possible electromagnetic energy absorption by the user's head.

Chapter 4 presents recent advances in base station antenna designs for GSM, DCS, PCS, and UMTS cellular communication systems. Many advanced designs reported recently are described in detail, and experimental results of the constructed prototypes for single-band, dual-band, and multi-band operations are presented. The designs for obtaining dual-polarized radiation, especially for dual-band dual-polarized radiation, are also shown. Chapter 5 is devoted to the planar antenna designs for WLAN or Bluetooth applications. Designs suitable for WLAN base station or access point antennas and capable of circular polarization (CP) radiation, dual-polarized radiation, and dual-band operation are first presented. Design examples of surface-mountable antennas such as the ceramic chip antenna, FR4 chip antenna, and strip monopole wrapped into a rectangular-disk-like structure are also described. Novel printed monopole and printed inverted-F antennas designed for WLAN operations in the 2.4 and 5.2 GHz bands are also reviewed, and design examples are given.

The advances in DR antennas for wireless communications are considered in Chapter 6. Recent advanced designs for achieving CP radiation, dual-polarized radiation, and broadband operation are introduced. With the use of the DR element with a very high relative permittivity (for example, larger than 80), the antenna can have a very low profile (less than 2 mm for operating in the 5.2 GHz WLAN band), which makes it very attractive for practical applications. Design examples of the very-high-permittivity DR antenna are also presented. Because of their low profile and no metallic loss, such DR antennas are expected to find many promising applications in wireless communications at higher operating frequencies, such as in the 5.2 GHz bands for WLAN operation.

Finally, in Chapter 7, the integration of antennas for different operating bands, for example, integration of two separate GPS and DCS antennas, two separate DCS and WLAN antennas, and so on, is demonstrated. The GPS patch antenna constructed with a very-high-relative-permittivity (about 90) ceramic material is also addressed, which has the attractive feature of compact size (can be about $12 \times 12\ \text{mm}^2$ or smaller in area) and is very suitable for integration with antennas with different operating bands. Details of some practical integration designs and the experimental results are shown.

This book is intended to organize new advanced planar antennas designs, mainly those reported since 2000, for wireless communications. Over 70 advanced planar antenna designs and their detailed experimental results are included. It is believed that this book can be a very useful design reference on mobile and WLAN antennas for antenna scientists and engineers.

Kin-Lu Wong
Kaohsiung, Taiwan

Planar Antennas for Wireless Communications

CHAPTER ONE

Introduction and Overview

1.1 INTRODUCTION

Wireless communications have progressed very rapidly in recent years, and many mobile units are becoming smaller and smaller. To meet the miniaturization requirement, the antennas employed in mobile terminals must have their dimensions reduced accordingly. Planar antennas, such as microstrip and printed antennas, have the attractive features of low profile, small size, and conformability to mounting hosts [1] and are very promising candidates for satisfying this design consideration. For this reason, compact and broadband design techniques for planar antennas [2] have attracted much attention from antenna researchers. Very recently, especially after the year 2000, many novel planar antenna designs to satisfy specific bandwidth specifications of present-day mobile cellular communication systems, including the global system for mobile communication (GSM; 890–960 MHz), the digital communication system (DCS; 1710–1880 MHz), the personal communication system (PCS; 1850–1990 MHz), and the universal mobile telecommunication system (UMTS; 1920–2170 MHz), have been developed and published in the open literature. This book organizes those novel designs for applications as internal mobile phone antennas and base station antennas, and detailed design considerations and experimental results are presented.

Planar antennas are also very attractive for applications in communication devices for wireless local area network (WLAN) systems in the 2.4 GHz (2400–2484 MHz) and 5.2 GHz (5150–5350 MHz) bands. Novel planar antenna designs for achieving broadband circular polarization (CP) and dual-polarized radiation in the WLAN band for overcoming multipath fading problem to enhance system performance have been demonstrated recently. In addition, surface-mountable antennas that can be easily integrated on the circuit board of a communication device to reduce the packaging cost have also received much attention, and related new designs for WLAN operations have been reported recently. These newly developed WLAN antennas are presented and discussed.

The applications of dielectric resonator (DR) antennas [3] for WLAN operation are also addressed. Dielectric resonator antennas also have the features of low profile and small size, like the microstrip and printed antennas, and in addition, there is no metallic loss, leading to low loss for operating at higher frequencies. Furthermore, when the DR element with a very high relative permittivity is used (for example, larger than 80), the antenna can have a very low profile (less than 2 mm for operating in the 5.2 GHz WLAN band), which is very attractive for practical applications. Some related promising designs are introduced in this book. Finally, the integration of antennas for different operating bands, such as the GPS antenna, the WLAN antenna, and the DCS antenna, is discussed and details of some practical integration designs and experimental results are shown.

1.2 PIFAs FOR INTERNAL MOBILE PHONE ANTENNAS

A planar inverted-F antenna (PIFA) [4] is in general achieved by short-circuiting its radiating patch or wire to the antenna's ground plane with a shorting pin and can resonate at a much smaller antenna size for a fixed operating frequency. Owing to their compact size the designs of PIFAs have attracted much attention, and a variety of dual-band or multiband PIFAs suitable for applications in mobile phones have been demonstrated recently [5–24]. Figure 1.1 shows a typical configuration of a PIFA mounted on the top portion of a circuit board of a practical mobile phone, and some promising radiating top patches for achieving dual-band operation of the PIFA are presented in Figure 1.2. These PIFA designs usually occupy a compact volume and can be integrated within the mobile phone housing, leading to concealed or internal mobile phone antennas. Because this kind of internal antenna can avoid the damages caused by catching on things and will not break, compared with the conventional protruded whip or rod antennas used for mobile phones, it is now becoming one of the major design considerations for mobile phones. In addition, in comparison to the conventional whip antennas showing omnidirectional radiation, such PIFAs have the advantage of relatively smaller backward radiation toward the mobile phone user. This suggests that the possible electromagnetic energy absorption by the user's head can be reduced. These advantageous characteristics have led many novel PIFA designs, most of them capable of dual-band operation, to be applied in mobile phones in the market.

For the promising radiating top patch designs for the dual-frequency PIFA shown in Figure 1.2, there are two major design concepts: One design uses two different resonant paths to generate two separate resonant modes for dual-frequency operation, and the other uses the first two resonant frequencies of a single resonant path for dual-frequency operation. In the former case, the design techniques include the use of an inserted L-shaped slit or a folded slit [8, 20] (Fig. 1.2a and b), or a U-shaped slot [10, 19] (Fig. 1.2c), or inductor and capacitor (LC) resonators [12]

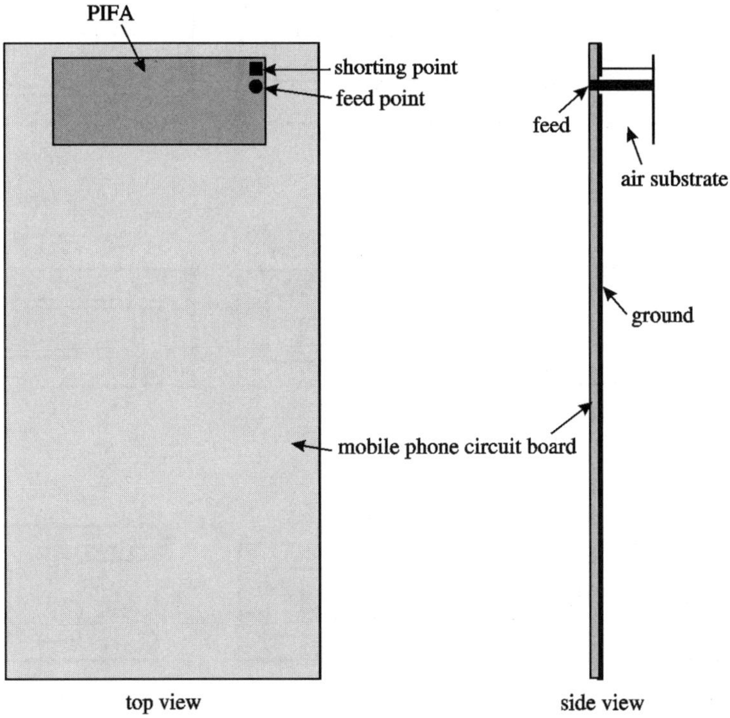

FIGURE 1.1 Geometry of a PIFA mounted on the top portion of a mobile phone circuit board.

(Fig. 1.2d), or a chip-inductor loading [14] (Fig. 1.2e) to separate the radiating top patch into two subpatches of different sizes, which provide two resonant paths of different lengths to achieve the 900/1800 MHz operation. In the latter case, the major design consideration is to adjust the antenna's first two resonant frequencies to be about 2. To achieve this goal, the techniques of using a branch-line slit to achieve a lengthened resonant path [13] (Fig. 1.2f), inserting several linear slits to meander the top patch [7, 22] (Fig. 1.2g), using a shorted spiral strip with various strip widths (Fig. 1.2h), folding a meandered top patch (Fig. 1.2i), and so on, have been successfully applied.

To include the WLAN operation in the 2.4 GHz band, designs of triple-frequency PIFAs for operations in the 900, 1800, and 2450 MHz bands have been reported, which include the use of two or three separate top patches [21], a meandered top patch [22], and a branch-line strip with a folded end [23]. The techniques of using a parasitic shorted patch placed coplanar with the driven patch [7, 11] or stacked on top of the driven patch [25–27] to enhance the operating bandwidth of the PIFA have also been reported. Recently, it was reported [28] that, by using an L-shaped ground plane (see two possible designs shown in Fig. 1.3) in place of the conventional flat ground plane, more reduction in the antenna's backward

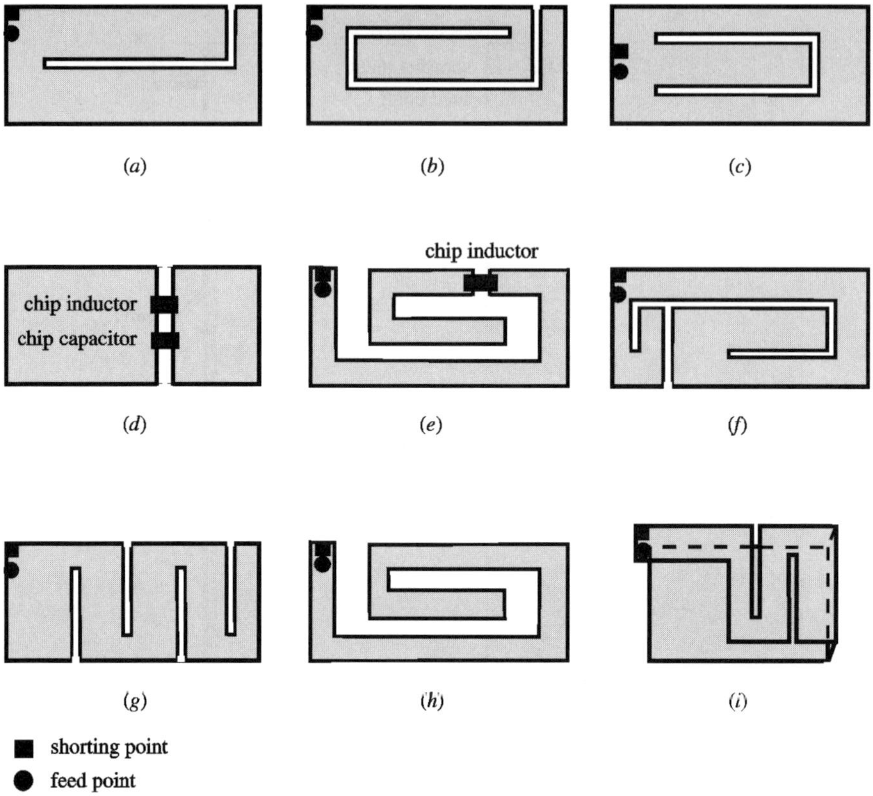

FIGURE 1.2 Some promising radiating top patches for dual-frequency PIFAs.

radiation is obtained and antenna performances are enhanced. The effects of ground plane size on the PIFA's impedance bandwidth characteristics have also been studied [29]. These recently reported results for the applications of PIFAs as internal mobile phones antennas are presented in Chapter 2.

1.3 VERY-LOW-PROFILE MONOPOLES FOR INTERNAL MOBILE PHONE ANTENNAS

Conventional monopole antennas in the form of a straight rod are usually operated with one-quarter wavelength; this means that, when operated at 900 MHz for mobile communications, the monopole antenna requires a height of about 83 mm. Because of this large antenna height, it is impossible to integrate such monopole antennas within the mobile phone housing. To reduce the monopole height, which makes the antenna less prone to breaking off, monopoles in the form of a helix or wound coil [30] or a folded loop [31] for mobile phone applications have been used. Many related designs for achieving 900/1800 MHz dual-frequency operations

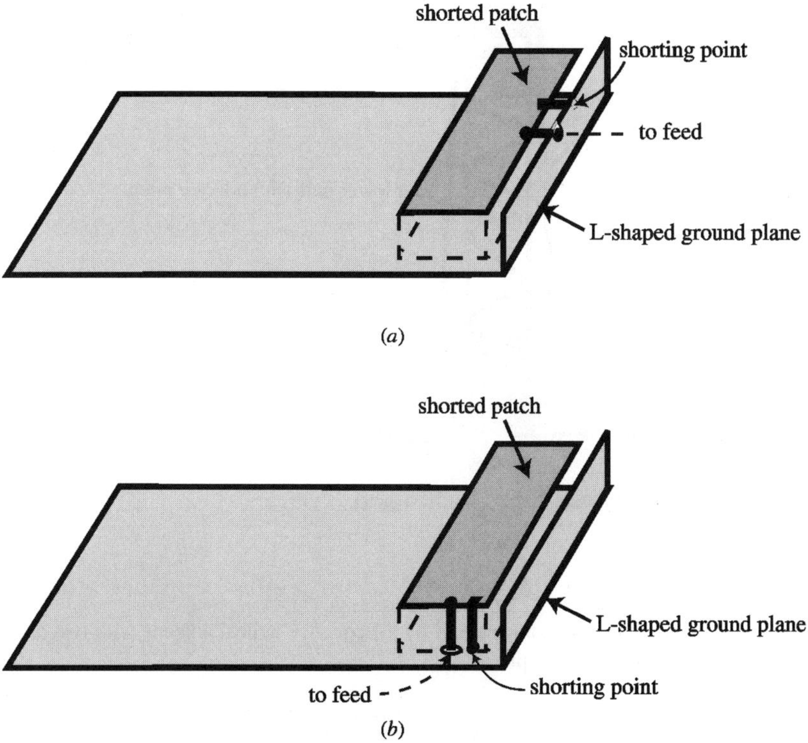

FIGURE 1.3 Geometries of a PIFA with an L-shaped ground plane. (a) The feed is in parallel with the radiating patch; (b) the feed is in perpendicular to the radiating patch.

have been reported [32–36]. A typical dual-frequency design is shown in Figure 1.4, which is a dual-frequency helical antenna with two different pitches for a GSM/DCS mobile phone [32, 33]. Figure 1.5 shows two other possible dual-frequency designs: The first design uses two helixes of different radii (the helix with a smaller radius for 1800 MHz operation is placed within the helix with a larger radius for 900 MHz operation) [32]. The second design uses a straight rod for 1800 MHz operation placed inside a uniform helix having the 900 MHz resonance [34, 35]. The design technique of configuring a folded loop to achieve two desired separate resonant frequencies for the 900 and 1800 MHz operations has also been used [36]. Although the antenna heights of these dual-frequency designs have been greatly reduced compared with that of a straight monopole, they are still larger than 10% of the wavelength of the lowest operating frequency.

To further reduce antenna height, a variety of novel dual-frequency monopole designs have been reported very recently [37–42]. These designs are mainly associated with bending, folding, or wrapping two-dimensional planar monopoles into three-dimensional structures. This technique greatly reduces the total antenna height from the ground plane of a mobile phone. Typical designs that have been reported have an antenna height of less than 15 mm (about 4% of the wavelength

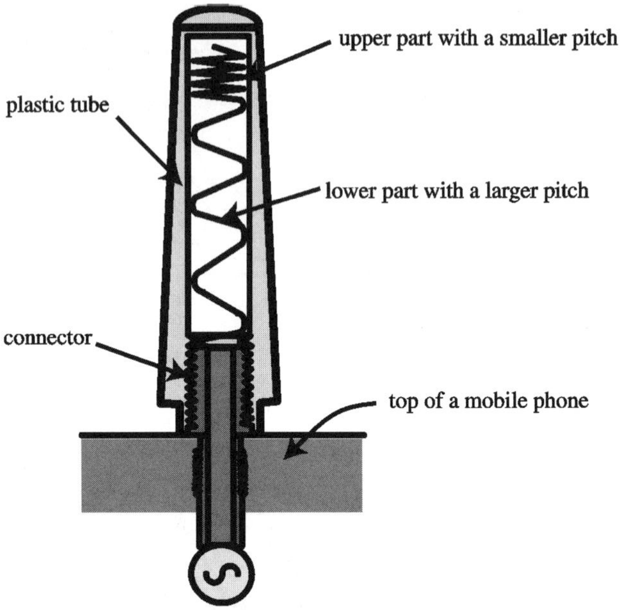

FIGURE 1.4 Geometry of a conventional dual-frequency helical antenna with two different pitches for a GSM/DCS mobile phone.

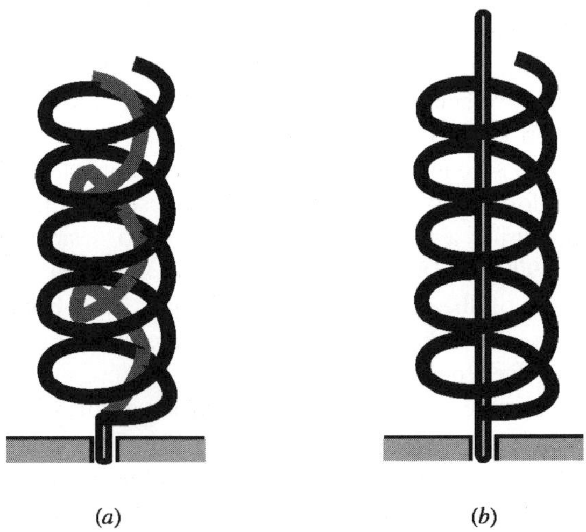

FIGURE 1.5 Geometries of two types of dual-frequency helical antennas. (a) Two helixes with different radii; (b) one helix and one straight rod.

at 900 MHz). Some designs can even have an antenna height as low as 7 mm from the ground plane (about 2% of the wavelength at 900 MHz). Such antennas are very promising for placement within the mobile phone housing, that is, a concealed antenna for the mobile phone can be obtained. This kind of very-low-profile monopole designs for dual-frequency internal mobile phone antennas are described in detail in Chapter 3.

Some newly developed designs for achieving very-low-profile GSM/DCS dual-frequency monopoles for mobile phones are depicted in Figures 1.6–1.9. In Figure 1.6, dual-frequency operation is obtained by using a branch-line planar monopole comprising two printed meandered strips resonating at different frequencies [37, 39]. Because of the meandering, the antenna height from the mobile phone ground plane can be greatly reduced. Furthermore, it has been shown that, by wrapping a similar branch-line planar monopole into a rectangular boxlike structure as shown in Figure 1.7, a much reduced antenna height is obtained [37]. It should be noted that, in this design example, there are three branch lines. Among these, one is for the GSM operation, and the other two are for the DCS operation, whose impedance bandwidth also covers the operating band of the PCS system; that is, a GSM/DCS/PCS multiband operation is achieved [37].

Dual-frequency operation obtained by loading a meandered wire, which is for generating the 900 MHz resonance, to an inverted-L wire having the 1800 MHz resonance, has also been reported [40], and a typical configuration of the design is shown in Figure 1.8. Note that, in Figures 1.6–1.8, dual-frequency operations

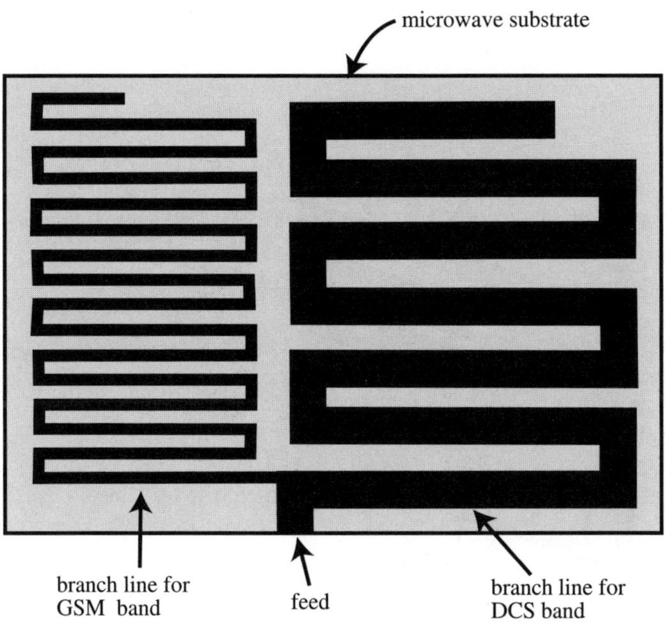

FIGURE 1.6 Geometry of a GSM/DCS dual-frequency planar branch-line monopole printed on a substrate.

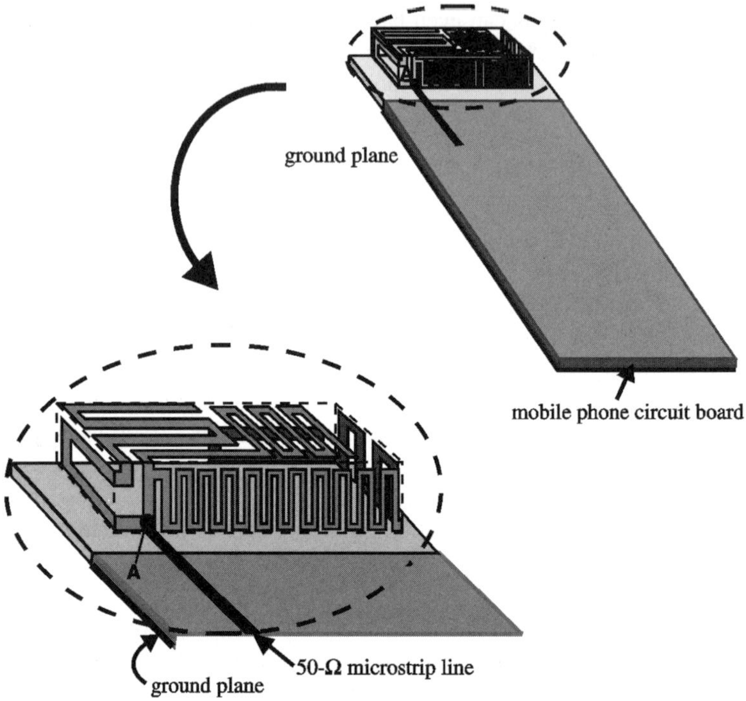

FIGURE 1.7 Geometry of a GSM/DCS/PCS multiband monopole obtained by wrapping a branch-line planar monopole and mounted on the top portion of a mobile phone circuit board.

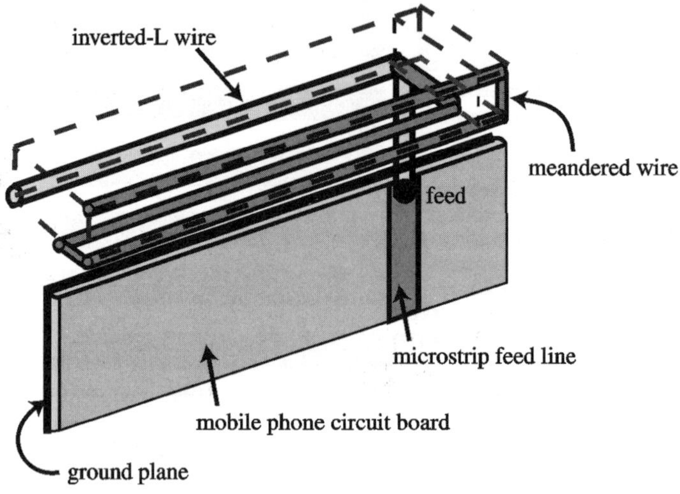

FIGURE 1.8 Geometry of a dual-frequency inverted-L wire antenna loaded with a meandered wire for a GSM/DCS mobile phone.

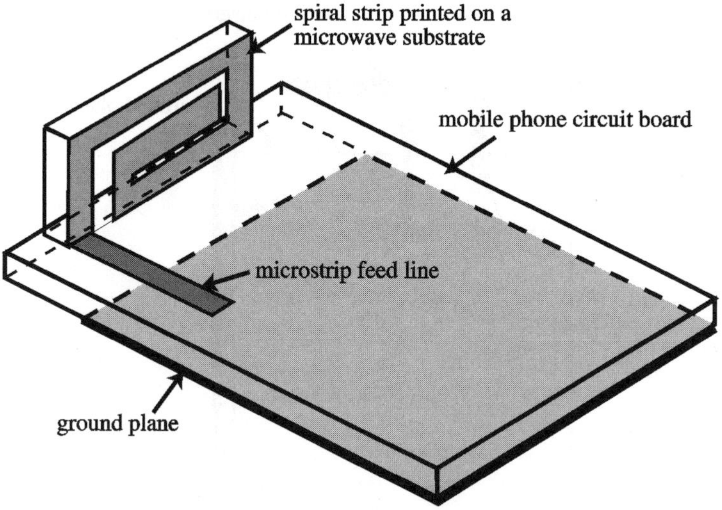

FIGURE 1.9 Geometry of a dual-frequency rectangular spiral monopole antenna mounted on the top portion of a mobile phone circuit board and positioned in perpendicular to the circuit board.

are obtained by using two separate resonant paths to generate the 900- and 1800-MHz-resonant frequencies (in Fig. 1.8, an additional resonant path is added to obtain an enhanced bandwidth for the 1800 MHz operation to cover the PCS band). This design concept has also been used in the designs shown in References 38 and 41. In Reference 38 a branch-line slit is inserted within a planar monopole to create two desired different resonant paths for the 900 and 1800 MHz resonances, and in Reference 41 a planar monopole comprising two subpatches of different sizes is wrapped into a rectangular boxlike structure to achieve 900 and 1800 MHz operations with a much reduced antenna height.

One can also use the antenna's first two resonant frequencies to achieve the 900 and 1800 MHz operations. A typical design using a rectangular spiral strip monopole is presented in Figure 1.9. The rectangular spiral strip monopole printed on a microwave substrate is mounted on the top portion of a mobile phone circuit board and positioned in perpendicular to the circuit board. With this arrangement, the distance of the rectangular spiral strip monopole from the mobile phone ground plane is greatly reduced; that is, a much reduced antenna height for the mobile phone is obtained. Also note that the spiral strip has different widths in different sections, which greatly affect the antenna's first two resonant frequencies. By utilizing this characteristic, the desired dual-frequency operation at 900 and 1800 MHz can be obtained by tuning the widths in different sections. This kind of dual-frequency rectangular spiral strip can also be folded onto a plastic chip as a surface-mountable antenna, as demonstrated in Reference 42.

The planar inverted-F antenna can also be directly printed on a mobile phone circuit board to operate as an integrated or on-board monopole antenna. A typical

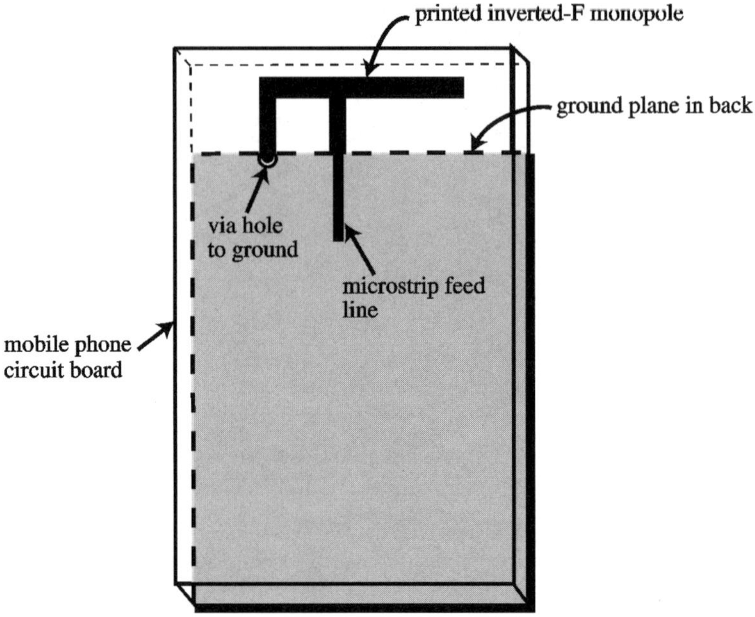

FIGURE 1.10 Geometry of an inverted-F monopole printed on a mobile phone circuit board.

design has been recently reported [43], and its geometry is shown in Figure 1.10. Monopole designs with reduced backward radiation have also been shown [44, 45] that have the advantages of enhancing antenna performance and reducing possible electromagnetic energy absorption by the user's head. These newly reported designs are discussed in detail in Chapter 3.

1.4 BASE STATION ANTENNAS FOR CELLULAR COMMUNICATION SYSTEMS

Planar antennas have also been widely applied in cellular system base stations. To achieve a broadband operation, conventional designs using a thick air-substrate patch antenna incorporating a U-slotted patch, an E-shaped patch, a wedge-shaped patch, an L-shaped probe feed, a three-dimensional transition feed, and so on, have been used [2, 46–52]. However, although wide impedance bandwidths are obtained for these designs, they usually do not cover the specific operating bandwidths of present-day cellular communication systems such as the GSM (890–960 MHz), DCS (1710–1880 MHz), PCS (1850–1990 MHz), and UMTS (1920–2170 MHz) systems.

To satisfy the specific operating bands, defined by 1.5:1 VSWR (voltage standing wave ratio), of the practical communication systems, many advanced designs have been demonstrated recently. To achieve single-band operation, a variety of designs

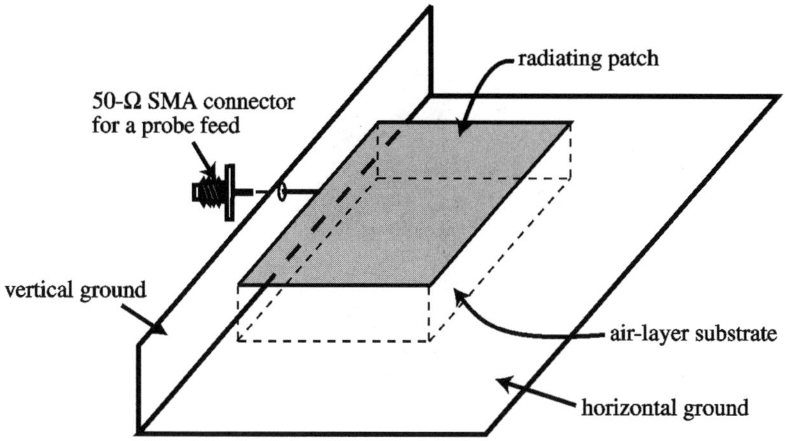

FIGURE 1.11 Geometry of a coplanar-probe-fed patch antenna for broadband operations.

using the slot-coupled feed [53], probe feed [54–59], coplanar probe feed [60], microstrip-line feed [61], and capacitively coupled feed [62, 63] have been used. Among these designs, many interesting novel design techniques are applied; for example, to avoid the large probe pin inductance introduced by a long probe pin required for a thick air-layer substrate, a very effective method of using a coplanar probe feed has been developed [60], in which the vertical ground plane added is for accommodating the coplanar probe feed (see Fig. 1.11). This design has a simple geometry, and good broadside radiation characteristics have also been obtained.

Other successful methods of using a conducting cylinder transition [54], a triangular transition patch [55], a ground plane with a small elevated portion [56], a W-shaped ground plane [57], an E-shaped patch incorporating a U-shaped ground plane [58, 59], and so on, have also been reported. In addition to the obtained wide impedance bandwidths covering the required bandwidths of the GSM, DCS, PCS, or UMTS systems, excellent radiation characteristics with much reduced cross-polarization radiation have been obtained in the designs shown in References 57 and 58. Moreover, enhanced antenna gain for the design with a W-shaped ground plane has been observed [57]. The peak antenna gain can reach about 10 dBi for a single-patch case in the DCS band. These designs are also very suitable for application in $1\times N$ linear arrays for applications in cellular-system base stations to achieve a much larger antenna gain and a narrower beam width in the elevation plane. Two promising linear array designs are shown in Figures 1.12 and 1.13. Figure 1.12 shows the geometry of 1×2 array antenna using two W-shaped radiating patches. Note that the two feeds for the two W-shaped patches are out of phase, which can result in the cancellation of the waves that are related to the excitation of the higher-order modes and thus can lead to the reduction in cross-polarization radiation [57]. The geometry shown in Figure 1.13 is a 1×4 array antenna using four E-shaped patches, and for each E-shaped patch there are two vertical ground

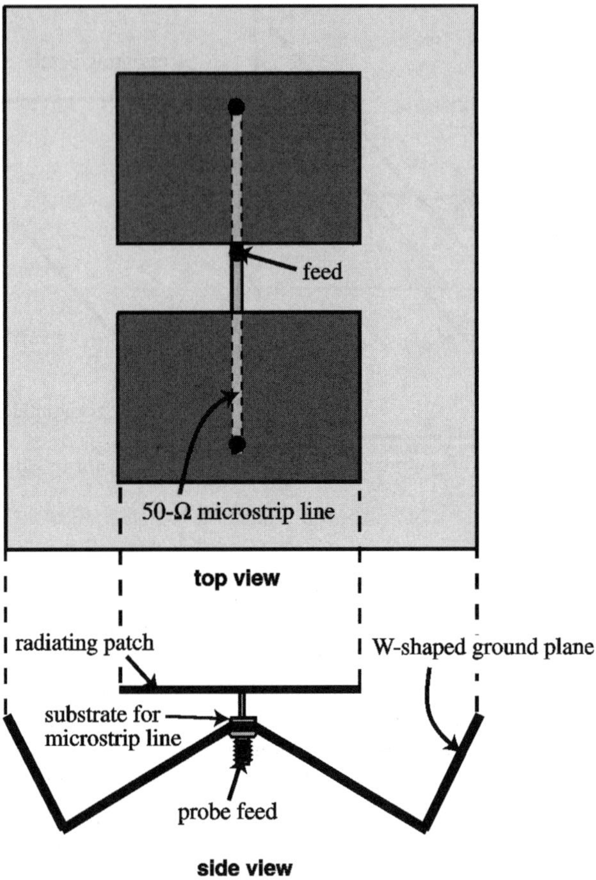

FIGURE 1.12 Geometry of a 1×2 W-shaped patch array antenna for applications in cellular system base stations.

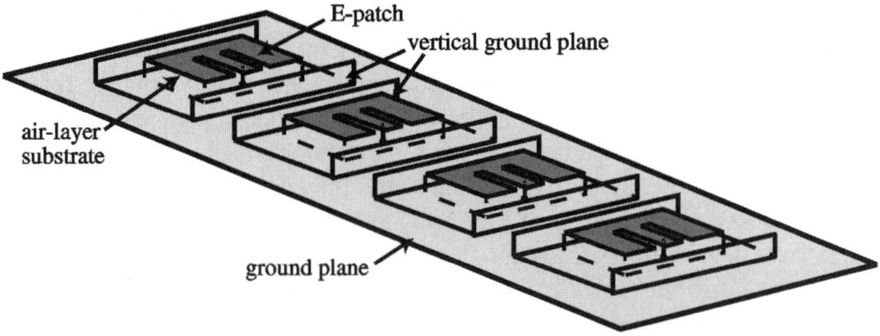

FIGURE 1.13 Geometry of a 1×4 E-patch array antenna for applications in cellular system base stations; feed network is not shown in the figure.

planes added, which can suppress the cross-polarization fields in the near fields, leading to reduced cross-polarization radiation [59].

Some novel dual-band (GSM/DCS) or multiband (GSM/DCS/PCS) designs have also been developed. The recent designs include a radiating patch incorporating a U-shaped ground plane [64] and a radiating patch incorporating a parasitic shorted patch. These designs have a more compact antenna size in comparison to the conventional design using two separate radiating patches [65]. Broadband dual-polarized patch antennas have also attracted much attention [66–72], owing to their capability of overcoming the multipath fading problem to enhance system performance. Several promising designs using two same feeds [66–69] or hybrid feeds [70, 71] to achieve high isolation (S_{21} less than -30 dB) between the two feeding ports have been reported. Much higher port decoupling (S_{21} less than -40 dB) for operating frequencies in the DCS band has been achieved in the design studied in Reference 71.

Promising designs to achieve dual-band dual-polarized radiation have also been reported. Figure 1.14 shows the geometry of a GSM/DCS dual-band dual-polarized patch antenna. This design, studied in Reference 72, consists of a rectangular ring patch for the GSM operation and a notched rectangular patch for the DCS operation, which are printed on the same layer, and each patch is aperture-coupled through two H-shaped coupling slots to generate two orthogonal linearly polarized waves. In addition to the compact dimensions obtained, this antenna shows good port decou-

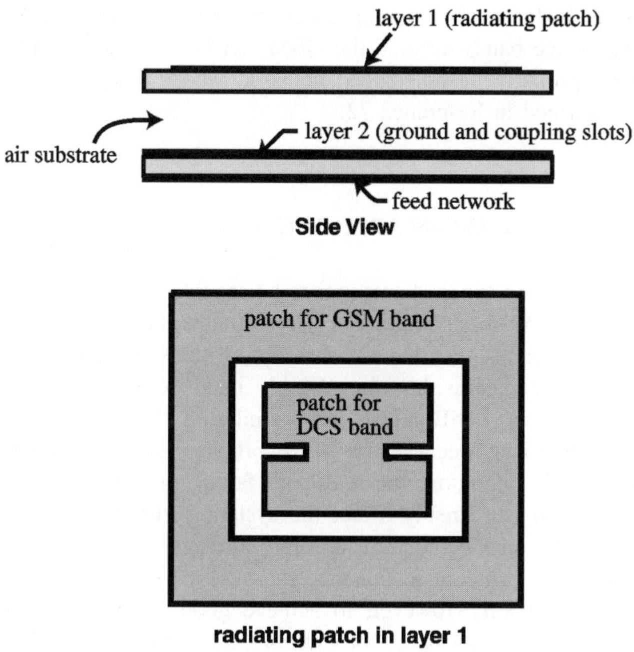

FIGURE 1.14 Geometry of a slot-coupled dual-band dual-polarized patch antenna for GSM/DCS operations; feed network and coupling slots are not shown in the figure.

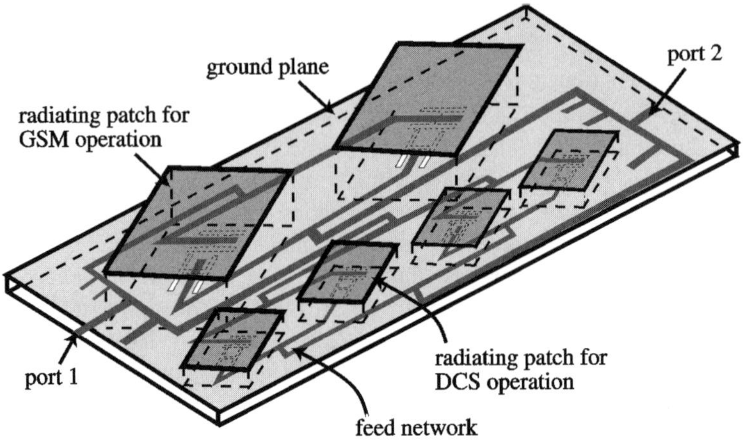

FIGURE 1.15 Geometry of an array antenna for GSM/DCS dual-band dual-polarized operations.

pling of less than −39 and −34 dB for dual linear polarizations in the GSM and DCS bands, respectively [72]. Figure 1.15 shows the promising design of a dual-polarized slot-coupled patch array for GSM/DCS base station antenna. Note that the 1×2 patch array with larger patches is for the GSM operation, and the 1×4 patch array with smaller patches is for the DCS operation. This design uses two slot-coupled feeds for each radiating patch to achieve dual-polarized radiation, and good impedance bandwidths in the GSM and DCS bands can be obtained by incorporating a properly designed feed network, whose design considerations are the same as described in Reference 72.

1.5 ANTENNAS FOR WLAN APPLICATIONS

There are in general two types of antennas to be considered for WLAN applications: one is for fixed WLAN base stations or access points, and the other is for mobile communication terminals. For base station applications, the impedance matching for operating frequencies within the WLAN bandwidth should be better than 1.5:1 VSWR or about 14 dB return loss, similar to that of the cellular system base stations. Moreover, because the wave propagation environment is usually complex for WLAN operation, the multipath fading problem can thus become a serious problem and can greatly affect the system performance. For this reason, the antennas capable of CP radiation are very attractive, because it can overcome the multipath fading problem to enhance the system performance, especially for indoor WLAN operation. However, to achieve good CP radiation for operating frequencies across a WLAN band, for example, the 2.4-GHz WLAN band (2400–2484 MHz), a two-feed design method [2] is usually required, which increases the complexity of the antenna design. Recently, a simple broadband CP design using a single-feed method has been developed. This design uses a coplanar probe feed and has a simple geometry similar to that shown in Figure 1.11. Experimental results of a constructed prototype for operating in the 2.4-GHz band showed that the antenna can have an impedance bandwidth (1.5:1 VSWR) of about 30% and a 3-dB axial-ratio CP bandwidth of about 10.4%, covering the frequency range of 2400–2484 MHz. Large antenna gain level (about 8.5 dBi) for frequencies within the CP bandwidth has also been obtained. This antenna design can be implemented with very low cost and is very suitable for WLAN access point applications. Other promising designs suitable for WLAN access-point applications, such as the use of an inverted-L patch antenna to achieve dual-band WLAN operations in the 2.4- and 5.2-GHz bands [73], have also been addressed. Details of these antenna designs and their experimental results are presented in Chapter 5.

On the other hand, the antennas for mobile communication units must have the features of small size, low cost, and good performance. For some cases, it is further required that the antennas should be surface mountable to reduce the packaging cost. For this purpose, the ceramic chip antennas [74–77] have been reported. Such chip antennas usually comprise a ceramic base and metal line patterns printed on or embedded within the base. In addition, they are compact in size and can be directly mounted on a circuit board in a communication device. The geometry of a typical ceramic chip antenna is shown in Figure 1.16. Note that, in this design, a gap is cut in the feed line for impedance tuning, and some small protruded grounds are also printed in the ceramic chip's side surfaces for easy connection to the system ground.

However, the ceramic chip antennas have a major disadvantage; that is, they tend to break easily, because the ceramic materials are fragile in nature. Furthermore, the fabrication cost of this kind of antennas is usually high. To overcome these problems, a low-cost surface-mountable antenna constructed by folding a planar monopole made of a copper plate into a rectangular disklike structure has

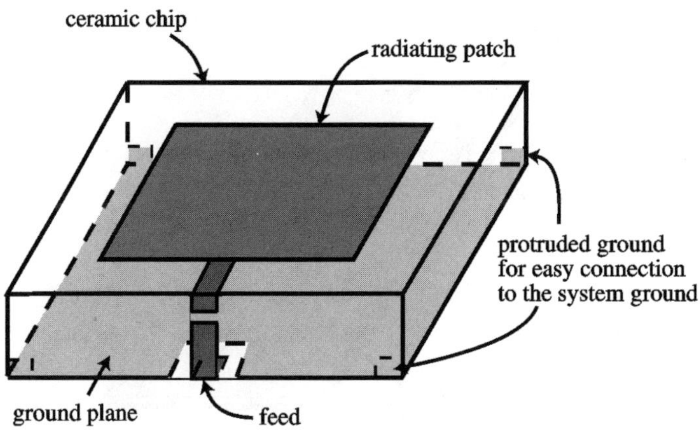

FIGURE 1.16 Geometry of a ceramic chip antenna as a surface-mountable antenna for WLAN applications.

been used [78]. This antenna uses no ceramic base and has a rigid structure of a rectangular disklike shape. In addition, the antenna can be easily fed by using a 50-Ω microstrip line, which makes it surface mountable like the ceramic chip antennas. Moreover, dual-frequency operation can also be obtained. A design example for achieving dual-band WLAN operations in the 2.4- and 5.2-GHz bands has been successfully implemented.

Several designs of the integrated monopole antenna have also been demonstrated that print and integrate a monopole on a small portion of the circuit board of a communication device [79-82]. A typical integrated monopole antenna design is shown in Figure 1.17, in which the monopole has an F shape and can perform dual-WLAN operations in the 2.4 and 5.2 GHz bands [82]. Such antennas have the advantages of occupying a very small volume of the system, decreasing the total fabrication cost, and easily integrating with the associated circuitry on the circuit board.

The printed diversity monopole antennas suitable for combating the multipath fading problem have also been available in the open literature. For this purpose, a printed dual orthogonal monopole antenna for diversity operation in the 2.4 GHz WLAN band has been devised [83], and the antenna geometry is shown in Figure 1.18. Note that the dual orthogonal straight monopoles for providing polarization diversity are printed and integrated on the same microwave substrate, and a protruded ground plane of suitable dimensions is added between the two monopoles to enhance the port decoupling between the two feeding ports. In this case, a constructed prototype for WLAN operation in the 2.4 GHz band [83] showed that the measured reflection coefficient (S_{11}) of the operating frequencies within the WLAN band is less than −20 dB, and the measured isolation (S_{21}) is less than −27 dB, which is much better than that without a protruded ground plane (less than about −12 dB). A design of the diversity monopole antenna capable of dual-band operations in the 2.4 and 5.2 GHz bands is also available and is described in Chapter 5.

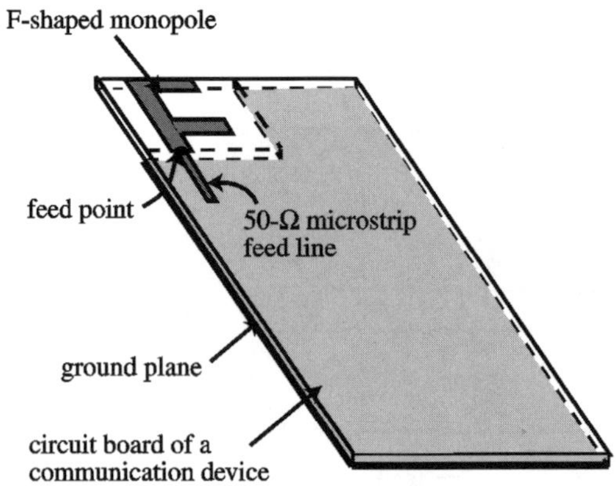

FIGURE 1.17 Geometry of an F-shaped monopole antenna integrated on the circuit board of a communication device for dual-band WLAN applications.

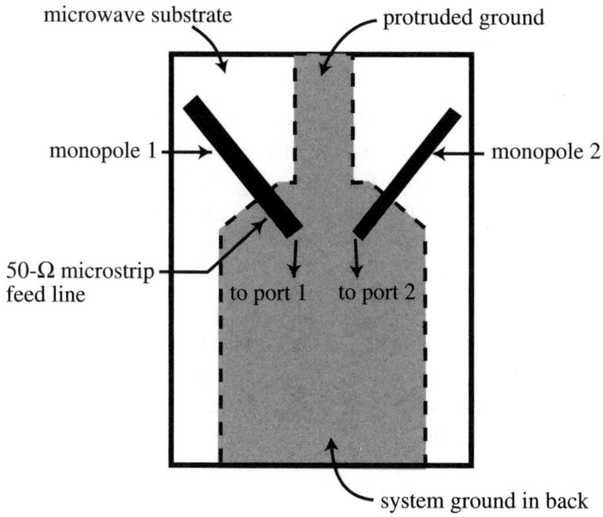

FIGURE 1.18 Geometry of an integrated diversity monopole antenna for WLAN applications.

The dual-band PIFA designs described in Section 1.2 are also very suitable for achieving dual-band WLAN operations in the 2.4 and 5.2 GHz bands. Several successful designs have been reported [10, 84, 85]. In contrast to the monopole antennas, these dual-band PIFAs can be placed directly on the ground plane of a communication device, which increases their design freedom and possible applications.

To obtain dual-polarized radiation for applications in a mobile communication unit, an interesting low-cost compact antenna design has been developed. This antenna design is obtained by applying the technique of using a slotted ground plane [86]. By embedding four linear slots, oriented in perpendicular to each other, in the ground plane of a microstrip antenna with an inexpensive FR4 substrate (thickness 1.6 mm and relative permittivity 4.4), and feeding the antenna using two probe feeds at two orthogonal positions, $\pm 45^\circ$ slanted linear polarizations can be generated. Owing to the embedded slots in the antenna's ground plane, which can lead to antenna size reduction without a sacrifice in the impedance bandwidth and antenna gain [86], this antenna provides a small size ($35 \times 35\,\text{mm}^2$) for operating in the 2.4 GHz band, which makes it suitable for integration into a restricted area as a part of a WLAN personal computer memory card international association (PCMCIA) card for wireless Internet access for notebook computers. In addition, good dual-polarized performance is obtained, and the antenna has an impedance bandwidth of about 3.5% covering the 2.4 GHz WLAN band. A peak antenna gain of about 2.5 dBi for the two orthogonal linear polarizations has been obtained, and an isolation between the two feeding ports of less than −20 dB has been achieved.

1.6 DIELECTRIC RESONATOR ANTENNAS FOR WIRELESS COMMUNICATIONS

Dielectric resonator antennas have been shown to be efficient radiators [3, 87]. They require no radiating metal patch and thus have relatively lower loss compared with microstrip antennas, especially for operating at higher frequencies. Dielectric resonator antennas also have the advantage of easy excitation through the use of transmission lines such as a microstrip line [88] or a coplanar waveguide [89] (see Fig. 1.19). Many studies have also been conducted and are available in the open literature [3].

Recently, it has been shown that, when a very-high-permittivity DR element is used, the DR antenna can have a very low profile. For operating in the 5.2 GHz WLAN band, the antenna has a low profile of only 1.9 mm and occupies a small area of $10 \times 10\,\text{mm}^2$, with the use of a DR element having a relative permittivity

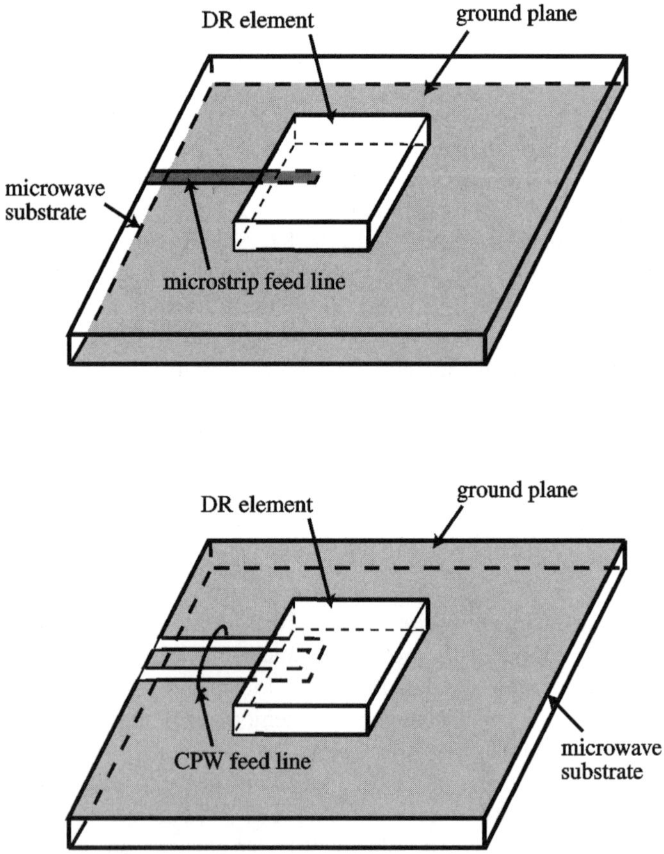

FIGURE 1.19 Low-profile, very-high-permittivity DR antennas fed by a microstrip line and a coplanar waveguide (CPW) feed.

of 90.5 [90]. Owing to its low profile and compact size, this kind of very-high-permittivity DR antenna is very suitable for WLAN applications. For this reason, many advanced designs for achieving CP radiation, broadband operation, and dual-polarized operation have been developed.

For the case of achieving CP radiation using a single feed, conventional designs are mainly based on the use of DR elements with special configurations, such as the nearly cubic DR [91], the rectangular DR with a special length-to-width ratio [92], or the cross-shaped DR [93]. In these designs, the requirement for special configurations of the DR elements increases the complexity in antenna fabrication, which makes them less attractive for practical applications. Recently, several interesting designs have been reported [88, 94–96]. These CP designs are applicable to cases with regular DR elements and are promising for practical applications. The design techniques include the use of the regular DR element fed by a cross-slot-coupled feed [94] and the use of parasitic conformal strips [95–97] and a loading patch [88].

A CP design using two DR elements has also been shown [98]. In this design, the technique of using two linearly polarized radiating elements for obtaining CP radiation was applied. Advances in very-high-permittivity DR antennas for achieving broadband operation [90] and dual-polarized radiation [99] have also been achieved. These recent advanced designs are addressed in Chapter 6, and their possible applications for WLAN operations are discussed.

1.7 INTEGRATION OF ANTENNAS FOR DIFFERENT OPERATING BANDS

For some applications, it is required that a global positioning system (GPS) antenna be integrated into a communication device to provide GPS signal reception. For this purpose, several integration designs for integrating a GPS antenna with a DCS antenna or a WLAN antenna have been devised recently [100]. Similarly, the integration of a WLAN antenna with a DCS antenna has also been shown [101]. These integration designs and the experimental results are described in detail in Chapter 7.

REFERENCES

1. K. L. Wong, Design of Nonplanar Microstrip Antennas and Transmission Lines, John Wiley & Sons, New York, NY, 1999.
2. K. L. Wong, Compact and Broadband Microstrip Antennas, John Wiley & Sons, New York, USA, 2002.
3. K.M. Luk and K.W. Leung, Editors, Dielectric Resonator Antennas, Research Studies Press Limited, Hertfordshire, England, UK, 2002.
4. T. Taga and K. Tsunekawa, "Performance analysis of a built-in planar inverted-F antenna for 800 MHz hand portable radio units," IEEE J. Select. Areas Commun., vol. 5, pp. 921–929, June 1987.

5. Z. D. Liu, P. S. Hall, and D. Wake, "Dual-frequency planar inverted F antenna," IEEE Trans. Antennas Propagat., vol. 45, pp. 1451–1458, Oct. 1997.
6. C. R. Rowell and R. D. Murch, "A compact PIFA suitable for dual-frequency 900/1800-MHz operation," IEEE Trans. Antennas Propagat., vol. 46, pp. 596–598, April 1998.
7. J. Ollikainen, O. Lehmus, M. Fischer, and P. Vainikainen, "Internal dual-band patch antenna for mobile phones," in Proc. Millennium Conf. on Antennas Propagat., Davos, Switzerland, p. 364.
8. S. Tarvas and A. Isohatala, "An internal dual-band mobile phone antenna," in 2000 IEEE Antennas Propagat. Soc. Int. Symp. Dig., pp. 266–269.
9. D. Viratelle and R. J. Langley, "Dual-band printed antenna for mobile telephone applications," IEE Proc.-Microwave Antennas Propagat., vol. 147, pp. 381–384, Oct. 2000.
10. P. Salonen, M. Keskilammi, and M. Kivikoski, "New slot configurations for dual-band planar inverted-F antenna," Microwave Opt. Technol. Lett., vol. 28, pp. 293-298, March 5, 2001.
11. D. Manteuffel, A. Bahr, D. Heberling, and I. Wolff, "Design considerations for integrated mobile phone antennas," in Proc. 2001 Int. Conf. on Antennas Propagat. (ICAP), pp. 252–254.
12. G. H. K. Lui and R. D. Murch, "Compact dual-frequency PIFA designs using LC resonators," IEEE Trans. Antennas Propagat., vol. 49, pp. 1016–1019, July 2001.
13. F. R. Hsiao, H. T. Chen, G. Y. Lee, T. W. Chiou, and K. L. Wong, "A dual-band planar inverted-F patch antenna with a branch-line slit," Microwave Opt. Technol. Lett., vol. 32, pp. 310–312, Feb. 20, 2002.
14. S. H. Yeh and K. L. Wong, "Compact dual-frequency PIFA with a chip-inductor-loaded rectangular spiral strip," Microwave Opt. Technol. Lett., vol. 33, pp. 394–397, June 20, 2002.
15. I. A. Korisch, "Planar dual frequency band antenna," U.S. Patent No. 5926139, July 20, 1999.
16. F. Geerert, "Dual band antenna for a hand portable telephone and a corresponding hand portable telephone," European Patent No. 1018777, Dec. 10, 1999.
17. I. Pankinaho, "Double resonance antenna structure for several frequency ranges," U.S. Patent No. 6140966, Oct. 31, 2000.
18. Z. Ying, "Printed twin spiral dual band antenna," U.S. Patent No. 6166694, Dec. 26, 2000.
19. N. Chiba, T. Amano, and H. Iwasaki, "Multifrequency inverted F-type antenna," U.S. Patent No. 6195048, Feb. 27, 2001.
20. S. Tarvas, J. Mikkola, S. Kivela, and A. Isohatala, "Planar dual-frequency antenna and radio apparatus employing a planar antenna," U.S. Patent No. 6252552, June 26, 2001.
21. C. T. P. Song, P. S. Hall, H. Ghafouri-Shiraz, and D. Wake, "Triple band planar inverted F antennas for handheld devices," Electron. Lett., vol. 36, pp. 112-114, Jan. 20, 2000.
22. W. P. Dou and Y. W. M. Chia, "Novel meandered planar inverted-F antenna for triple-frequency operation," Microwave Opt. Technol. Lett., vol. 27, pp. 58–60, Oct. 5, 2000.
23. F. R. Hsiao and K. L. Wong, "Compact planar inverted-F patch antenna for triple-frequency operation," Microwave Opt. Technol. Lett., vol. 33, pp. 459–462, June 20, 2002.

24. M. Martinez-Vazquez, M. Geissler, and D. Heberling, "Volume considerations in the design of dual-band handset antennas," in 2001 IEEE Antennas Propagat. Soc. Int. Symp. Dig., pp. 112–115.

25. J. Ollikainen, M. Fischer, and P. Vainikainen, "Thin dual-resonant stacked shorted patch antenna for mobile communications," Electron. Lett., vol. 35, pp. 437–438, March 18, 1999.

26. J. Barreiros, P. Cameirao, and C. Peixeiro, "Microstrip patch antenna for GSM 1800 handsets," in 1999 IEEE Antennas Propagat. Soc. Int. Symp. Dig., pp. 2074–2077.

27. G. Y. Lee, T. W. Chiou, and K. L. Wong, "Broadband stacked shorted patch antenna for mobile communication handsets," in Proc. 2001 Asia-Pacific Microwave Conf., pp. 232–235.

28. F. R. Hsiao and K. L. Wong, "A shorted patch antenna with an L-shaped ground plane for internal mobile handset antenna," Microwave Opt. Technol. Lett., vol. 33, pp. 314–316, May 20, 2002.

29. T. Y. Wu and K. L. Wong, "On the impedance bandwidth of a planar inverted-F antenna for mobile handsets," Microwave Opt. Technol. Lett., vol. 32, pp. 249–251, Feb. 20, 2002.

30. J. D. Kraus, Antennas, McGraw-Hill Book Co., New York, NY, 1988.

31. K. D. Katsibas, C. A. Balanis, P. A. Tirkas, and C. R. Birtcher, "Folded loop antenna for mobile hand-held units," IEEE Trans. Antennas Propagat., vol. 46, pp. 260–266, Feb. 1998.

32. P. Haapala, P. Vainikainen, and P. Eratuuli, "Dual frequency helical antennas for handsets," in Proc. 1996 IEEE Vehicular Technol. Conf., pp. 336–338.

33. Z. Ying, "Multi-band non-uniform helical antennas," U.S. Patent No. 6112102, Aug. 29, 2000.

34. P. Haapala, "Antenna operating in two frequency ranges," U.S. Patent No. 6054966, April 25, 2000.

35. D. Maldonado, "Dual band antenna," U.S. Patent No. 5812097, Sept. 22, 1998.

36. E. Lee, P. S. Hall, and P. Gardner, "Dual band folded monopole/loop antenna for terrestrial communication system," Electron. Lett., vol. 36, pp. 1990–1991, Nov. 23, 2000.

37. P. L. Teng and K. L. Wong, "Planar monopole folded into a compact structure for very-low-profile multi-band mobile phone antenna," Microwave Opt. Technol. Lett., vol. 33, pp. 22–25, April 5, 2002.

38. G. Y. Lee and K. L. Wong, "Very-low-profile bent planar monopole antenna for GSM/DCS dual-band mobile phones," Microwave Opt. Technol. Lett., vol. 35, Sept. 20, 2002.

39. K. Rutkowski and G. J. Hayes, "Multiple frequency band branch antennas for wireless communications," U.S. Patent No. 6198442, March 6, 2001.

40. H. C. Tung, C. Y. Fang, and K. L. Wong, "An inverted-L monopole antenna loaded with a meandered wire for GSM/DCS dual-band mobile phones," Microwave Opt. Technol. Lett., vol. 33, pp. 212–214, May 5, 2002.

41. F. S. Chang, S. H. Yeh, and K. L. Wong, "Planar monopole in wrapped structure for low-profile GSM/DCS mobile phone antenna," Electron. Lett., vol. 38, pp. 499–500, May 23, 2002.

42. K. L. Wong, S. W. Su, T. W. Chiou, and Y. C. Lin, "Dual-band plastic chip antenna for GSM/DCS mobile phones," Microwave Opt. Technol. Lett., vol. 33, pp. 330–332, June 5, 2002.
43. Y. L. Kuo and K. L. Wong, "Coplanar waveguide-fed folded inverted-F antenna for UMTS application," Microwave Opt. Technol. Lett., vol. 32, pp. 364–366, March 5, 2002.
44. Y. H. Yeh, Y. Y. Chen, and K. L. Wong, "A low-profile, bent and shorted monopole antenna with reduced backward radiation for mobile handsets," Microwave Opt. Technol. Lett., vol. 33, pp. 146–147, April 20, 2002.
45. S. Eggleston, "Antenna assembly, and associated method, having parasitic element for altering antenna pattern characteristics," U.S. Patent No. 6249255, June 19, 2001.
46. K. F. Lee, K. M. Luk, K. F. Tong, S. M. Shum, T. Huynh, and R. Q. Lee, "Experimental and simulation studies of the coaxially fed U-slot rectangular patch antenna," IEE Proc.-Microwave Antennas Propagat., vol. 144, pp. 354–358, Oct. 1997.
47. Y. M. Jo, "Broad band patch antennas using a wedge-shaped air substrate structure," in 1999 IEEE Antennas Propagat. Soc. Int. Symp. Dig., pp. 932–935.
48. C. L. Mak, K. M. Luk, K. F. Lee, and Y. L. Chow, "Experimental study of a microstrip patch antenna with an L-shaped probe," IEEE Trans. Antennas Propagat., vol. 48, pp. 777–782, May 2000.
49. N. Herscovici, "A wide-band single-layer patch antenna," IEEE Trans. Antennas Propagat., vol. 46, pp. 471–474, April 1998.
50. K. L. Wong and W. H. Hsu, "A broadband rectangular patch antenna with a pair of wide slits," IEEE Trans. Antennas Propagat., vol. 49, pp. 1345–1347, Sept. 2001.
51. F. Yang, X. X. Zhang, X. Ye, and Y. Rahmat-Samii, "Wide-band E-shaped patch antennas for wireless communications," IEEE Trans. Antennas Propagat., vol. 49, pp. 1094–1100, July 2001.
52. B. L. Ooi and Q. Shen, "A novel E-shaped broadband microstrip antenna," Microwave Opt. Technol. Lett., vol. 27, pp. 348–352, Dec. 5, 2000.
53. H. C. Tung and K. L. Wong, "A compact dual-polarized patch antenna for 1800-MHz band operation," Microwave Opt. Technol. Lett., vol. 29, pp. 1–2, April 5, 2001.
54. F. S. Chang and K. L. Wong, "A broadband probe-fed planar patch antenna with a short probe pin and a conducting cylinder transition," Microwave Opt. Technol. Lett., vol. 31, pp. 282–284, Nov. 20, 2001.
55. F. S. Chang and K. L. Wong, "A broadband probe-fed patch antenna for a DCS base station," Microwave Opt. Technol. Lett., vol. 30, pp. 341–343, Sept. 5, 2001.
56. P. L. Teng, C. L. Tang, and K. L. Wong, "A broadband planar patch antenna fed by a short probe feed," in Proc. 2001 Asia-Pacific Microwave Conf., pp. 1243–1246.
57. K. L. Wong, C. L. Tang, and J. Y. Chiou, "Broadband probe-fed patch antenna with a W-shaped ground plane," IEEE Trans. Antennas Propagat., vol. 50, June 2002.
58. W. H. Hsu and K. L. Wong, "Broadband probe-fed patch antenna with reduced cross-polarization radiation," in Proc. 2001 Int. Conf. on Antennas Propagat. (ICAP), pp. 525–528.
59. W. H. Hsu and K. L. Wong, "Broadband probe-fed patch antenna with a U-shaped ground plane for cross-polarization reduction," IEEE Trans. Antennas Propagat., vol. 50, pp. 352–355, March 2002.

60. F. S. Chang and K. L. Wong, "Broadband patch antenna edge-fed by a coplanar probe feed," Microwave Opt. Technol. Lett., vol. 31, pp. 287–289, Nov. 20, 2001.
61. J. S. Kuo and K. L. Wong, "A low-cost microstrip-line-fed shorted-patch antenna for a PCS base station," Microwave Opt. Technol. Lett., vol. 29, pp. 146–148, May 5, 2001.
62. F. S. Chang and K. L. Wong, "Compact L-strip capacitively coupled patch antenna with a folded radiating patch for a DCS base station," Microwave Opt. Technol. Lett., vol. 31, pp. 60–62, Oct. 5, 2001.
63. W. H. Hsu and K. L. Wong, "A dual-capacitively-fed broadband patch antenna with reduced cross-polarization radiation," Microwave Opt. Technol. Lett., vol. 26, pp. 169–171, Aug. 5, 2000.
64. F. S. Chang and K. L. Wong, "Broadband dual-frequency coplanar probe-fed patch antenna for GSM/DCS/PCS base stations," Microwave Opt. Technol. Lett., vol. 33, pp. 370–372, June 5, 2002.
65. K. M. Luk, C. H. Lai, and K. F. Lee, "Wideband L-probe-feed patch antenna with dual-band operation for GSM/DCS base stations," Electron. Lett., vol. 35, pp. 1123–1124, July 8, 1999.
66. S. Hienonen, A. Lehto, and A. V. Raisanen, "Simple broadband dual-polarized aperture-coupled microstrip antenna," in 1999 IEEE Antennas Propagat. Soc. Int. Symp. Dig., pp. 1228–1231.
67. B. Lindmark, "A novel dual polarized aperture coupled patch element with a single layer feed network and high isolation," in 1997 IEEE Antennas Propagat. Soc. Int. Symp. Dig., pp. 2190–2193.
68. I. Nystrom and D. Karlsson, "Reduction of back radiation and cross-coupling in dual polarized aperture coupled patch antennas," in 1997 IEEE Antennas Propagat. Soc. Int. Symp. Dig., pp. 2222–2225.
69. K. L. Wong, H.C. Tung, and T.W. Chiou, "Broadband dual-polarized aperture-coupled patch antennas with modified H-shaped coupling slots," IEEE Trans. Antennas Propagat., vol. 50, pp. 188–191, Feb. 2002.
70. K. L. Wong and T. W. Chiou, "Broadband dual-polarized patch antennas fed by capacitively coupled feed and slot-coupled feed," IEEE Trans. Antennas Propagat., vol. 50, pp. 346–351, March 2002.
71. T. W. Chiou and K. L. Wong, "Broadband dual-polarized single microstrip patch antenna with high isolation and low cross polarization," IEEE Trans. Antennas Propagat., vol. 50, pp. 399–401, March 2002.
72. T. W. Chiou and K. L. Wong, "A compact dual-polarized aperture-coupled patch antenna for GSM 900/1800-MHz systems," in Proc. 2001 Asia-Pacific Microwave Conf., pp. 95–98.
73. Y. L. Kuo and K. L. Wong, "A planar inverted-L patch antenna for 2.4/5.2-GHz dual-band operation," Microwave Opt. Technol. Lett., vol. 31, pp. 394–396, Dec. 5, 2001.
74. H. Tanidokoro, N. Konishi, E. Hirose, Y. Shimphara, H. Arai, and N. Goto, "1-wavelength loop type dielectric chip antennas," in 1998 Antennas Propagat. Soc. Int. Symp. Dig., pp. 1950–1953.
75. H. Matsushima, E. Hirose, Y. Shinohara, H. Arai, and N. Goto, "Electromagnetically coupled dielectric chip antenna," in 1998 Antennas Propagat. Soc. Int. Symp. Dig., pp. 1954–1957.

76. Y. Dakeya, T. Suesada, K. Asakura, N. Nakajima, and H. Mandai, "Chip multilayer antenna for 2.45 GHz-band application using LTCC technology," in 2000 IEEE MTT-S Int. Microwave Symp. Dig., pp. 1693–1696.
77. W. Choi, S. Kwon, and B. Lee, "Ceramic chip antenna using meander conductor line," Electron. Lett., vol. 37, pp. 933–934, July 19, 2001.
78. F. S. Chang and K. L. Wong, "Planar monopole folded into a rectangular-disk-like structure as surface-mountable antenna for 2.4/5.2 GHz dual-band operation," Microwave Opt. Technol. Lett., vol. 34, pp. 166–169, Aug. 5, 2002.
79. M. Ali and G. J. Hayes, "Analysis of integrated inverted-F antennas for Bluetooth applications," in Proc. 2000 IEEE-APS Conf. on Antennas and Propagation for wireless communications, pp. 21–24.
80. Y. H. Suh and K. Chang, "Low cost microstrip-fed dual frequency printed dipole antenna for wireless communications," Electron. Lett., vol. 36, pp. 1177–1179, July 6, 2000.
81. Y. L. Kuo, T. W. Chiou, and K. L. Wong, "A novel dual-band printed inverted-F antenna," Microwave Opt. Technol. Lett., vol. 31, pp. 353–355, Dec. 5, 2001.
82. S. H. Yeh and K. L. Wong, "Integrated F-shaped monopole antenna for 2.4/5.2 GHz dual-band operation," Microwave Opt. Technol. Lett., vol. 34, pp. 24–26, July 5, 2002.
83. Y. L. Kuo and K. L. Wong, "Dual-polarized monopole antenna for WLAN application," in 2002 IEEE Antennas Propagat. Soc. Int. Symp. Dig., vol. 4, pp. 80–83.
84. H. C. Tung and K. L. Wong, "A shorted microstrip antenna for 2.4/5.2-GHz dual-band operation," Microwave Opt. Technol. Lett., vol. 30, pp. 401–402, Sept. 20, 2001.
85. S. H. Yeh, S. T. Fang, and K. L. Wong, "Dual-band shorted patch antenna for dual ISM-band application," Microwave Opt. Technol. Lett., vol. 32, pp. 79–80, Jan. 5, 2001.
86. T. W. Chiou and K. L. Wong, "Designs of compact microstrip antennas with a slotted ground plane," in 2001 IEEE Antennas Propagat. Soc. Int. Symp. Dig., pp. 732–735.
87. S. A. Long, M. W. McAllister, and L. C. Shen, "The resonant cylindrical dielectric cavity antennas," IEEE Trans. Antennas Propagat., vol. 31, pp. 406–412, May 1983.
88. F. R. Hsiao, T. W. Chiou, and K. L. Wong, "Circularly polarized low-profile square dielectric resonator antenna with a loading patch," Microwave Opt. Technol. Lett., vol. 31, pp. 157–159, Nov. 5, 2001.
89. J. Y. Wu, C. Y. Huang, and K. L. Wong, "Low-profile, very-high-permittivity DR antenna excited by a coplanar waveguide," Microwave Opt. Technol. Lett., vol. 22, pp. 96–97, July 20, 1999.
90. F. S. Hsiao, J. S. Kuo, T. W. Chiou, and K. L. Wong, "A very-high-permittivity broadband dielectric resonator antenna for WLAN application in the 5.2 GHz Band," Microwave Opt. Technol. Lett., vol. 32, pp. 426–427, March 20, 2002.
91. M. B. Oliver, Y. M. M. Antar, R. K. Mongia, and A. Ittipiboon, "Circularly polarized rectangular dielectric resonator antenna," Electron. Lett., vol. 31, pp. 418–419, March 16, 1995.
92. K. P. Esselle, "Circularly polarized higher-order rectangular dielectric-resonator antenna," Electron. Lett., vol. 32, pp. 150–151, Feb. 1, 1996.
93. G. Drossos, Z. Wu, and L. E. Davis, "Circularly polarized cylindrical dielectric resonator antenna," Electron. Lett., vol. 32, pp. 282–283, Feb. 15, 1996.

94. C. Y. Huang, J. Y. Wu, and K. L. Wong, "Cross-slot-coupled microstrip antenna and dielectric resonator antenna for circular polarization," IEEE Trans. Antennas Propagat., vol. 47, 605–609, April 1999.

95. M. T. Lee, K. M. Luk, E. K. N. Yung, and K.W. Leung, "Circularly polarized dielectric resonator antenna with a microstrip feed," in Proc. 1999 Asia-Pacific Microwave Conf., pp. 722–723.

96. K. W. Leung, W. C. Wong, K. M. Luk, and E. K. N. Yung, "Circular-polarized dielectric resonator antenna excited by dual conformal strips," Electron. Lett., vol. 37, pp. 484–486, March 16, 2000.

97. R. T. Long, R. J. Dorris, S. A. Long, M. A. Khayat, and J. T. Williams, "Use of parasitic strip to produce circular polarization and increased bandwidth for cylindrical dielectric resonator antenna," Electron. Lett., vol. 37, pp. 406–408, March 29, 2001.

98. C. Y. Huang, K. L. Wong , C. F. Yang, and J. Y. Wu, "Planar array composed of two linearly polarized dielectric resonator antennas for circular polarization," Microwave Opt. Technol. Lett., vol. 21, pp. 323–324, June 5, 1999.

99. C. Y. Huang, T. W. Chiou, and K. L. Wong, "Dual-polarized dielectric resonator antennas," Microwave Opt. Technol. Lett., vol. 31, pp. 222–223, Nov. 5, 2001.

100. J. Y. Wu, J. S. Row, and K. L. Wong, "A compact dual-band microstrip patch antenna suitable for DCS/GPS operations," Microwave Opt. Technol. Lett., vol. 29, pp. 410–412, June 20, 2001.

101. J. Y. Wu and K. L. Wong, "Two integrated stacked shorted patch antennas for DCS/WLAN dual-band operations," Microwave Opt. Technol. Lett., vol. 30, pp. 134–136, July 20, 2001.

CHAPTER TWO

PIFAs for Internal Mobile Phone Antennas

2.1 INTRODUCTION

In comparison to the conventional protruded whip or rod antennas for applications in mobile phones, there are two major advantages of PIFAs: One advantage is that PIFAs are concealable within the housing of the mobile phones, and the other is that they capable of having reduced backward radiation toward the user's head to reduce electromagnetic wave power absorption and enhance antenna performance. Owing to these advantages, many PIFA designs have been devised, especially for achieving dual-frequency operations for applications in GSM/DCS dual-band or GSM/DCS/PCS triple-band mobile phones [1–14]. Design examples of the dual-frequency PIFAs using various promising techniques are described in Section 2.2, and detailed design considerations and experimental results are presented.

PIFA designs capable of triple-frequency operations in the 900, 1800, and 2450 MHz bands for GSM, DCS, and WLAN applications are also available in the open literature [15–17], in which the techniques of using two or three separate patches, a meandered patch, and a branch-line strip with a folded end have been applied. The related designs are addressed in Section 2.3. In Section 2.4, an interesting design of using an L-shaped ground plane to replace the conventional flat ground plane for PIFAs is described [18]. Such PIFAs can have further reduced backward radiation and enhanced antenna gain.

The techniques of using stacked PIFAs to achieve wider impedance bandwidths [19, 20] are discussed in Section 2.5. Finally, significant effects of ground plane dimensions on the impedance bandwidth of the PIFA are demonstrated. Experimental results of a constructed prototype for a UMTS mobile phone showed that the PIFA can have an impedance bandwidth varying from about 150 to 490 MHz, a variation of more than 300% [21]. The detailed results are presented and discussed in Section 2.6.

2.2 DUAL-FREQUENCY PIFA

2.2.1 Two Separate Patches with a Dual-Feed Design

Figure 2.1 shows the top patch of the PIFA comprising two separate patches of different sizes for achieving 900 and 1800 MHz operations. Note that the larger and smaller patches have separate shorting points and are operated, respectively, in the 900 and 1800 MHz bands as quarter-wavelength resonant structures. A dual feed (feed point 1 for the larger patch and feed point 2 for the smaller patch) is also utilized in the design, which can find application where mobile phones have receivers with dual front ends. In addition, to achieve a compact antenna size, a portion of the larger patch is removed to accommodate the smaller patch. Related studies have been reported in References 1 and 2]. A compact 900/1800 MHz dual-frequency PIFA with an area of $25 \times 26\,mm^2$ and a height of 4 mm has been studied in Reference 2, in which the PIFA has a capacitive load and is fed by a capacitive feed and in addition, a dielectric substrate with a relative permittivity of 2.1 is used between the two top patches and the ground plane.

2.2.2 Patch with an L-Shaped or Folded Slit

Figure 2.2 shows two promising designs of the top patch of the dual-frequency PIFA using a single probe feed and a common shorting pin. In Figure 2.2a, with the use of an L-shaped slit inserted at the patch boundary [3], the rectangular top patch can be separated into two subpatches of different sizes. The dimensions of the larger and the smaller subpatches can be designed to roughly resonate as a quarter-wavelength structure to operate in the 900 and 1800 MHz bands, respectively. Two subpatches can also be obtained by using a folded slit [4, 5], as shown in Figure 2.2b. Note that, in this case, the resonant path provided by the smaller subpatch starts from the feed point and then extends into the center portion of the rectangular patch. A large portion of this resonant path is thus surrounded by the outer larger subpatch,

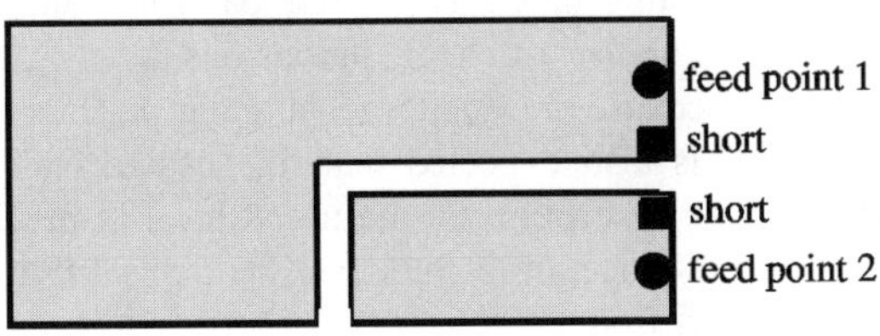

FIGURE 2.1 Top patch of the dual-frequency PIFA comprising two separate patches for achieving 900 and 1800 MHz operations.

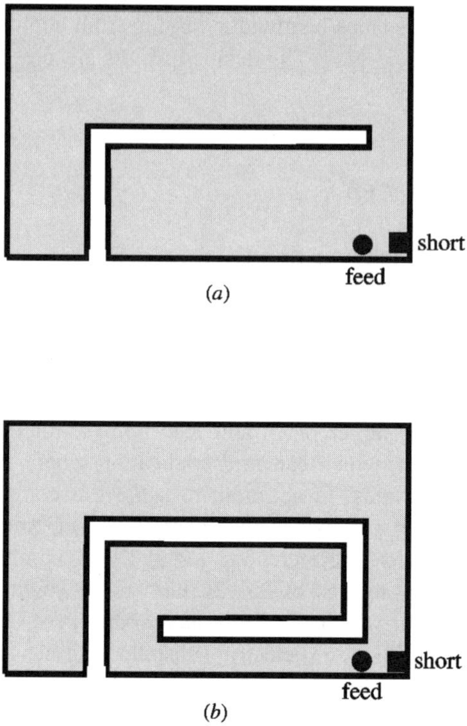

FIGURE 2.2 Top patch of the dual-frequency PIFA with (a) an L-shaped slit and (b) a folded slit.

which has the advantage of obtaining reduced backward radiation, leading to a smaller electromagnetic wave power absorption by the user's head in the 1800 MHz band operation.

2.2.3 Patch with a U-Shaped Slot

Recently, instead of using an L-shaped slit or a folded slit to obtain two separate subpatches in the radiating top patch, a design using an embedded U-shaped slot (see Fig. 2.3) has been used [6, 7]. In this design, a smaller rectangular patch of dimensions $L_2 \times W_2$ for operating in the 1800 MHz band is obtained in the central portion of the original rectangular patch of dimensions $L_1 \times W_1$, which is for the 900 MHz band operation. Because the smaller rectangular patch is embedded within the radiating top patch, it is also expected that this design can provide reduced electromagnetic wave power absorption by the user's head in the 1800 MHz band. The lower (f_1) and upper (f_2) operating frequencies of this PIFA design can also be approximately determined from

$$f_1 \cong \frac{c}{4(L_1 + W_1)}, \tag{2.1}$$

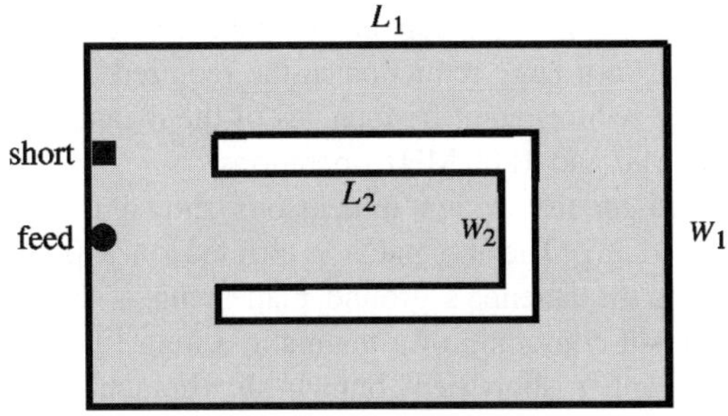

FIGURE 2.3 Top patch of the dual-frequency PIFA with an embedded U-shaped slot.

$$f_2 \cong \frac{c}{4(L_2 + W_2)}. \tag{2.2}$$

The above two equations make the design of such a dual-frequency PIFA easy to achieve, and c in the equations is the speed of light in free space.

2.2.4 Patch with a Branch Line Slit

The designs shown in Sections 2.2.1–2.2.3 mainly utilize two different resonant paths for dual-frequency operation (the smaller one for the 1800 MHz band and the larger one for the 900 MHz band). Figure 2.4 shows another promising PIFA design to achieve 900/1800 MHz dual-frequency operations. This design utilizes

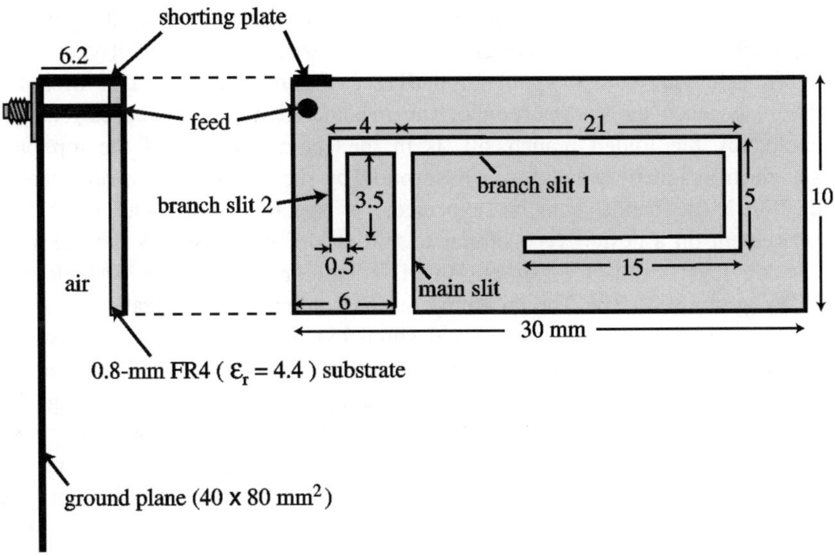

FIGURE 2.4 Geometry of the dual-frequency PIFA with a branch-line slit; dimensions given in the figure are in millimeters. (From Ref. 8, © 2002 John Wiley & Sons, Inc.)

an asymmetric branch-line slit to meander the excited patch surface currents in the top patch, which leads to a large reduction in the required top patch dimensions of the PIFA, and the first two resonant frequencies of the meandered resonant path are also tuned to obtain 900 and 1800 MHz operations.

A prototype based on the design dimensions shown in Figure 2.4 has been constructed and tested [8]. The top patch is printed on a 0.8-mm FR4 substrate and short-circuited to the antenna's ground plane with a shorting plate of length 2 mm. A branch-line slit consisting of a main slit, a long folded branch slit (branch slit 1 in the figure), and a short bent branch slit (branch slit 2 in the figure) is embedded in the top patch. The main slit, which has an open end at the patch boundary, and the long folded branch slit in the top patch are mainly for effectively meandering the excited patch surface currents. On the other hand, the short bent branch slit is mainly for achieving enhanced impedance matching of the first two excited resonant frequencies to obtain wider operating bandwidths. Also note that the required length of the short bent branch slit is much smaller than that of the long folded branch slit. The feed point is located close to the shorting plate, and both the feed point and the shorting plate are also required to be placed at the patch edge near the short bent branch slit. In this case, a large meandering of the excited patch surface currents starting from the feed point to the portion of the patch encircled by the long folded branch slit can be obtained. This large meandering leads to a large reduction in the required patch dimensions for the proposed 900/1800 MHz band operations.

It should also be noted that the wavelengths of the first two resonant frequencies in this PIFA design correspond, respectively, to about 4.0 and 2.0 times the length of the meandered patch surface current path, starting from the feed point to the portion of the patch encircled by the long folded branch slit. That is, this meandered patch portion is operated as a radiator with resonant lengths of about 0.25 and 0.5 wavelengths in the proposed design. This characteristic also makes it very convenient in achieving the 900 and 1800 MHz excitation for this PIFA design. In addition, because the end portion of the meandered patch surface current path, encircled by the folded branch slit, is in the central portion of the top patch, lower electromagnetic wave power absorption by the user's head in both the 900 and 1800 MHz bands can be expected. A specific absorption rate (SAR) measurement on a design very similar to that shown in Figure 2.4 showed a very low SAR level of 0.21 W/kg at 900 MHz and an even lower SAR level of 0.11 W/kg at 1800 MHz. The measured results confirm this expectation.

Figure 2.5 shows the measured and simulated return loss for the constructed prototype based on the design dimensions shown in Figure 2.4. The simulation results, which agree well with the measured data, are obtained by using IE3D simulation software. From the results obtained, the first two resonant frequencies at about 900 and 1800 MHz are excited with good impedance matching. The measured bandwidths, determined from 2.5:1 VSWR or about 7.3 dB return loss, are about 24 and 123 MHz in the 900 and 1800 MHz bands, respectively, which are comparable to those obtained in Reference 9, although the constructed prototype has a much smaller antenna volume of only $30\times10\times7\ \text{mm}^3$. However, it should be

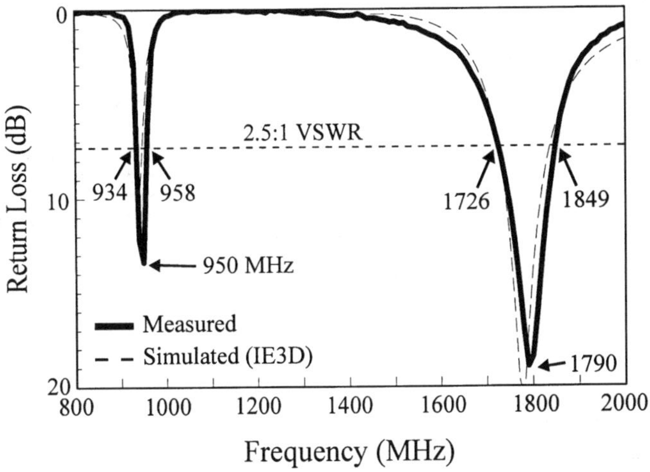

FIGURE 2.5 Measured and simulated return loss of the PIFA shown in Figure 2.4. (From Ref. 8, © 2002 John Wiley & Sons, Inc.)

noted that the obtained bandwidths are still not large enough to meet the bandwidth requirements of the GSM/DCS cellular communications systems [70 MHz (890–960 MHz) and 170 MHz (1710–1880 MHz) for the GSM and DCS systems, respectively]. Further enhancement in the impedance bandwidths is required to obtain optimal performance for GSM/DCS mobile phones.

The excited patch surface current distributions in the top patch for the two resonant frequencies have also been studied. Figure 2.6 shows the simulation results of the excited patch surface currents at resonance of the two excited resonant modes. It is seen that the current path, which starts from the feed point to the portion of the patch encircled by the long folded branch slit, determines the excitation of the two desired resonant frequencies. The patch portion with this current path thus acts as a main radiator in this PIFA design to resonate at 0.25 and 0.5 wavelengths. In this design, the length (ℓ) of this meandered current path is about 80 mm. That is, the first or lower frequency for the proposed antenna has a wavelength corresponding to be about 4.0ℓ, and the second or upper frequency has a wavelength of about 2.0ℓ.

Figures 2.7 and 2.8 plot the measured radiation patterns at two resonances of the antenna. For both resonant frequencies, good omnidirectional radiation in the azimuthal plane (x–z plane) is seen. The antenna gain is also measured to be about 0.5 and 1.5 dBi for the lower and upper resonant frequencies, respectively.

2.2.5 Meandered Patch With/Without a Parasitic Shorted Patch

By properly meandering the top patch of the PIFA, the frequency ratio of the PIFA's first two resonant frequencies can be tuned to be about 2.0, which makes it very promising for obtaining the 900 and 1800 MHz operations [9, 10]. By further incor-

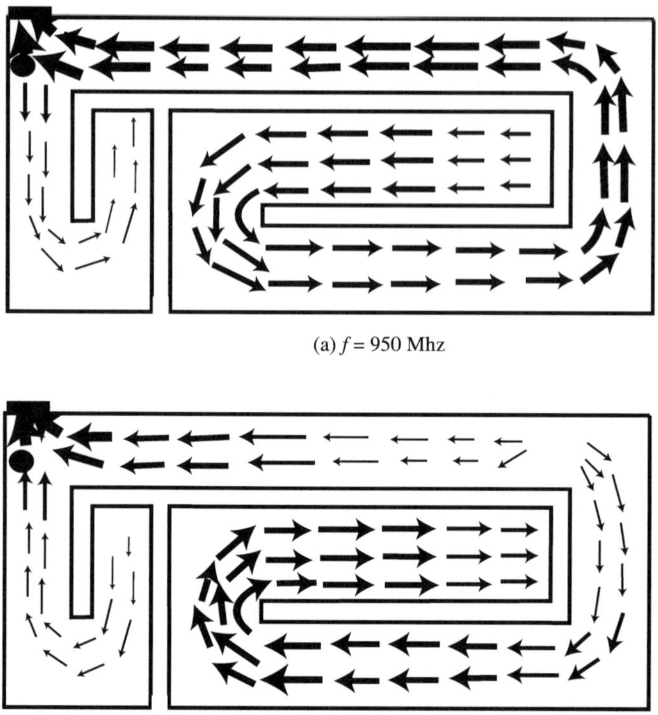

(a) $f = 950$ Mhz

(b) $f = 1790$ Mhz

FIGURE 2.6 Simulated patch surface current distributions (IE3D simulation results) at (a) $f = 950$ MHz and (b) $f = 1790$ MHz for the PIFA shown in Figure 2.4. (From Ref. 8, © 2002 John Wiley & Sons, Inc.)

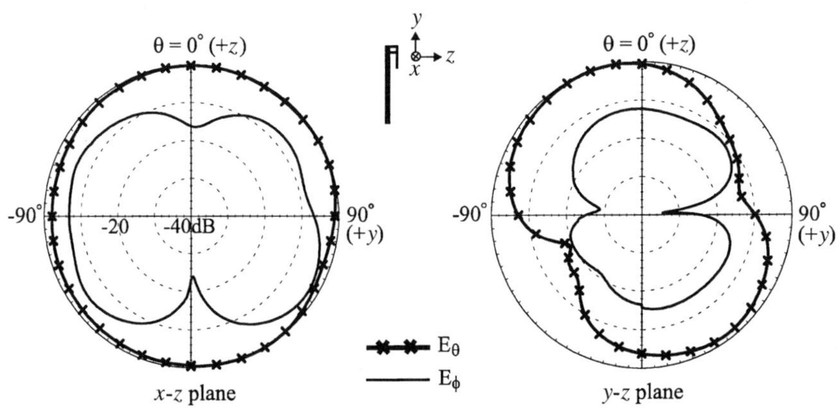

FIGURE 2.7 Measured radiation patterns at 950 MHz for the PIFA shown in Figure 2.4. (From Ref. 8, © 2002 John Wiley & Sons, Inc.)

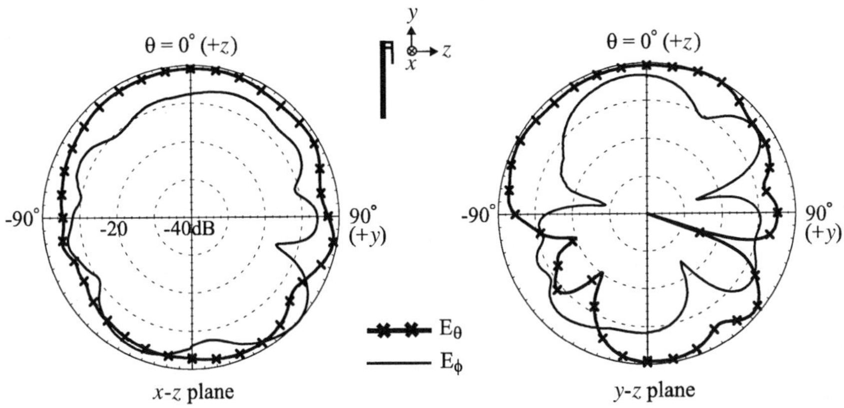

FIGURE 2.8 Measured radiation patterns at 1790 MHz for the PIFA shown in Figure 2.4. (From Ref. 8, © 2002 John Wiley & Sons, Inc.)

porating the use of a parasitic shorted patch (see the geometry shown in Fig. 2.9), a much enhanced bandwidth in the upper band of the PIFA has also been demonstrated [10]. In this design, with the PIFA occupying a volume of $40\times30.4\times7.2\,mm^3$ and mounted on the top portion of a ground plane having an area of $40\times110\,mm^2$, the obtained impedance bandwidth, determined by 6 dB return loss (3:1 VSWR), reaches 535 MHz in the upper band, covering the required bandwidth of the DCS and PCS operations. This relatively larger upper band bandwidth is mainly due to the added parasitic shorted patch, which introduces an additional resonant mode to the upper band and makes the upper band become a dual-resonance band. This condition leads to an enhanced impedance bandwidth. In addition, this design has a sufficient bandwidth in the lower band (about

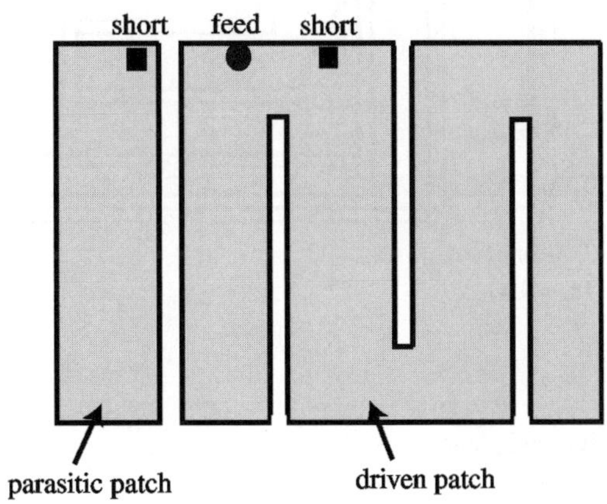

FIGURE 2.9 Top patch of the dual-frequency PIFA with a meandered patch and a parasitic shorted patch.

115 MHz) to cover the required bandwidth of the GSM operation. In this case, the obtained PIFA design is suitable for applications in GSM/DCS/PCS triple-band mobile phones. A similar design using a parasitic shorted patch to obtaining a PIFA capable of GSM/DCS/PCS triple-band operations has also been shown in Reference 11.

2.2.6 Two or More Shorted Patches or Strips

By using two or more shorted patches with a common shorting pin, bandwidth enhancement in both the lower and upper bands of the PIFA has been obtained. A promising design is shown in Figure 2.10. This design utilizes a folded slit-loaded patch (patch 1) and a folded patch (patch 2), both sharing a common shorting pin. The two patches are printed on the same FR4 supporting substrate of thickness 0.8 mm and relative permittivity 4.4. Patch 2 is also configured to be around patch 1 to result in a compact size for the PIFA. In this design, the two patches occupy a small area of $18 \times 32\ \mathrm{mm}^2$ and an air layer of thickness 6.4 mm is between the ground plane and the supporting FR4 substrate; that is, the total antenna height is 7.2 mm. The two patches are excited by a probe feed located close to the common shorting pin.

Notice that a folded slit, as discussed in Section 2.2.2, is embedded in patch 1. This folded slit separates patch 1 into two subpatches of different sizes, which generate a resonant mode in each of the 900 and 1800 MHz bands. On the other

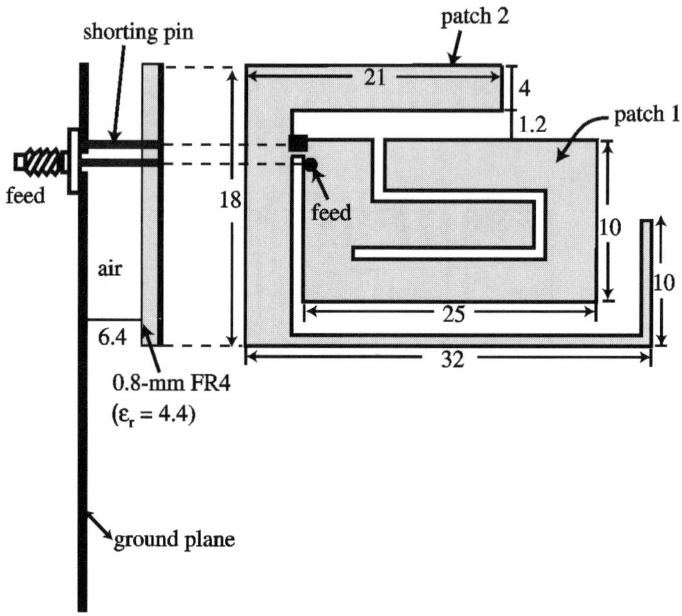

FIGURE 2.10 Geometry of the dual-frequency PIFA with a folded-slit-loaded patch and a folded patch.

hand, because of the shorting pin loading, it is possible for patch 2 to operate as either a quarter-wavelength or a half-wavelength antenna [22, 23]. This behavior has been confirmed in experiments, and through adjustments of the dimensions of patch 2 this PIFA design can have two additional resonant modes excited, one close to the 900 MHz band and one close to the 1800 MHz band. Thus, by incorporating the two resonant modes excited by patch 1, the proposed PIFA can have two resonant modes excited in each of its lower and upper bands. This behavior not only gives the proposed PIFA sufficient operating bandwidths for the GSM and DCS systems but also makes it possible to cover the additional frequency band of the PCS system, leading to a GSM/DCS/PCS triple-band PIFA.

Figure 2.11 shows the geometry of a related promising PIFA design. In this case, three resonant elements are used: two meandered metallic strips of slightly different lengths (strips 1 and 2) and one nearly rectangular patch (patch 1), which are printed

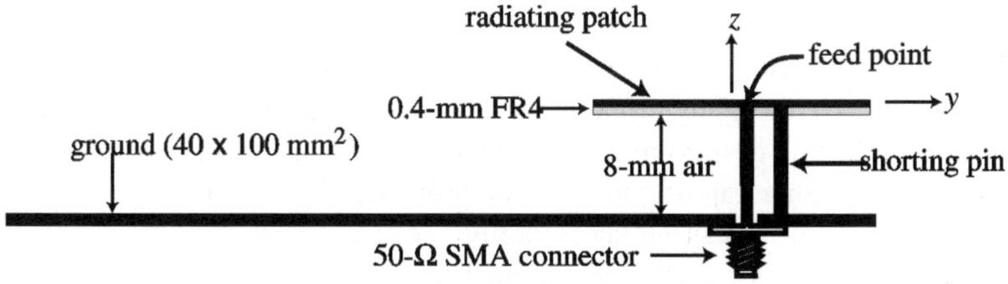

(*a*)

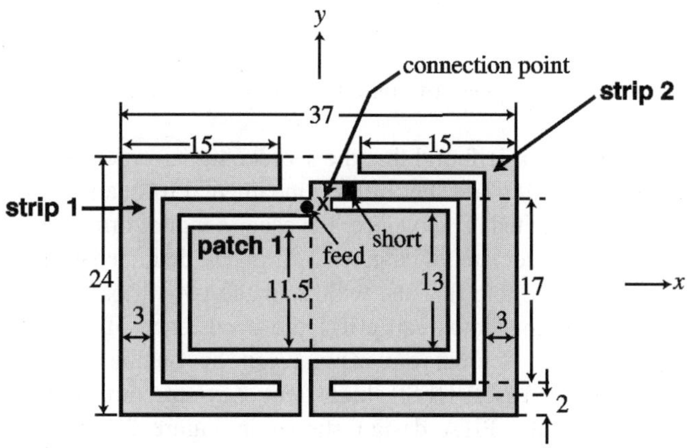

(*b*)

FIGURE 2.11 Geometry of the dual-frequency PIFA with two shorted strips and one shorted patch. (a) Side view of the PIFA mounted on the top portion of a ground plane of dimensions $40 \times 100\,\text{mm}^2$; (b) top patch of the PIFA. (From Ref. 12, © 2002 IEEE, reprinted with permission)

on a same supporting FR4 substrate (thickness 0.4 mm and relative permittivity 4.4) and arranged in a compact configuration. These three resonant elements share a common shorting pin, and for the GSM operation in the lower band, the proposed PIFA is operated with the two meandered strips both resonated as quarter-wavelength structures, leading to a wide bandwidth formed by two resonant modes. For the DCS operation in the upper band, three resonant modes are generated, two from the second higher-order modes of the two meandered strips and one from the nearly rectangular patch, leading to a wide bandwidth covering the DCS band.

Also note that the three resonant elements are excited by a common probe feed at a position close to the connection point. Patch 1 is a nearly rectangular patch and is placed in the center. Note that the use of a nearly rectangular patch, instead of a simple rectangular patch, in this design is for fine-tuning the PIFA's lower and upper bands to exactly cover the GSM and DCS bands, respectively. The dimensions of patch 1 are selected such that it will generate a resonant mode at about 1900 MHz to widen the operating bandwidth of the upper band. The dimensions of patch 1 can also be roughly determined from Equation 2.1 or Equation 2.2.

Strips 1 and 2 have slightly different lengths (about 84 and 86 mm for strips 1 and 2, respectively, in this design) and are meandered around the left and right boundaries of patch 1, with a gap of 1 mm. In the proposed design, the first two resonant modes of strips 1 and 2 can be successfully excited. The first or the fundamental resonant mode of strips 1 and 2 is designed to occur at about 930 and 900 MHz, respectively. In this case, strips 1 and 2 are both operated as quarter-wavelength structures; that is, the lengths (ℓ) of strips 1 and 2 can be roughly determined from

$$f_0 = \frac{c}{4\ell}, \tag{2.3}$$

where c is again the speed of light in free space and f_0 is the desired resonant frequency.

At the same time, it is found that the second higher-order mode, whose resonant frequency is about 1.9 times that of the fundamental resonant frequency, of strips 1 and 2 can also be excited. The two excited second higher-order modes incorporating the resonant mode associated with patch 1 form a wide impedance bandwidth for the DCS operation. Also note that the widths of the two strips are not constants. The variations in the strip widths can affect the occurrence of the resonant modes of the PIFA, and their optimal dimensions can be obtained with the aid of the simulation software Ansoft HFSS (high-frequency structure simulator).

A prototype of the PIFA design shown in Figure 2.11 was constructed and studied. The PIFA is mounted on the top portion of a ground plane of width 40 mm and length 100 mm, which can be considered to be the ground plane of a practical mobile phone. Figure 2.12 shows the measured and simulated return loss for the constructed prototype. Reasonable agreement between the measured data and the simulated results is obtained. For the lower band, two resonant modes are

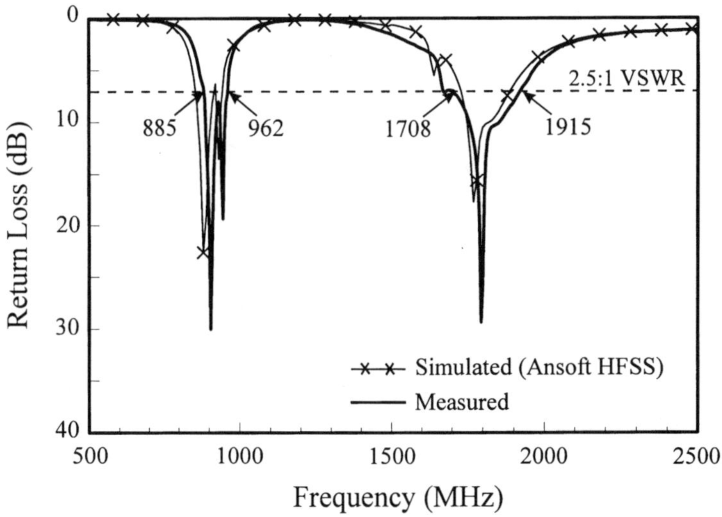

FIGURE 2.12 Measured and simulated return loss of the PIFA shown in Figure 2.11. (From Ref. 12, © 2002 IEEE, reprinted with permission)

successfully excited as expected, and an impedance bandwidth, determined from 2.5:1 VSWR or about 7.3 dB return loss, of 77 MHz (885–962 MHz) is obtained, which meets the required bandwidth of the GSM system. For the upper band, three resonant modes are seen to be excited, also as expected. Also note that the first two resonances at about 1.7 and 1.8 GHz in the upper band are greatly controlled by the lengths of strips 2 and 1, respectively. As for the third resonance at about 1.9 GHz, it is greatly affected by the dimensions of patch 1. The impedance bandwidth obtained is as large as 207 MHz (1708–1915 MHz), which covers the required bandwidth of the DCS system.

The radiation characteristics of the proposed PIFA have also been studied. Figure 2.13 plots the measured radiation patterns at the center frequencies of the GSM and DCS bands. The measured patterns show a better front-to-back ratio at 1795 MHz than at 925 MHz. The measured antenna gain for operating frequencies in the GSM and DCS bands is presented in Figure 2.14. For the GSM band, the peak antenna gain reaches about 2.7 dBi, with gain variations within 1.7 dBi. As for the DCS band, the peak antenna gain is about 2.1 dBi and the gain variations are less than 1.1 dBi.

2.2.7 Meandered Patch with a Folded End

To reduce the size of the dual-frequency PIFA, a novel design of using a meandered patch with a folded end has recently been devised. The geometry of such a compact dual-frequency PIFA is shown in Figure 2.15. In this design, the meandered patch is printed on both sides of an FR4 substrate of area $10 \times 24\ mm^2$, thickness 0.8 mm, and

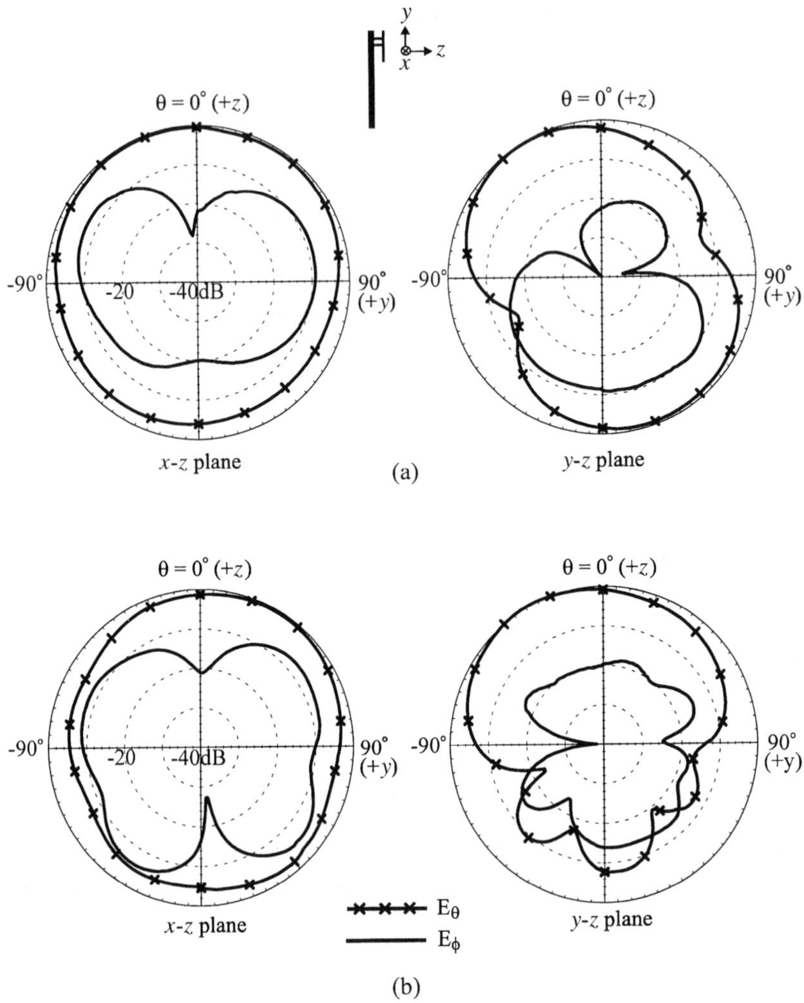

FIGURE 2.13 Measured radiation patterns at (a) 925 MHz and (b) 1795 MHz for the PIFA shown in Figure 2.11. (From Ref. 12, © 2002 IEEE, reprinted with permission)

relative permittivity 4.4. For the patch on the upper side of the FR4 substrate, two linear slits are embedded to meander the excited patch surface currents. By incorporating the patch on the lower side of the FR4 substrate, the total length of the effective patch surface current path can be much larger than the physical length of the FR4 substrate, which significantly reduces the required length of the PIFA to obtain the 900 and 1800 MHz operations.

Also note that, in this design, a rectangular notch is removed near the feed and shorting points on the patch printed on the upper side of the substrate. By varying the aspect ratio (ratio of the length and width) of this rectangular notch, the

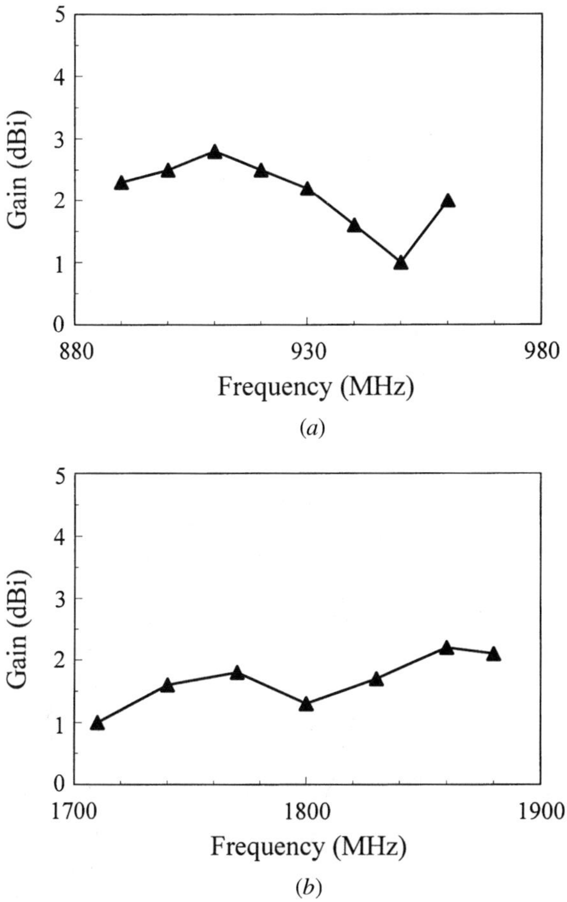

FIGURE 2.14 Measured antenna gain in (a) the GSM band and (b) the DCS band for the PIFA shown in Figure 2.11. (From Ref. 12, © 2002 IEEE, reprinted with permission)

frequency ratio of the first two resonant frequencies of the PIFA is varied in a wide range. This behavior makes the tuning of the first two resonant frequencies to be within the desired operating bands easy to achieve. A prototype of this PIFA design shown in Figure 2.15 has been successfully implemented. The prototype has an antenna height of 7.2 mm and occupies an area of $10 \times 24\,\text{mm}^2$. Measured return loss is shown in Figure 2.16. Although the constructed prototype has a compact size, the lower and upper bands still show an impedance bandwidth (2.5:1 VSWR) of about 30 and 155 MHz, respectively.

2.2.8 Patch with LC Resonators

The technique of using LC resonators to achieve dual-frequency operation has been applied to monopole antennas. This technique is also applicable for the PIFA to

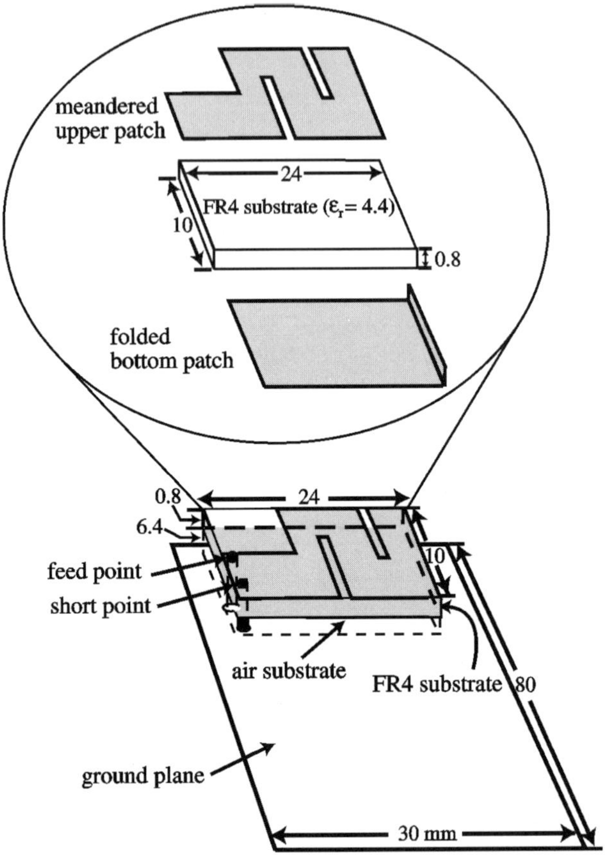

FIGURE 2.15 Geometry of the dual-frequency PIFA using a meandered patch with a folded end.

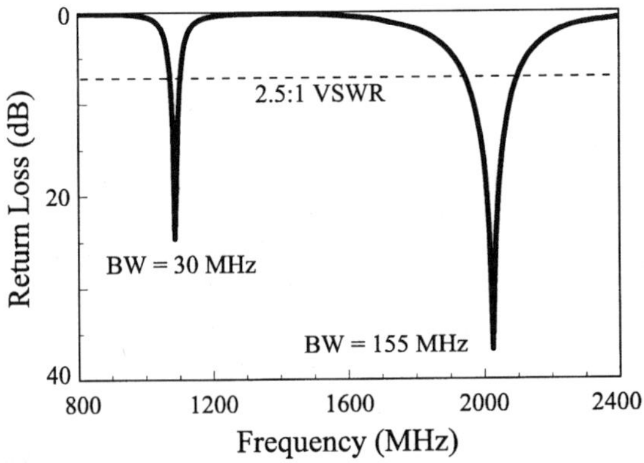

FIGURE 2.16 Measured return loss of the PIFA shown in Figure 2.15.

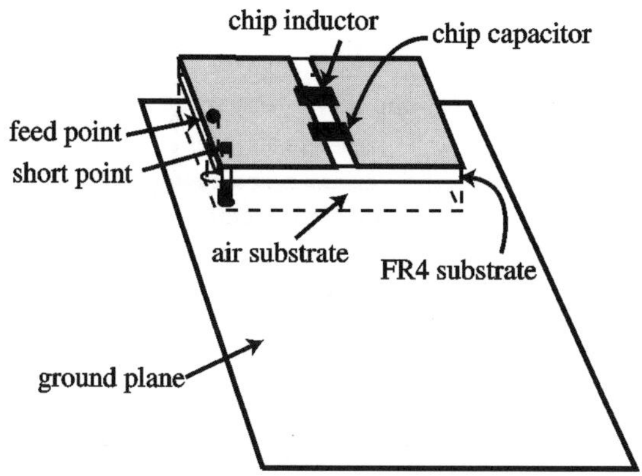

FIGURE 2.17 Geometry of the dual-frequency PIFA using a rectangular patch loaded with a chip inductor and a chip capacitor.

obtain dual-frequency operation in the 900 and 1800 MHz bands [13]. Figure 2.17 shows a typical design of the dual-frequency PIFA using a rectangular patch loaded with a chip inductor and a chip capacitor. By selecting suitable values of the chip inductor and chip capacitor, the fundamental resonant mode of the PIFA can be split into two separate resonant modes and dual-frequency operation in the 900 and 1800 MHz bands can be obtained. With a design example shown in Reference 13 using an 11-nH chip inductor and a 0.5-pF chip capacitor, the 900 and 1800 MHz dual-frequency operation has been obtained. The PIFA has a volume of about $6 \times 20 \times 30\,\text{mm}^3$, and the obtained impedance bandwidths (10 dB return loss) are about 35 and 140 MHz for the 900 and 1800 MHz bands, respectively.

2.2.9 Rectangular Spiral Strip with a Chip Inductor

Instead of using a chip inductor and a chip capacitor as described in Section 2.2.8, a simpler method of using a single chip inductor has been successfully applied to obtain dual-frequency operation of the PIFA [14]. It has been shown that, by loading a chip inductor of suitable inductance (about 12 nH or less) to the PIFA with a rectangular spiral strip (see the geometry shown in Fig. 2.18), the frequency ratio of its first two resonant frequencies can be controlled, which makes the dual-frequency operation with a desired frequency ratio easy to achieve. In addition, owing to the use of a rectangular spiral strip instead of the conventional rectangular patch, the PIFA's top patch can occupy a much reduced area for a fixed operating frequency. That is, a compact dual-frequency operation for the PIFA can be obtained.

Prototypes of the proposed dual-frequency PIFA to achieve the 900 and 1800 MHz operations have been constructed and tested. The design dimensions of the constructed prototypes, obtained with the aid of the simulation software Ansoft HFSS, are shown in Figure 2.18. The radiating element is a shorted rectangular spiral strip of constant width 2 mm and occupies an area of $10 \times 26\,\text{mm}^2$. The spiral

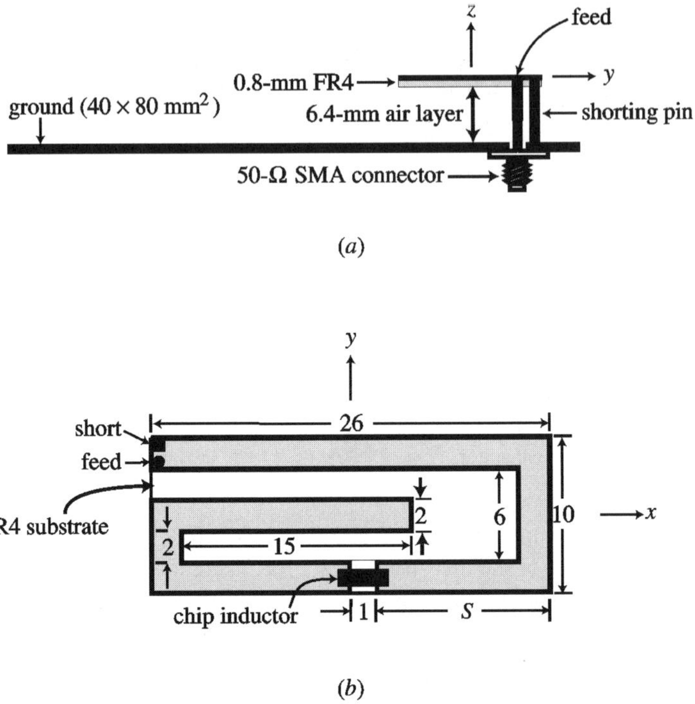

FIGURE 2.18 Geometry of the dual-frequency PIFA using a rectangular spiral strip with a chip inductor. (a) Side view of the PIFA mounted on the top portion of a ground plane of dimensions $40 \times 80\,mm^2$; (b) top patch of the PIFA. (From Ref. 14, © 2002 John Wiley & Sons, Inc.)

strip was printed on an FR4 substrate (thickness 0.8 mm and relative permittivity 4.4) and supported by plastic posts (not shown in the figure) of length 6.4 mm above the ground plane. Thus the total height of the proposed PIFA is 7.2 mm from the ground plane. Also note that the proposed PIFA occupies a compact area of $10 \times 26\,mm^2$.

At the outer end of the spiral strip are the feed and the shorting pin, and the total length from the open end to the inner end is about 77 mm ($= \ell_1$) in the proposed design. This length is close to one-quarter wavelength at 900 MHz, and generates a resonant mode for the lower operating frequency of the proposed dual-frequency operation. To control the upper operating frequency, a gap of 1 mm is cut in the spiral strip and a chip inductor is added as a connecting bridge across the gap. It is then found that, by varying the position of the chip inductor or the gap (S shown in Fig. 2.18b), the upper operating frequency is varied and can be controlled. In the proposed PIFA, the length from the strip's open end to the chip inductor position is about $(32 + S)$ mm ($= \ell_2$), which is also designed to be about one-quarter wavelength of the desired upper operating frequency. However, it should be noted that the inductance of the loading chip inductor also has a large effect on generating the upper resonance of the proposed PIFA. In general, when a chip inductor of suitable inductance is used, it will perform like a short circuit for lower operating frequencies and, conversely, like an open circuit for higher operating frequencies. This behavior makes the lower and upper operating frequencies controlled, respectively, by the lengths 77 mm (the total strip length in this design) and $(32+S)$ mm (the length from the feed to the chip inductor or the gap) in the proposed PIFA.

Figure 2.19 shows the measured and simulated return loss of the proposed PIFA with various chip inductor positions of $S=4$, 13, and 23 mm. The loading chip

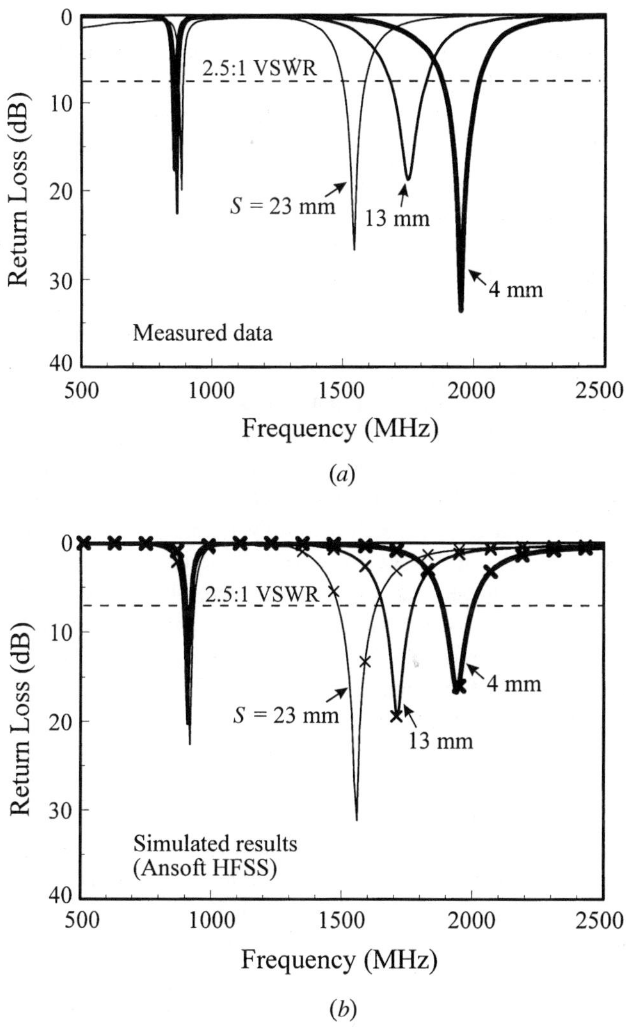

FIGURE 2.19 (a) Measured and (b) simulated return loss as a function of chip inductor position for the PIFA shown in Figure 2.18. (From Ref. 14, © 2002 John Wiley & Sons, Inc.)

TABLE 2.1 Performance of the Proposed Dual-band PIFA Shown in Figure 2.18 with a 12 nH Chip Inductor [14]. Impedance Bandwidth (BW) is Determined From 2.5:1 VSWR.

	S (mm)	ℓ_1 (mm)	ℓ_2 (mm)	f_1, BW (MHz, MHz)	f_2, BW (MHz, MHz)	f_2/f_1
Antenna A	4	77	36	860, 14	1952, 116	2.27
Antenna B	13	77	45	865, 14	1747, 145	2.02
Antenna C	23	77	55	884, 18	1542, 115	1.74

inductor has an inductance of 12 nH. It is first seen that the measured data in general agree with the simulated results. The measured data are also given in Table 2.1 for comparison. It is seen that the lower operating frequencies (f_1) are all close to 900 MHz, as designed, and very slight variations are observed for the three cases studied. Conversely, the upper operating frequency (f_2) decreases with an increase in S. This is because increasing the length S increases the length ℓ_2 from the feed to the gap, and thus the resonant frequency controlled by this length decreases. In this study, as S is varied from 4 to 23 mm, the obtained frequency ratios (f_2/f_1) of the two operating frequencies are in a range from about 2.27 to 1.74. Also note that antenna B ($S = 13$ mm) has a frequency ratio of about 2.0 and its upper operating frequency is close to 1800 MHz. The obtained impedance bandwidths, determined from 2.5:1 VSWR or about 7.3 dB return loss, are, respectively, 14 and 145 MHz for the lower and upper bands of antenna B.

Figures 2.20 and 2.21 plot the measured radiation patterns of antenna B at the lower and upper resonant frequencies, respectively. In the azimuthal plane (x–z plane), the radiation patterns for both operating frequencies are close to omnidirectional, especially for the 900 MHz operation. Radiation patterns at other operating frequencies across the lower and upper bands were also measured and are similar to those shown here; that is, the radiation patterns are stable across the respective operating bands. Measured antenna gain for operating frequencies across the lower and upper bands of antenna B is shown in Figure 2.22. For the lower band, the antenna gain level is about 1.3 dBi. For the upper band, the peak antenna gain is about 1.7 dBi and gain variations are less than 1 dBi across the band.

Effects of the loading chip inductors with various inductances on the performance of the proposed PIFA were also experimentally studied. Table 2.2 lists the measured lower and upper resonant frequencies for antenna B as a function of the chip inductor with various inductances. First note that the position of the loading chip inductor is fixed ($S = 13$ mm) in this study. For the case with $L = 0$, it represents the short-circuiting condition; that is, there is no gap cut in the spiral strip. The obtained results also show relatively small variations in the lower operating frequency. On the other hand, the upper operating frequency or the antenna's second resonant frequency decreases from 2222 to 1454 MHz, when the inductance L increases from 0 to 22 nH. Note that, for the case with $L = 22$ nH, although there is a resonance at 758 MHz, the obtained return loss is worse than 7.3 dB, and thus

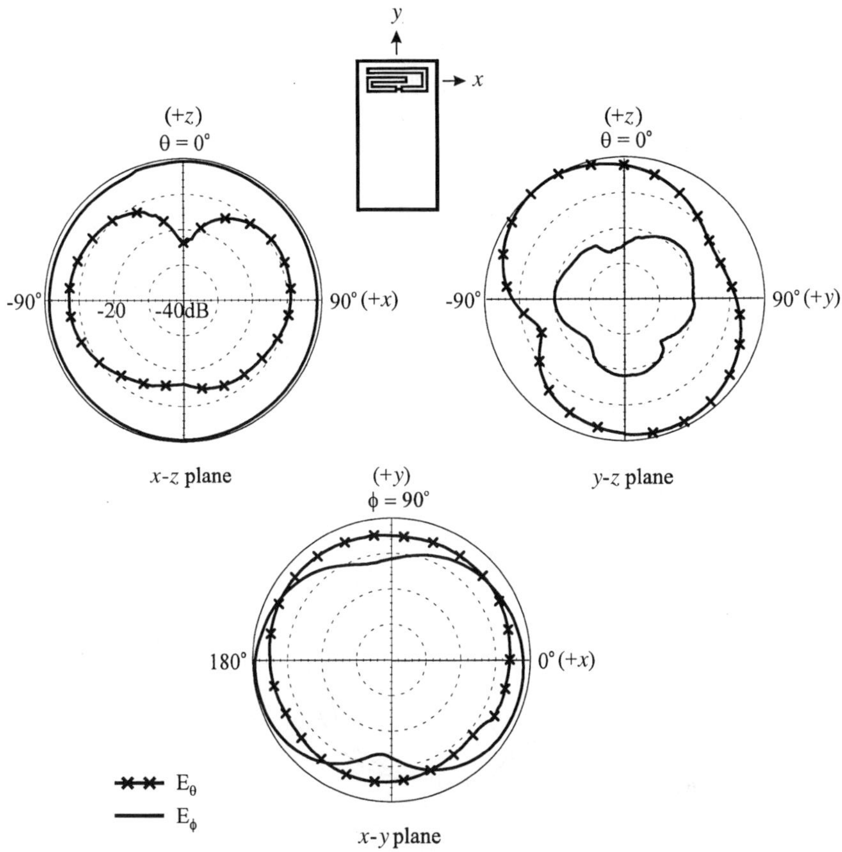

FIGURE 2.20 Measured radiation patterns at 865 MHz for the PIFA with $S = 13$ mm studied in Figure 2.19. (From Ref. 14, © 2002 John Wiley & Sons, Inc.)

no impedance bandwidth is given in Table 2.2. When L is larger than 22 nH even worse impedance matching is obtained for f_1, and the results are therefore not shown. The results obtained in Tables 2.1 and 2.2 suggest that both the position and the inductance of the loading chip inductor can effectively control the occurrence of the upper operating frequency of the proposed PIFA, with the lower operating frequency relatively very slightly affected. Thus, by suitably choosing the position and inductance of the loading chip inductor, dual-frequency operation with a wide frequency ratio range can be obtained.

2.2.10 Rectangular Spiral Strip with Nonuniform Widths

It has also been found that, by selecting different widths in different sections of a shorted rectangular spiral strip, dual-frequency operation in the 900 and 1800 MHz bands can be obtained. Figure 2.23 shows the geometry of such a

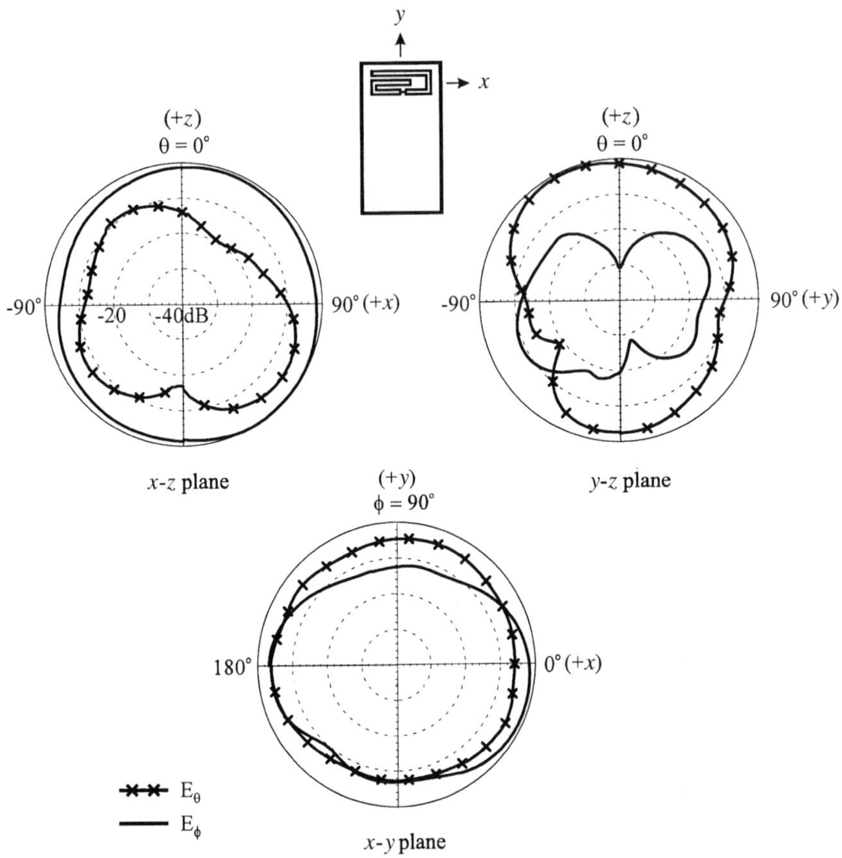

FIGURE 2.21 Measured radiation patterns at 1747 MHz for the PIFA with $S = 13$ mm studied in Figure 2.19. (From Ref. 14, © 2002 John Wiley & Sons, Inc.)

dual-frequency PIFA. The rectangular spiral strip used in this design has a structure similar to that shown in Figure 2.18, in which the strip width is a constant and dual-frequency operation is obtained by loading a chip inductor. In this design, however, the strip width is nonuniform, and each straight section (there are 6 straight sections in the design shown in Fig. 2.23) of the rectangular spiral strip has different strip widths. In general, to achieve better impedance bandwidth, it is preferred that the strip width increases from the first straight section at the outer open end, where the feed and shorting points are located, to the last straight section having the inner surrounded open end.

Also note that the two operating frequencies in this design utilize the first two resonant frequencies of the PIFA, which share a common resonant path starting from the feed point near the strip's outer open end to the strip's inner surrounded open end. In addition, the total length of the shorted rectangular spiral strip can be approximately determined from one-quarter wavelength of the lower frequency

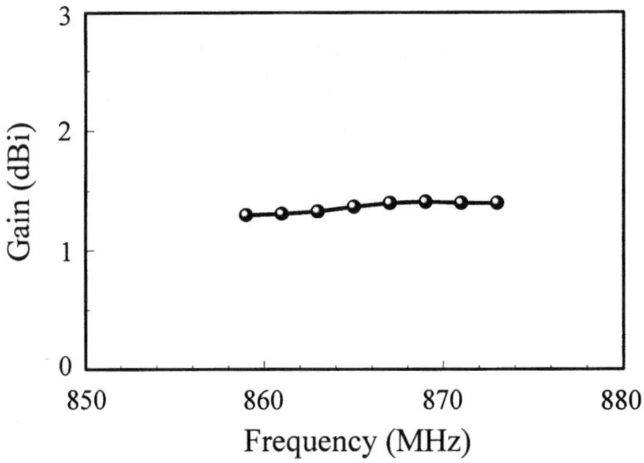

(*a*)

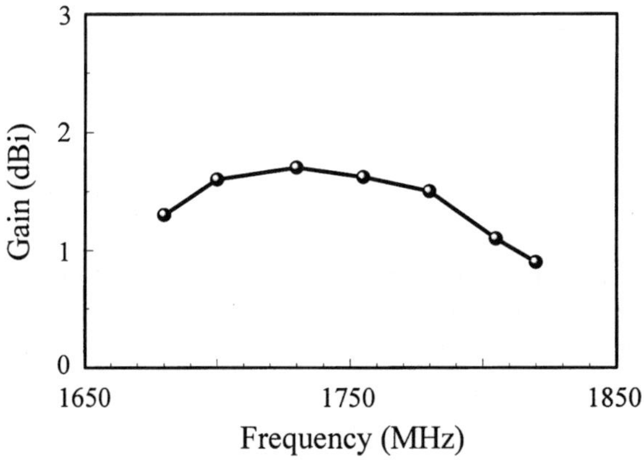

(*b*)

FIGURE 2.22 Measured antenna gain across (a) the lower band and (b) the upper band of the PIFA studied in Figure 2.19. (From Ref. 14, © 2002 John Wiley & Sons, Inc.)

TABLE 2.2 Performance of Antenna B Shown in Table 2.1 with Various Chip Inductors [14].

L (nH)	f_1, BW (MHz, MHz)	f_2, BW (MHz, MHz)	f_2/f_1
0(short-circuit)	992, 20	2222, 112	2.24
1.5	980, 22	2138, 116	2.18
3.3	962, 18	2048, 112	2.13
6.8	914, 18	1910, 128	2.09
12	865, 14	1747, 145	2.02
22	758, —	1454, 94	1.92

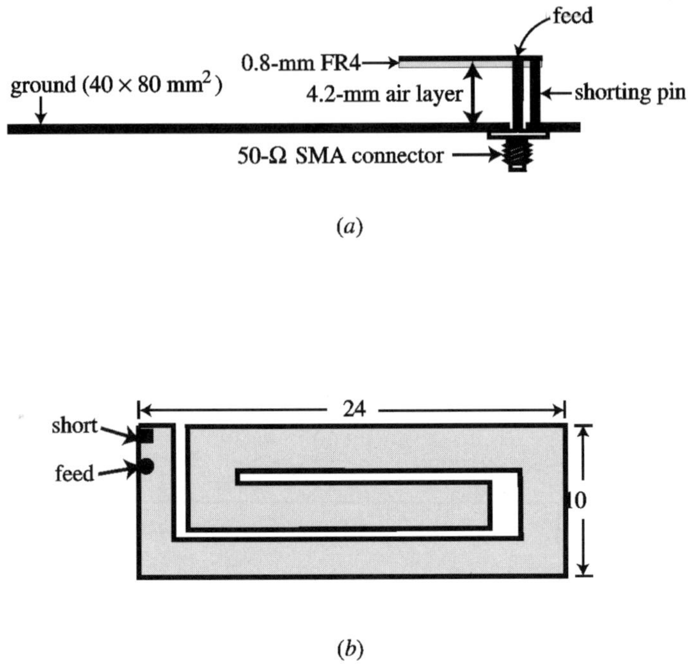

FIGURE 2.23 Geometry of the dual-frequency PIFA using a rectangular spiral strip of nonuniform widths. (a) Side view of the PIFA mounted on the top portion of a ground plane of dimensions $40 \times 80\,\text{mm}^2$; (b) top patch of the PIFA.

(900 MHz in this design) of the desired dual-frequency operation. A prototype of the proposed PIFA for 900 and 1800 MHz dual-frequency operations has also been implemented and tested. The constructed prototype has a height of 5 mm and occupies an area of only $10 \times 24\,\text{mm}^2$. Figure 2.24 shows the measured and simulated return loss. Reasonable agreement of the measurement and the simulation is obtained, and two separate resonant modes near 900 and 1800 MHz are successfully excited. Because the constructed prototype has compact dimensions ($5 \times 10 \times 24\,\text{mm}^3$), the proposed PIFA design is a very attractive candidate for practical applications in mobile phones that have a very limited realty space for the antenna.

2.3 TRIPLE-FREQUENCY PIFA

2.3.1 Two or Three Separate Patches

Figure 2.25 shows two promising configurations of the top patch of the triple-frequency PIFA for the 900, 1800, and 2450 MHz operations. In Figure 2.25a, the top patch comprises three separate shorted patches with a triple feed; patch 1 is

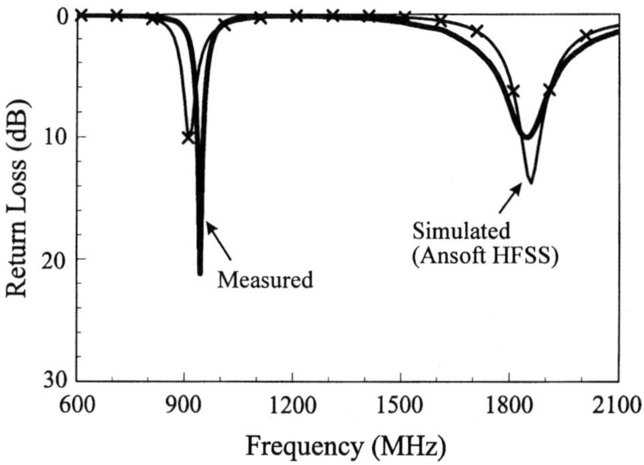

FIGURE 2.24 Measured and simulated return loss of the PIFA shown in Figure 2.23.

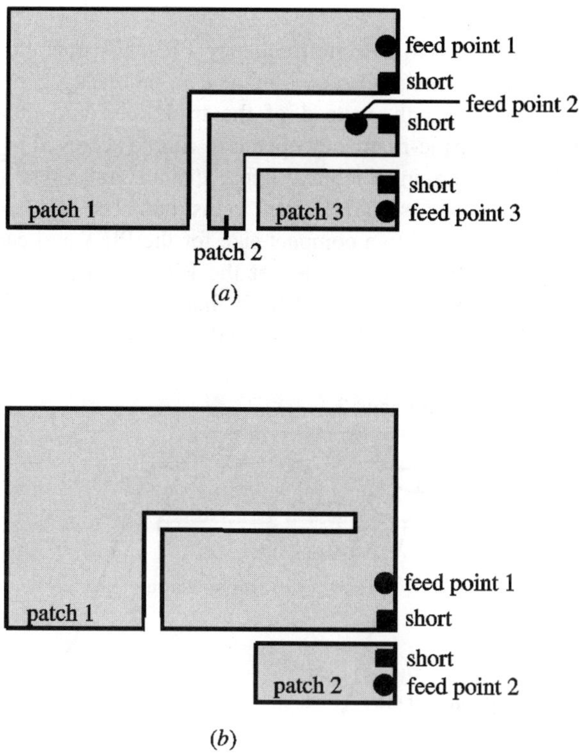

FIGURE 2.25 Top patch of the triple-frequency PIFA comprising (a) three separate patches with a triple feed and (b) two separate patches with a dual feed for achieving 900, 1800, and 2450 MHz operations.

for GSM operation in the 900 MHz band, patch 2 is for DCS operation in the 1800 MHz band, and patch 3 is for WLAN operation in the 2450 MHz band. Notice that these three patches are arranged in a compact configuration and have separate shorting and feed points. These three patches are all operated as quarter-wavelength resonant structures. Such a PIFA design with a triple feed can find applications where mobile phones require separate feeds for operations in GSM, DCS, and WLAN bands.

For the configuration shown in Figure 2.25b, the top patch consists of two separate patches, with the larger patch (patch 1 in the figure) loaded with an L-shaped slit for GSM and DCS dual-frequency operations (see the discussion in Section 2.2.2). In this design, the smaller patch (patch 2) is for WLAN operation in the 2045 MHz band and is also operated as a quarter-wavelength resonant structure. The two patches utilize different shorting and feed points, and such a design is suitable for mobile phones using the same feed for GSM and DCS operations and a second feed for WLAN operation.

2.3.2 A Meandered Patch

By using a meandered patch, a triple-frequency PIFA for operations in the 900, 1800, and 2450 MHz bands can also be obtained. A promising geometry is shown in Figure 2.26. In this design, instead of the triple-feed and dual-feed designs shown in Section 2.3.1, triple-frequency operation can be achieved by using a single feed. A related PIFA design, in which two linear slits are embedded in the top patch to meander the excited patch surface current paths, has been reported in Reference 16. This meandering can lead to a compact size for the PIFA and can also strongly affect the first three resonant frequencies of the PIFA. Thus, by choosing proper dimensions of the two linear slits, the PIFA's first three resonant frequencies can be adjusted to be about 900, 1800, and 2450 MHz, respectively. Also note that, to

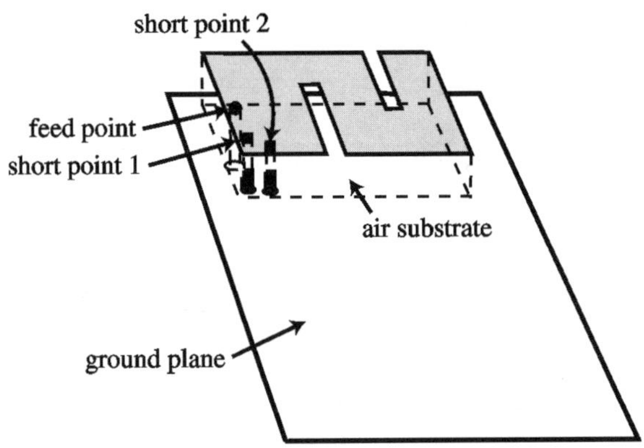

FIGURE 2.26 Geometry of the triple-frequency PIFA with a meandered patch for achieving 900, 1800, and 2450 MHz operations.

achieve good impedance matching of the three excited resonant modes, the use of two shorting pins, instead of only one, is suggested [16].

2.3.3 A Branch Line Strip with a Folded End

Recently, a triple-frequency PIFA with two shorted branch strips sharing a common shorting point and having a same feed point has been devised [17]. The geometry of this PIFA mounted on the top portion of a ground plane of dimensions $30 \times 80\,\text{mm}^2$ is shown in Figure 2.27. The PIFA consists of two shorted branch strips: the longer one (strip 1) controls the excitation of the resonant frequencies at about 900 and 1800 MHz, and the shorter one (strip 2) controls the excitation of the resonant

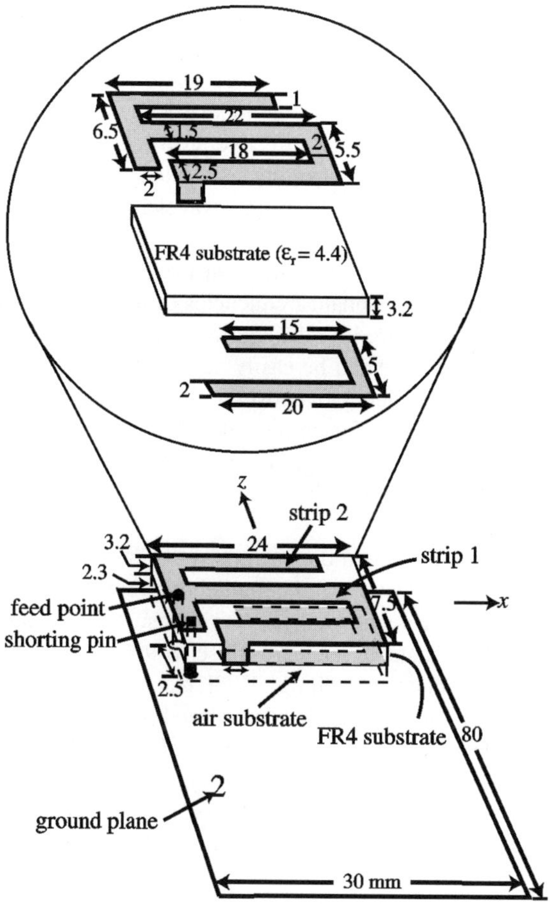

FIGURE 2.27 Geometry of the triple-frequency PIFA using a branch line strip with a folded end to achieve 900, 1800, and 2450 MHz operations. (From Ref. 17, © 2002 John Wiley & Sons, Inc.)

frequency at about 2450 MHz. The two branch strips are printed on an FR4 substrate of thickness 3.2 mm, area $7.5 \times 24\ mm^2$, and relative permittivity 4.4. Between the FR4 substrate and the ground plane is an air substrate of thickness 2.3 mm (plastic posts for supporting the FR4 substrate above the ground plane are not shown in the figure), which makes the total antenna height from the ground plane 5.5 mm. That is, the total antenna volume is $5.5 \times 7.5 \times 24\ mm^3$ or only about $1\ cm^3$. The design dimensions shown in the figure were obtained with the aid of the commercially available simulation software Ansoft HFSS.

Note that the shorter strip has a length of 24 mm (starting from the shorting pin), which corresponds to about 20% of the wavelength at 2450 MHz. This length is less than one-quarter wavelength, and this behavior is largely due to the effect of the thick FR4 substrate used, which results in a larger effective relative permittivity of the combined FR4 and air substrates between the shorted strip and the ground plane. The longer strip has a length of about 87 mm, corresponding to about 26% of the wavelength at 900 MHz. Note that the longer strip is meandered and printed on both sides of the FR4 substrate, which effectively reduces the PIFA. However, probably because of the meandering, which increases the possible coupling between two adjacent meandered sections, the effective strip length is decreased; that is, a larger strip length is required for operating at a fixed frequency. This is probably the reason that the strip length in operating wavelength is larger for the longer strip than for the shorter strip.

Also note that the proposed PIFA operates at the first resonant frequency of the shorted strip and the first two resonant frequencies of the longer strip. Because only one resonant frequency is required for the shorter strip, the adjusting of this resonant frequency to be the desired one can be easily achieved by fine-tuning the length of the shorter strip. On the other hand, to adjust the first two resonant frequencies to the desired frequencies, the widths of each meandered sections of the longer strip (see the design dimensions given in Fig. 2.27) are chosen to be different values, which are very effective in adjusting the frequency ratio of the longer strip's first two resonant frequencies.

A prototype based on the design dimensions shown was constructed and studied. Figure 2.28 shows the measured and simulated return loss of the constructed prototype. Very good agreement between the measured data and the simulated results is seen. Three resonant frequencies close to 900, 1800, and 2450 MHz are excited with good impedance matching. With respect to their respective center frequency, the three frequency bands, determined from 2.5:1 VSWR or about 7.3 dB return loss, have impedance bandwidths of about 18 MHz (about 1.9%), 85 MHz (about 4.8%), and 70 MHz (about 2.9%), respectively. Note that, although the proposed PIFA occupies a much smaller volume (less than 10%) than that in References 15 and 16, no large decreases in the obtained impedance bandwidths are seen.

The radiation characteristics were also studied. Figures 2.29–2.31 plot the measured radiation patterns at the center frequency of the three frequency bands. For 950 MHz (Fig. 2.29), the measured radiation pattern in the azimuthal plane (x–y plane) is close to omnidirectional radiation and the measured antenna gain is

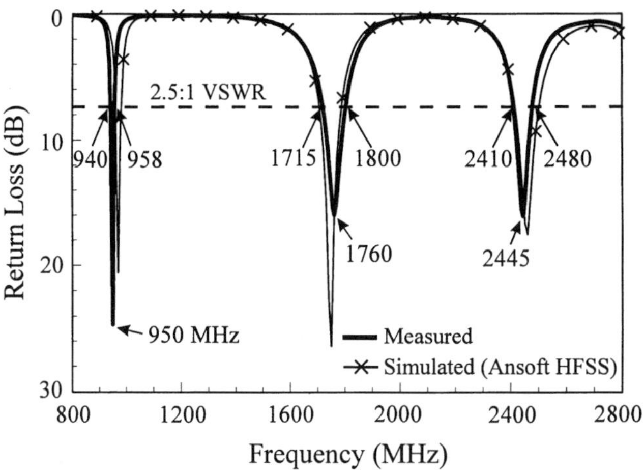

FIGURE 2.28 Measured and simulated return loss for the PIFA shown in Figure 2.27. (From Ref. 17, © 2002 John Wiley & Sons, Inc.)

about 0 dBi. For 1760 MHz (Fig. 2.30), the measured radiation pattern in the azimuthal plane (x–y plane) is less close to omnidirectional radiation than that for 950 MHz. However, the measured antenna gain is higher and is about 2.2 dBi. In the elevation plane of y–z plane, the forward radiation ($-y$ direction) is seen to be much greater than the backward radiation ($+y$ direction or behind the ground plane). The measured radiation patterns for 2450 MHz are plotted in Figure 2.31, and the measured antenna gain is about 1.5 dBi. Also note that the radiation characteristics of the operating frequencies across each frequency band were also measured, and the radiation patterns are stable in each frequency band.

2.4 PIFA WITH AN L-SHAPED GROUND

2.4.1 Patch PIFA Design

Instead of using a simple flat ground plane, a PIFA with an L-shaped ground plane has also been studied. It has been found that such a PIFA can have decreased backward radiation and enhanced antenna gain. A related study is reported in Reference 18. Figure 2.32a shows the geometry of a conventional PIFA with a simple flat ground plane for an internal mobile phone antenna, and the proposed PIFA with an L-shaped ground plane is presented in Figure 2.32b. In both geometries, the simple flat ground plane in Figure 2.32a and the horizontal ground plane in Figure 2.32b have the same dimensions of $60 \times 40\,\text{mm}^2$. An additional vertical ground is added to the edge of the horizontal ground plane in the proposed design to enhance the antenna performance. In this case, the vertical ground and the horizontal ground plane form an L-shaped ground plane for the proposed design.

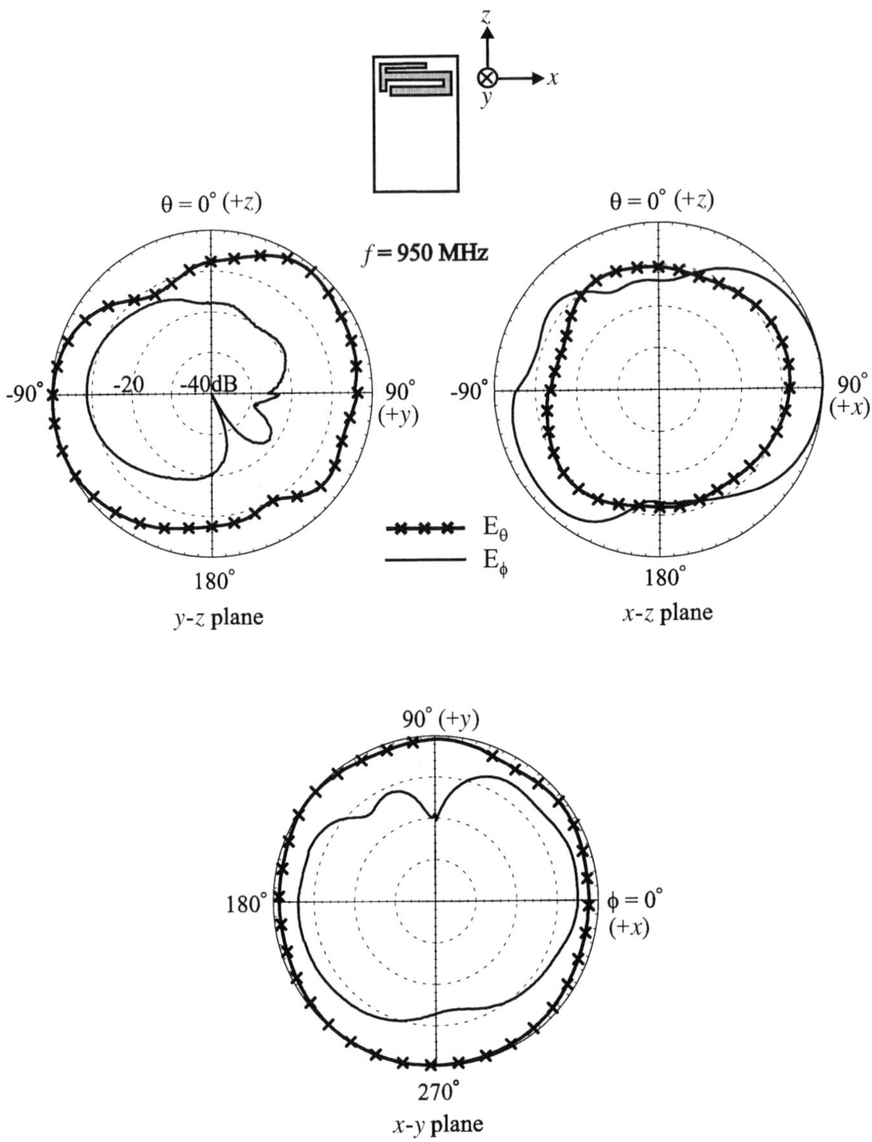

FIGURE 2.29 Measured radiation patterns at 950 MHz for the PIFA studied in Figure 2.28. (From Ref. 17, © 2002 John Wiley & Sons, Inc.)

In addition, because of the additional vertical ground, the coax feed can be placed in the same plane as the radiating patch as seen in Figure 2.32b or below the horizontal ground plane as seen in Figure 2.32a, which gives the proposed design flexible feed arrangements.

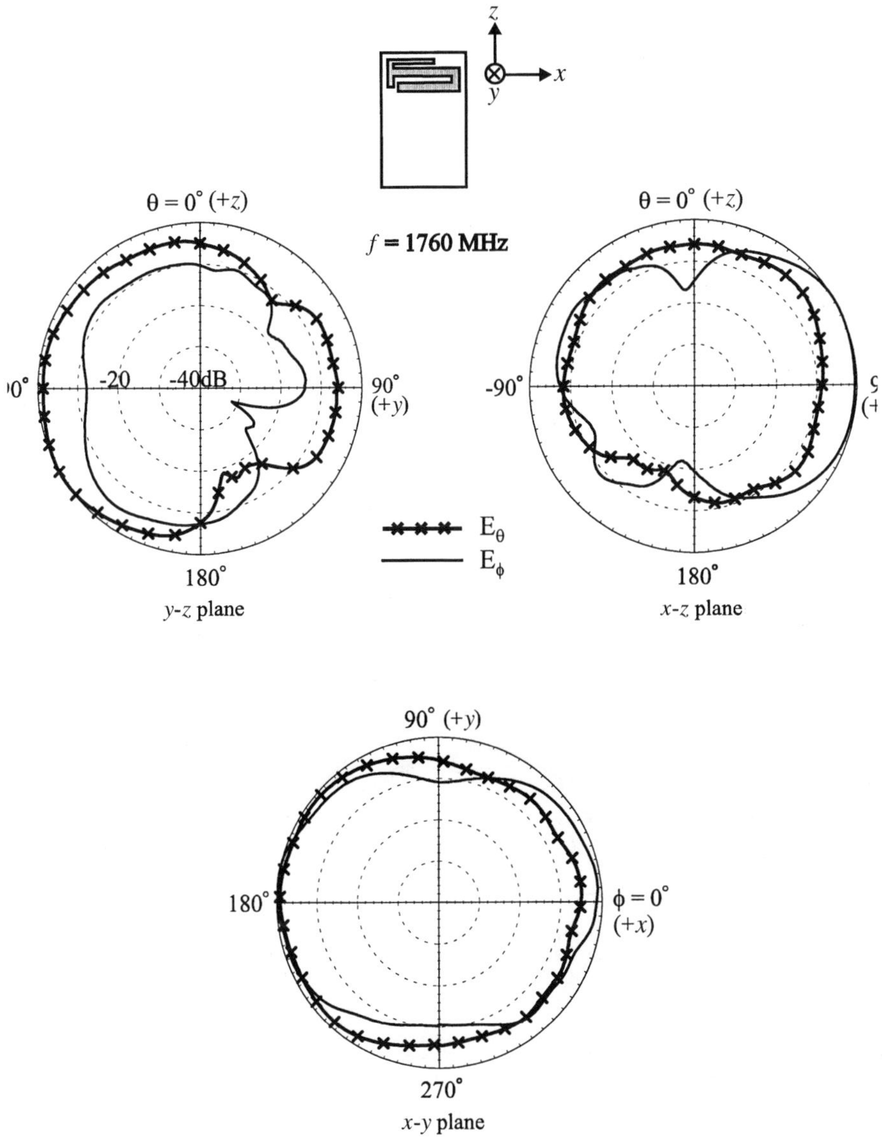

FIGURE 2.30 Measured radiation patterns at 1760 MHz for the PIFA studied in Figure 2.28. (From Ref. 17, © 2002 John Wiley & Sons, Inc.)

It should be noted that the dimensions given in Figure 2.32b are for achieving UMTS band operation. The PIFA is fed at the center of the edge of the top patch, which has dimensions $30 \times 15\,\text{mm}^2$ and is placed above the horizontal ground plane with a height of 7 mm. The top patch is also placed close to the vertical ground to reduce the required probe pin length of the coplanar coax feed, which makes

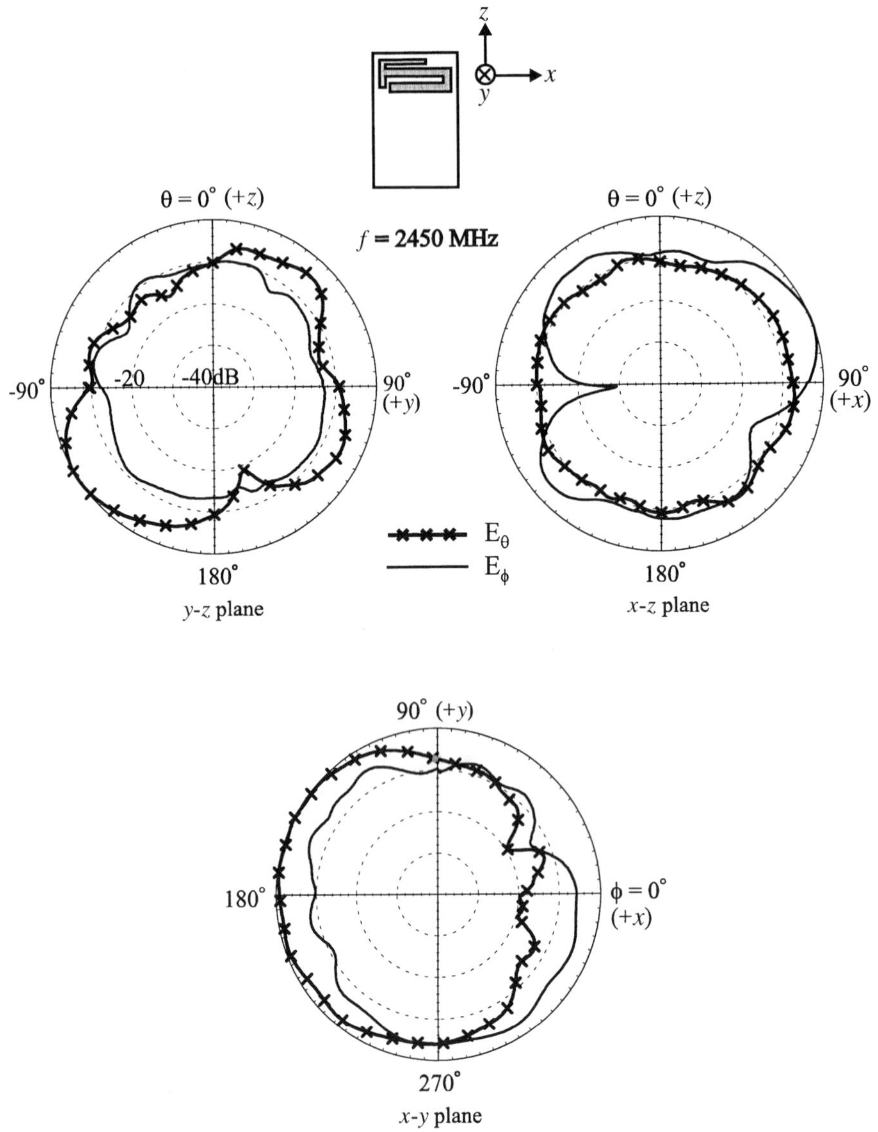

FIGURE 2.31 Measured radiation patterns at 2450 MHz for the PIFA studied in Figure 2.28. (From Ref. 17, © 2002 John Wiley & Sons, Inc.)

impedance matching of the proposed PIFA over a wide bandwidth easy to achieve. In this design, the distance between the vertical ground and the patch edge is 2 mm and the shorting point is 1 mm away from the feed point. Also note that, for the corresponding conventional PIFA shown in Figure 2.32a, all the dimensions are

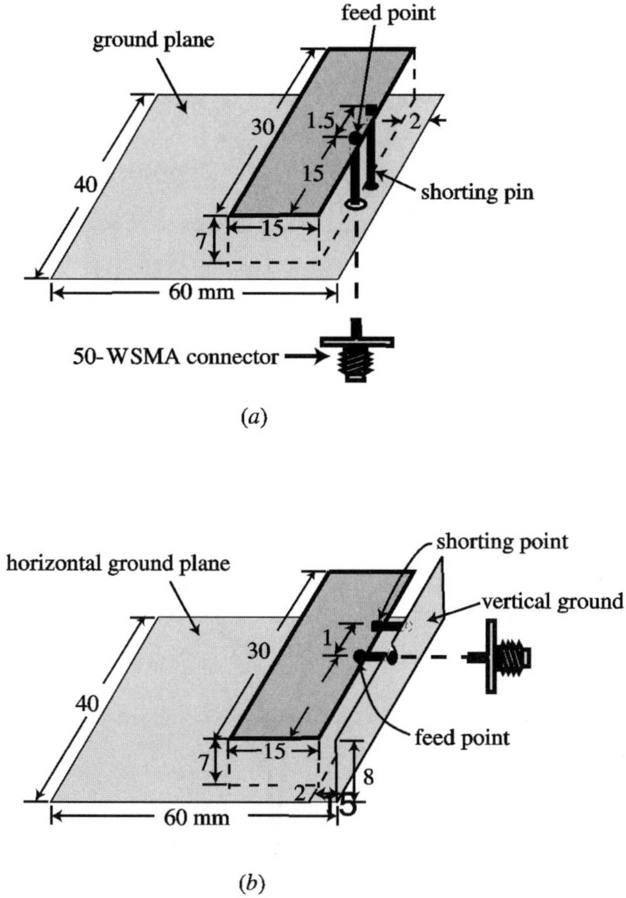

FIGURE 2.32 Geometries of the PIFA with (a) a simple ground plane (reference 1) and (b) an L-shaped ground plane. (From Ref. 18, © 2002 John Wiley & Sons, Inc.)

the same as those given in Figure 2.32b except that the distance between the shorting point and the feed point is 1.5 mm to achieve good impedance matching.

Figure 2.33 shows the measured and simulated return loss for the corresponding conventional PIFA (references 1 and 2) and the proposed PIFA. Good agreement between the measurement and the simulation for references 1 and 2 shown in Figure 2.33a is obtained. For the results of the proposed PIFA in Figure 2.33b, reasonable agreement between the measured data and the simulated results is also obtained. It is seen that the proposed PIFA has a wide impedance bandwidth (10 dB return loss) of 310 MHz (1920–2230 MHz), which covers the required operating bandwidth of the UMTS band. It should also be noted that, although the patch size is the same, the obtained bandwidth of the proposed PIFA has higher frequencies than those of the corresponding conventional design of reference 1. By decreasing the patch

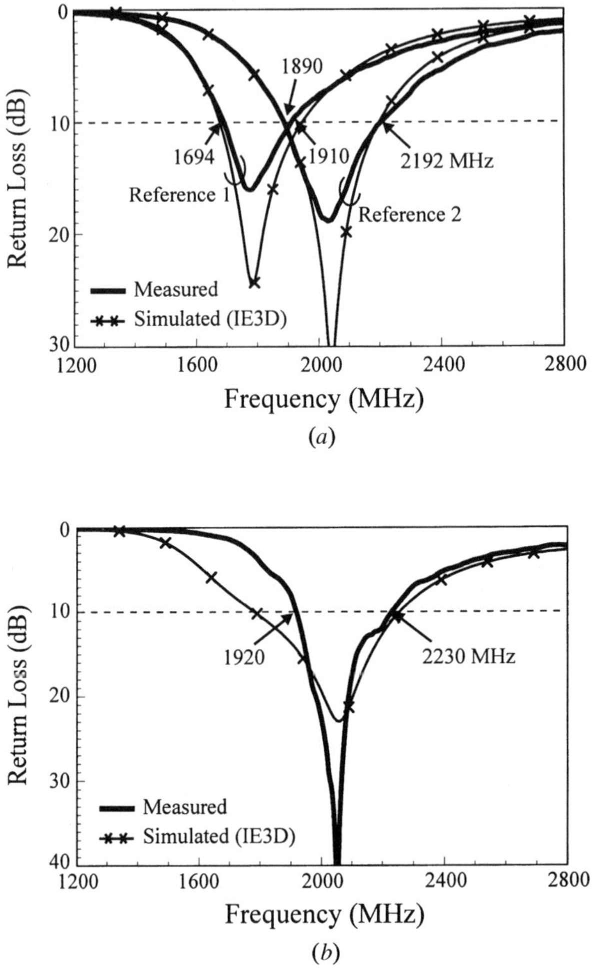

FIGURE 2.33 Measured and simulated return loss against frequency. (a) The conventional design [reference 1: patch size $15 \times 30\,\text{mm}^2$ as shown in Figure 2.32a; reference 2: patch size $13 \times 26\,\text{mm}^2$; other parameters for both references 1 and 2 are the same and are given in Figure 2.32a]; (b) the proposed design with an L-shaped ground plane [dimensions given in Figure 2.32b]. (From Ref. 18, © 2002 John Wiley & Sons, Inc.)

size from $15 \times 30\,\text{mm}^2$ to $13 \times 26\,\text{mm}^2$, the conventional design of reference 2 has an operating bandwidth at frequencies about the same as those of the proposed PIFA.

Figure 2.34 shows the measured radiation patterns for references 1 and 2 at resonance. No prominent difference in the radiation patterns is observed, although references 1 and 2 have different ground plane size in wavelength. However, with comparison to the measured radiation patterns of the proposed design shown in Figure 2.35, it is clearly seen that, in the azimuthal plane (*x*–*y* plane), the backward

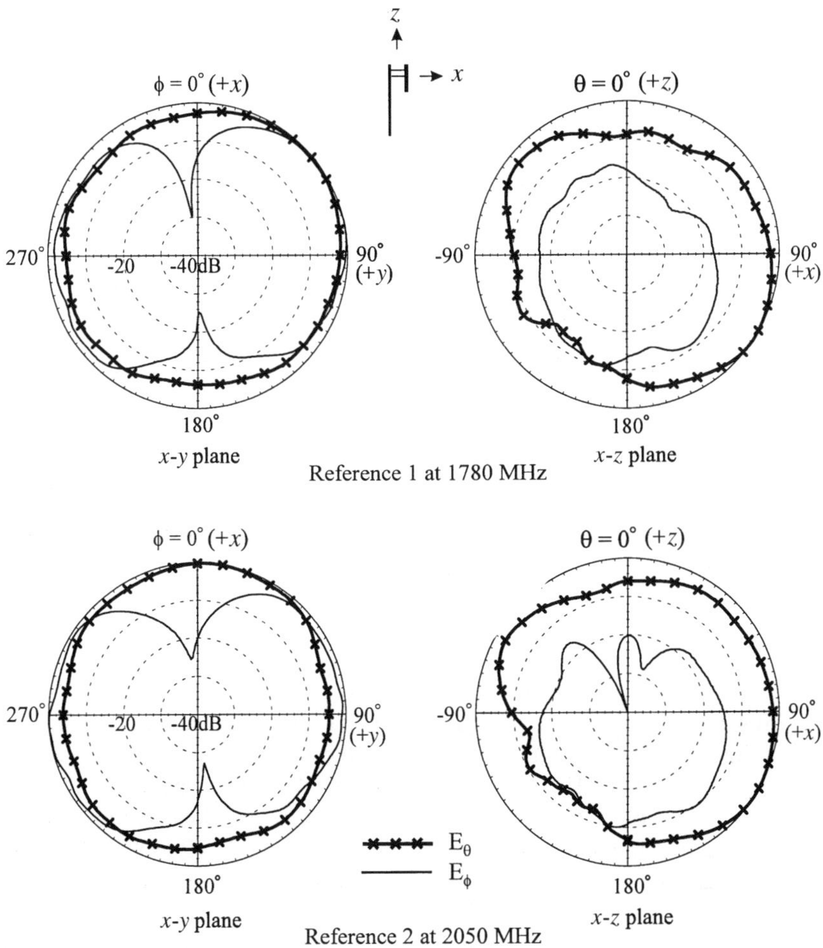

FIGURE 2.34 Measured radiation patterns at resonance for the reference PIFAs studied in Figure 2.33. (From Ref. 18, © 2002 John Wiley & Sons, Inc.)

radiation in the direction of $\varphi = 180°$ is smaller for the proposed PIFA than for the conventional PIFA. For the vertical plane (x–z plane), it is also observed that the radiation behind the ground plane is much smaller for the proposed PIFA than for the conventional PIFA. Figure 2.36 also shows the measured antenna gain for frequencies within the operating bandwidths for the reference PIFAs and the proposed PIFA. References 1 and 2 show about the same antenna gain level, which is lower than that of the proposed PIFA by about 1 dBi. The proposed PIFA has a peak antenna gain of about 3.5 dBi, and the gain variations are within 1.2 dBi for frequencies within the UMTS band.

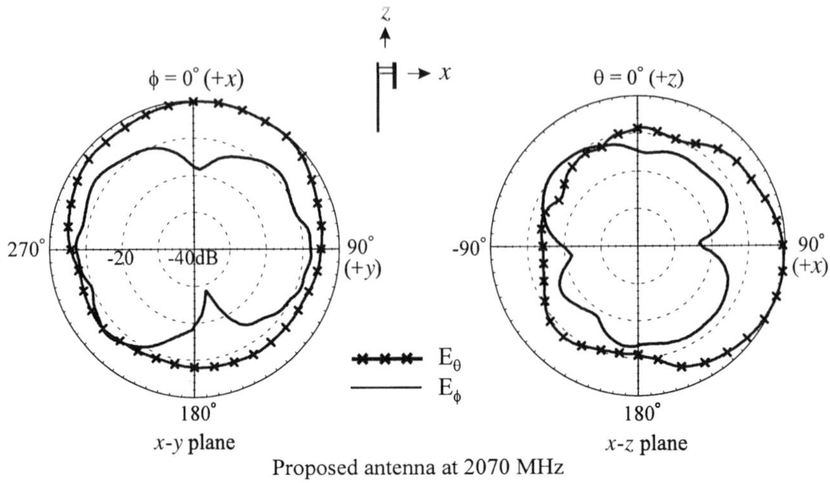

FIGURE 2.35 Measured radiation patterns at resonance for the proposed PIFA studied in Figure 2.33. (From Ref. 18, © 2002 John Wiley & Sons, Inc.)

2.4.2 Wire PIFA Design

A wire PIFA with an L-shaped ground plane has also been studied. A prototype of such a PIFA mounted on the top portion of a UMTS mobile phone ground plane is shown in Figure 2.37. The dimensions given in the figure are for achieving UMTS operation in the 2050 MHz band. The wire PIFA was constructed by using conducting wires with a diameter of 1 mm, and there are three sections in this wire PIFA: one shorting arm, one feed arm, and an inverted-L wire and an L-shaped ground plane comprising a vertical ground and a horizontal ground. The vertical ground, with dimensions $15 \times 30\,\text{mm}^2$, is for reducing the PIFA's backward radiation toward the user's head and the horizontal ground, with dimensions $10 \times 30\,\text{mm}^2$, can reduce the PIFA's radiation toward the associated microwave circuitry of the mobile phone, thereby reducing the electromagnetic interference and enhancing the system performance.

The shorting arm has a length of 9 mm and short-circuits the wire PIFA to the vertical ground. This arrangement can achieve enhanced impedance matching for the proposed wire PIFA. The feed arm has a length of 10 mm and is connected to the feed through a via-hole in the horizontal ground. The inverted-L wire has a total length of 30 mm, and its open end is bent to reduce its linear dimension to 25 mm. Note that it is possible, by further bending the open end of the inverted-L wire toward the feed arm, to obtain a much smaller linear dimension.

A prototype based on the design dimensions given in Figure 2.37 has been constructed and tested. Figure 2.38 shows the measured and simulated return loss for the prototype. A resonant mode is excited with good impedance matching at about 2 GHz, and excellent agreement between the measurement and the simulation is observed. However, the obtained impedance bandwidth (2.5:1 VSWR) is about

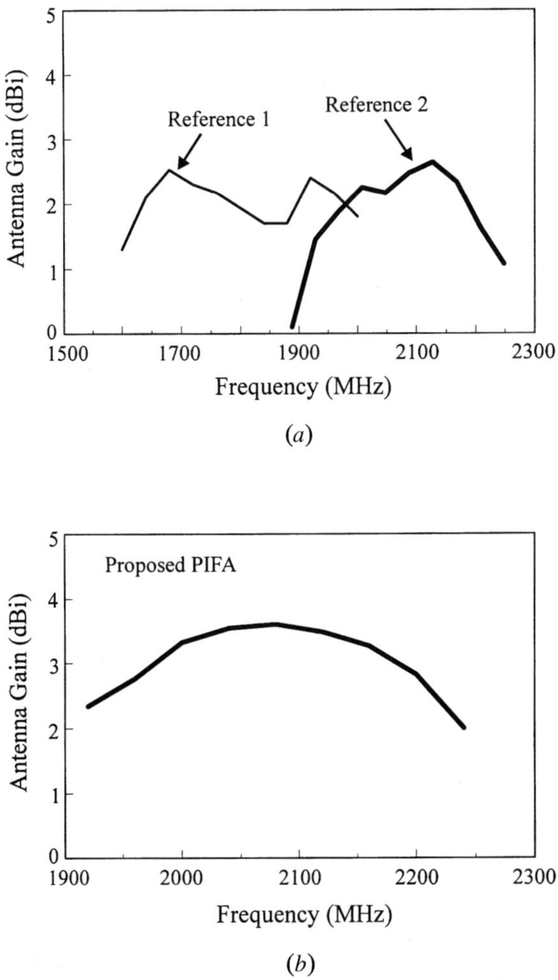

FIGURE 2.36 Measured antenna gain for (a) the reference PIFAs and (b) the proposed PIFA studied in Figure 2.33. (From Ref. 18, © 2002 John Wiley & Sons, Inc.)

150 MHz only, and wider impedance bandwidth is required for UMTS operation (250 MHz). Measured radiation patterns at 2000 MHz are also plotted in Figure 2.39. In the elevation plane (*x–z* plane), it is clearly seen that the forward radiation is much larger than the backward radiation, and the radiation in the upper half plane is lower than that in the lower half plane. In addition, the E_θ and E_φ components are comparable in the three principal planes, which can provide polarization diversity for mobile phones and is an advantage for wireless communications where their wave propagation environment is complex. In this case, the effect of the multipath fading problem is decreased and the system performance can be enhanced. Antenna gain

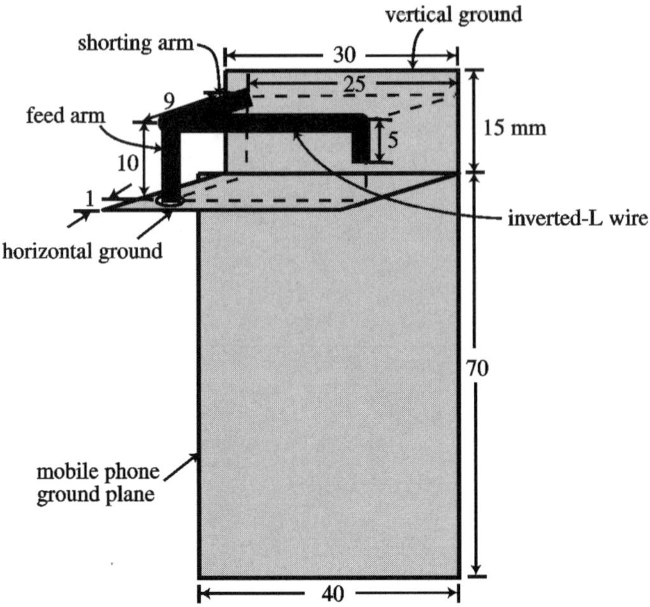

FIGURE 2.37 Geometry of the wire PIFA with an L-shaped ground plane for a UMTS mobile phone.

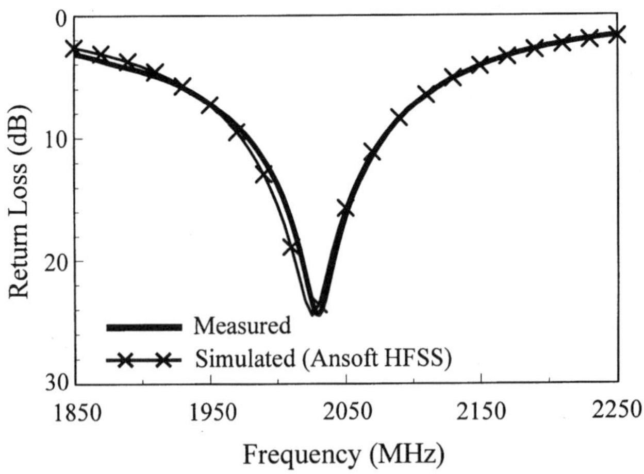

FIGURE 2.38 Measured and simulated return loss for the PIFA shown in Figure 2.37.

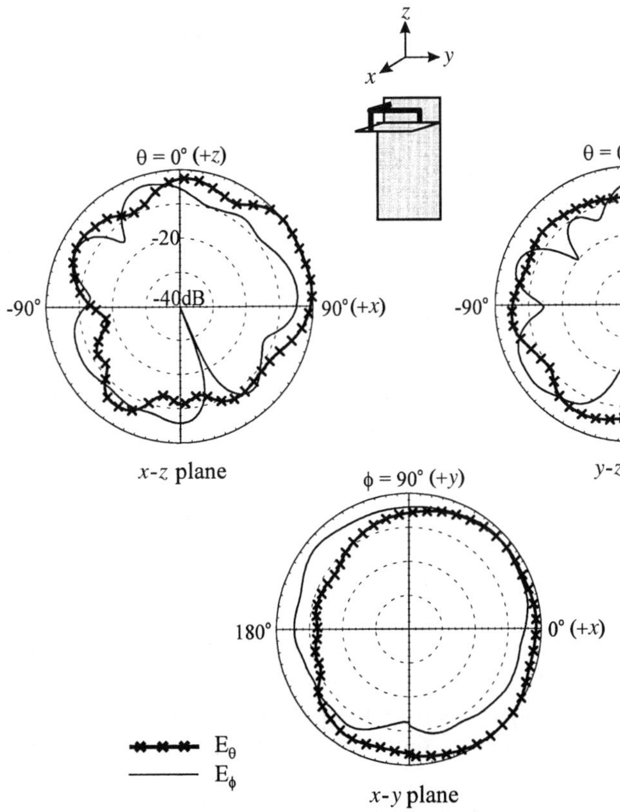

FIGURE 2.39 Measured radiation patterns at 2000 MHz for the PIFA shown in Figure 2.37.

for the constructed prototype was also measured, and a gain level of about 2.5 dBi is obtained.

2.5 STACKED PIFA

By stacking two PIFAs, enhanced impedance bandwidth can be obtained for a fixed antenna volume [19, 20]. This design method is very attractive for applications where realty space is very limited but wide operating bandwidth for the antenna is required. The bandwidth enhancement for this kind of stacked PIFAs is obtained by adjusting the lengths of the two stacked shorted patches to be slightly different, which makes it possible for the excitation of two resonant modes at close frequencies, thereby leading to a wider impedance bandwidth formed by the two closely excited resonant modes.

Figure 2.40 shows the geometry of a stacked PIFA for DCS operation in the 1800 MHz band. To be suitable as a surface mountable device, the lower patch is

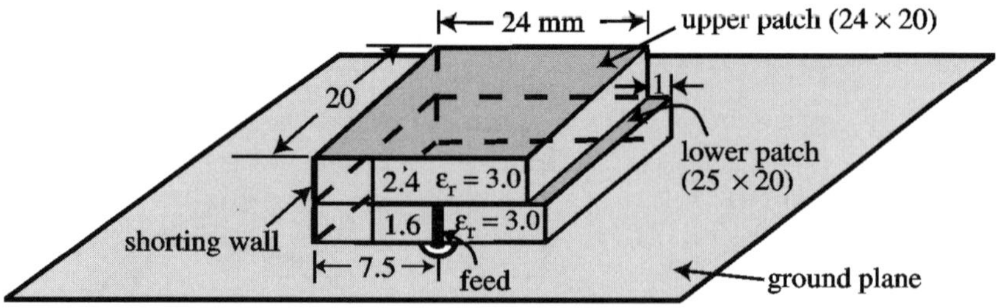

FIGURE 2.40 Geometry of a stacked PIFA for DCS operation in the 1800 MHz band. (From Ref. 20, © 2001 IEEE, reprinted with permission)

directly fed by a probe feed at the patch edge and the upper patch is parasitically coupled from the lower patch. Both patches have a common shorting wall and are printed on a low-loss dielectric substrate of relative permittivity 3.0. The substrate thicknesses for the upper and lower patches are 2.4 and 1.6 mm, respectively; that is, the total antenna height is only 4 mm. The length of the lower patch is 25 mm, whereas that of the upper patch is 24 mm. The feed point is 7.5 mm away from the common shorting wall. Figure 2.41 shows the measured return loss for the design shown in Figure 2.40. It is clearly seen that two resonant modes are excited at close frequencies, and the impedance bandwidth, determined from 10 dB return loss, reaches 202 MHz (1688–1890 MHz) or about 11.2% referenced to 1800 MHz. The obtained bandwidth covers the required bandwidth of the GSM system and is estimated to be about 1.5 times that of a single-layer PIFA occupying the same physical volume.

Figure 2.42 shows a design example of the stacked PIFA for GSM operation in the 900 MHz band. In this design, the lengths of the lower and upper patches are

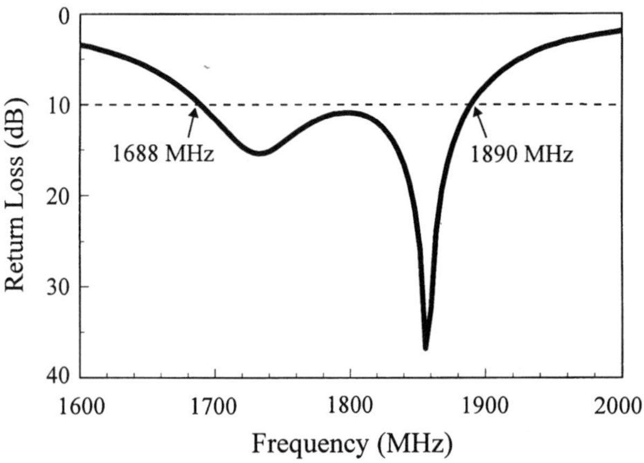

FIGURE 2.41 Measured return loss for the PIFA shown in Figure 2.40. (From Ref. 20, © 2001 IEEE, reprinted with permission)

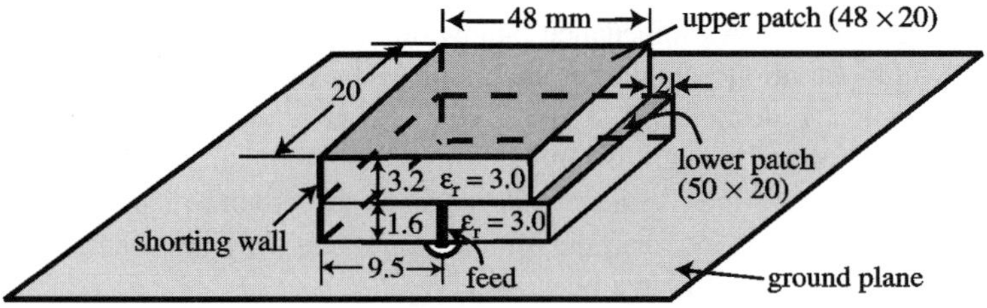

FIGURE 2.42 Geometry of the stacked PIFA for GSM operation in the 900 MHz band. (From Ref. 20, © 2001 IEEE, reprinted with permission)

adjusted to be 50 and 48 mm, respectively. The substrate thickness of the upper patch is also increased to be 3.2 mm, and the feed point is at a distance of 9.5 mm from the common shorting wall. Other design parameters are the same as described in Figure 2.40 for DCS operation. Measured return loss is shown in Figure 2.43, and the obtained impedance bandwidth reaches 94 MHz (868–962 MHz) or about 10.4% referenced to 900 MHz, which covers the required bandwidth of the GSM system. The obtained bandwidth is also about 1.5 times that of a single-layer PIFA occupying the same physical volume.

2.6 GROUND PLANE EFFECT ON IMPEDANCE BANDWIDTH OF THE PIFA

Effects of the ground plane dimensions on the impedance bandwidth of the PIFA have been studied, and it has been found that, by varying either the ground plane

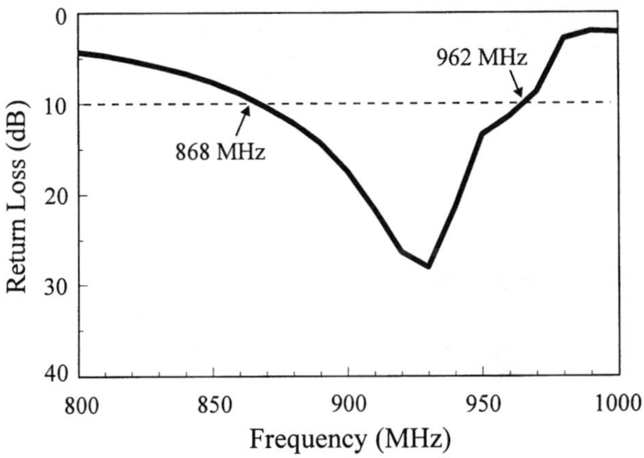

FIGURE 2.43 Measured return loss for the PIFA shown in Figure 2.42. (From Ref. 20, © 2001 IEEE, reprinted with permission)

length or width, the PIFA's impedance bandwidth can be significantly varied [21]. The bandwidth variations are larger than 300 MHz for the PIFA with a center operating frequency designed at about 2 GHz for a UMTS mobile phone, and there exists an optimal ground-plane length for obtaining maximum impedance bandwidth. In addition, there is a critical value for the ground plane width. When the ground plane width is less than the critical value, relatively very sharp variations in the impedance bandwidth have been observed.

Figure 2.44 shows the geometry of the PIFA mounted on the top portion of a ground plane of length L and width W. For applications in the UMTS band, whose center frequency is at 2050 MHz and whose required bandwidth is 250 MHz, the PIFA was chosen to have a length 24 mm and a width 18 mm and has an air-layer substrate of thickness 6.4 mm; that is, the PIFA occupies a volume of $24 \times 18 \times 6.4\ \mathrm{mm}^3$. To excite the antenna, a probe feed is placed at the center of the top patch (it can also be at the edge of the top patch), which is short-circuited to the ground plane with a shorting pin placed close to the feed point with a distance of d. It should be noted that, to achieve optimal impedance matching for the PIFA with

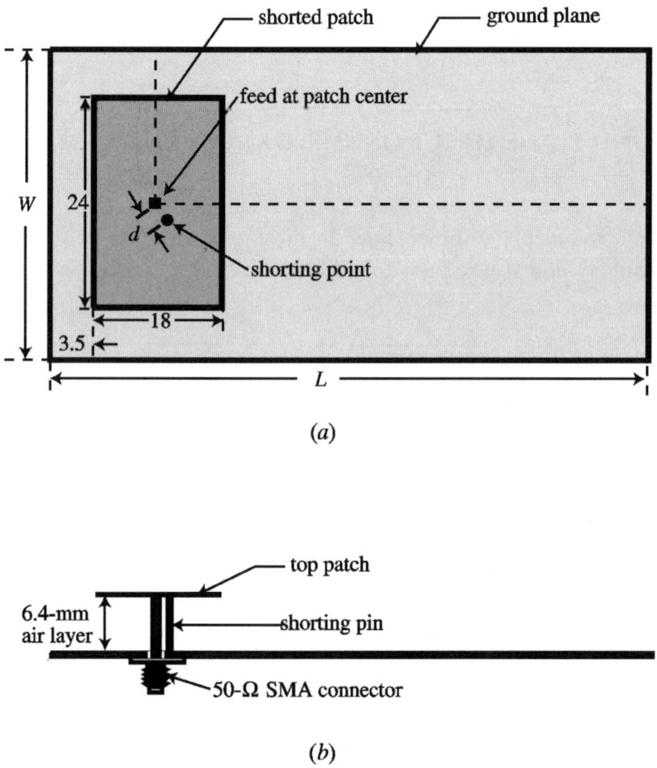

FIGURE 2.44 Geometry of the PIFA with a simple rectangular top patch for a UMTS mobile phone. (a) Top view; (b) side view. (From Ref. 21, © 2002 John Wiley & Sons, Inc.)

various ground plane dimensions, the distance d was not fixed and was slightly varied from about 0.75 to 2 mm in the experiment.

The PIFA shown in Figure 2.44 with various ground plane lengths and widths was studied experimentally. In the study the ground plane width was first fixed to be 30 mm, and Figure 2.45 shows the measured return loss against frequency for the PIFA with various ground plane lengths of 65–90 mm. It is clearly seen that the measured return loss is significantly affected by the variation in the ground plane length. The obtained impedance bandwidth, determined by 10 dB return loss, and the center operating frequency, defined to be the average of the lower and higher frequencies with 10 dB return loss, as a function of the possible ground plane length for a mobile phone are presented in Figure 2.46. It is observed that, for L varying from 50 to 105 mm, the obtained impedance bandwidths vary from about 150 to 490 MHz, a variation of over 300%. The optimal impedance bandwidth occurs at about $L = 65$ mm, which corresponds with about 45% of the free space wavelength at 2050 MHz, the center frequency of the UMTS band. It should also be noted that, for cases with $L = 51$–83 mm, the obtained impedance bandwidths are all greater than 250 MHz, the required bandwidth for UMTS operation. As for the center operating frequency shown in Figure 2.46b, the variations in the ground plane length also cause a variation of about 70 MHz, from about 2045 to 2115 MHz.

Effects of various ground plane widths on the measured impedance bandwidth and the center operating frequency have also been studied. Figure 2.47 shows the results for the case with the ground plane length fixed at 65 mm, which is the optimal length in the study shown in Figure 2.46. Measured results show that both the impedance bandwidth and the center operating frequency increase monotonically with decreasing ground plane width, which is quite different from the results of the ground plane length variations observed in Figure 2.46. It is also seen that,

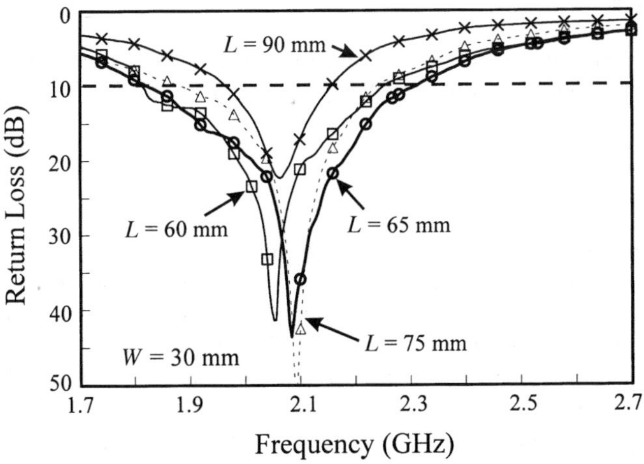

FIGURE 2.45 Measured return loss for the PIFA shown in Figure 2.44. (From Ref. 21, © 2002 John Wiley & Sons, Inc.)

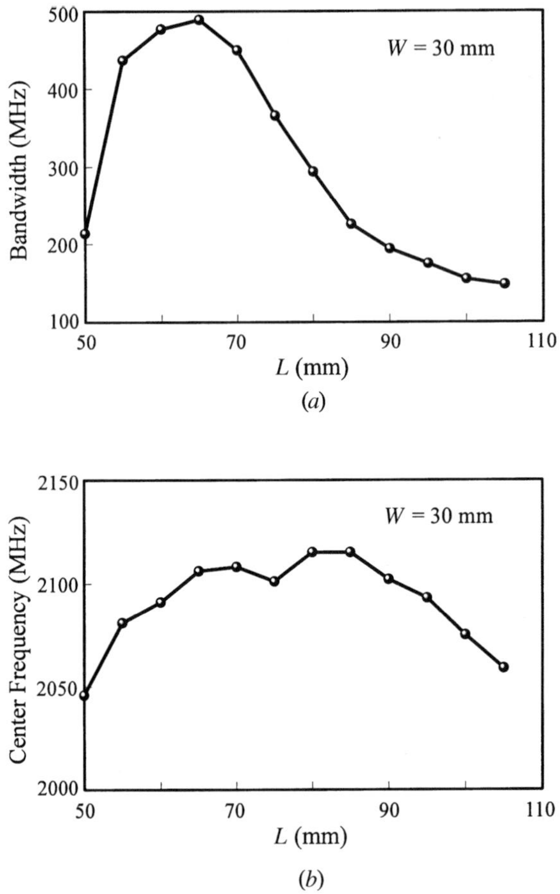

FIGURE 2.46 Measured (a) impedance bandwidth and (b) center operating frequency for the PIFA shown in Figure 2.44 with $W = 30$ mm. (From Ref. 21, © 2002 John Wiley & Sons, Inc.)

when the ground-plane width is smaller than 40 mm (about 27% of the free space wavelength at 2050 MHz), a relatively very sharp variation from about 1945 to 2105 MHz for the center operating frequency is observed for a small variation of 10 mm (W from 40 to 30 mm). Similarly, for W varied from 40 to 30 mm, the variations in the measured impedance bandwidth are as large as about 250 MHz (from about 240 to 490 MHz). This suggests that there exists a critical ground plane width for large variations in the impedance bandwidth and the center operating frequency of the PIFA. Also note that, because it becomes impractical for the ground plane width of a mobile phone to be smaller than 30 mm, the cases with W less than 30 mm are not studied.

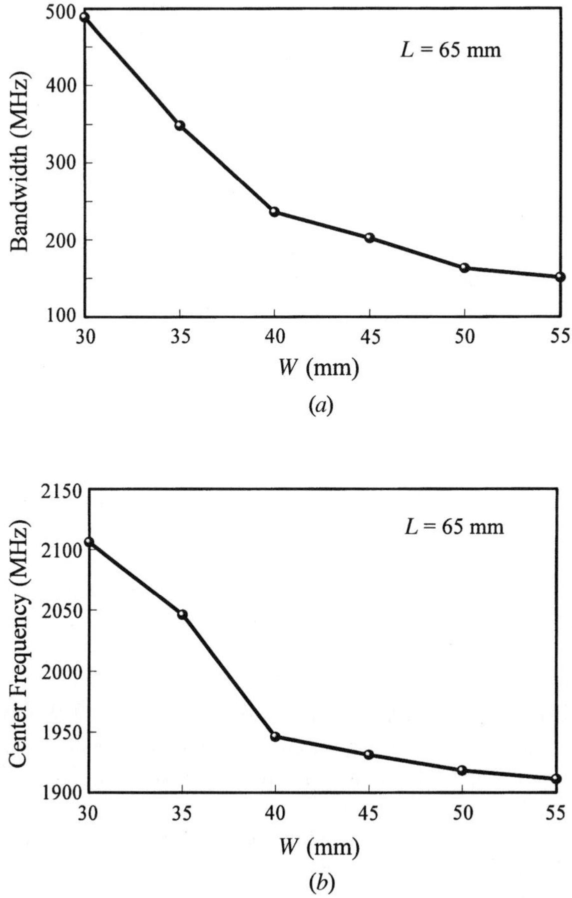

FIGURE 2.47 Measured (a) impedance bandwidth and (b) center operating frequency for the PIFA shown in Figure 2.44 with $L = 65$ mm. (From Ref. 21, © 2002 John Wiley & Sons, Inc.)

REFERENCES

1. Z. D. Liu, P. S. Hall, and D. Wake, "Dual-frequency planar inverted F antenna," IEEE Trans. Antennas Propagat., vol. 45, pp. 1451–1458, Oct. 1997.
2. C. R. Rowell and R. D. Murch, "A compact PIFA suitable for dual-frequency 900/1800-MHz operation," IEEE Trans. Antennas Propagat., vol. 46, pp. 596–598, April 1998.
3. S. Tarvas, J. Mikkola, S. Kivela, and A. Isohatala, "Planar dual-frequency antenna and radio apparatus employing a planar antenna," U.S. Patent No. 6252552, June 26, 2001.
4. S. Tarvas and A. Isohatala, "An internal dual-band mobile phone antenna," in 2000 IEEE Antennas Propagat. Soc. Int. Symp. Dig., pp. 266–269.

5. M. Martinez-Vazquez. M. Geissler, D. Heberling, A. Martinez-Gonzalez, and D. Sanchez-Hernandez, "Compact dual-band antenna for mobile handsets," Microwave Opt. Technol. Lett., vol. 32, pp. 87–88, Jan. 20, 2002.
6. P. Salonen, M. Keskilammi, and M. Kivikoski, "New slot configurations for dual-band planar inverted-F antenna," Microwave Opt. Technol. Lett., vol. 28, pp. 293–298, March 5, 2001.
7. N. Chiba, T. Amano, and H. Iwasaki, "Multifrequency inverted F-type antenna," U.S. Patent No. 6195048, Feb. 27, 2001.
8. F. R. Hsiao, H. T. Chen, G. Y. Lee, T. W. Chiou, and K. L. Wong, "A dual-band planar inverted-F patch antenna with a branch-line slit," Microwave Opt. Technol. Lett., vol. 32, pp. 310–312, Feb. 20, 2002.
9. W. Dou and W. Y. M. Chia, "Novel single-feed dual-band planar inverted-F antenna with a slot," in Proc. 2000 Int. Symp. Antennas Propagat. (ISAP). pp. 1115–1118
10. J. Ollikainen, O. Lehmus, M. Fischer, and P. Vainikainen, "Internal dual-band patch antenna for mobile phones," in Proc. Millennium Conf. on Antennas Propagat., Davos, Switzerland, p. 364.
11. D. Manteuffel, A. Bahr, D. Heberling, and I. Wolff, "Design considerations for integrated mobile phone antennas," in Proc. 2001 Int. Conf. on Antennas Propagat. (ICAP), pp. 252–254.
12. S. T. Fang, S. H. Yeh, and K. L. Wong, "Planar Inverted-F antennas for GSM/DCS mobile phones and dual ISM-band applications," in 2002 IEEE Antennas Propagat. Soc. Int. Symp. Dig., vol. 4, pp. 524–527.
13. G. H. K. Lui and R. D. Murch, "Compact dual-frequency PIFA designs using LC resonators," IEEE Trans. Antennas Propagat., vol. 49, pp. 1016–1019, July 2001.
14. S. H. Yeh and K. L. Wong, "Compact dual-frequency PIFA with a chip-inductor-loaded rectangular spiral strip," Microwave Opt. Technol. Lett., vol. 33, pp. 394–397, June 20, 2002.
15. C. T. P. Song, P. S. Hall, H. Ghafouri-Shiraz, and D. Wake, "Triple band planar inverted F antennas for handheld devices," Electron. Lett., vol. 36, pp. 112–114, Jan. 20, 2000.
16. W. P. Dou and Y. W. M. Chia, "Novel meandered planar inverted-F antenna for triple-frequency operation," Microwave Opt. Technol. Lett., vol. 27, pp. 58–60, Oct. 5, 2000.
17. F. R. Hsiao and K. L. Wong, "Compact planar inverted-F patch antenna for triple-frequency operation," Microwave Opt. Technol. Lett., vol. 33, pp. 459–462, June 20, 2002.
18. F. R. Hsiao and K. L. Wong, "A shorted patch antenna with an L-shaped ground plane for internal mobile handset antenna," Microwave Opt. Technol. Lett., vol. 33, pp. 314–316, May 20, 2002.
19. J. Ollikainen, M. Fischer, and P. Vainikainen, "Thin dual-resonant stacked shorted patch antenna for mobile communications," Electron. Lett., vol. 35, pp. 437–438, March 18, 1999.
20. G. Y. Lee, T. W. Chiou, and K. L. Wong, "Broadband stacked shorted patch antenna for mobile communication handsets," in Proc. 2001 Asia-Pacific Microwave Conf., pp. 232–235.
21. T. Y. Wu and K. L. Wong, "On the impedance bandwidth of a planar inverted-F antenna for mobile handsets," Microwave Opt. Technol. Lett., vol. 32, pp. 249–251, Feb. 20, 2002.

22. K. L. Wong and W. S. Chen, ‘‘Compact microstrip antenna with dual-frequency operation,’’ Electron. Lett., vol. 33, pp. 646–647, April 10, 1997.

23. C. L. Tang, H. T. Chen, and K. L. Wong, ‘‘Small circular microstrip antenna with dual-frequency operation,’’ Electron. Lett., vol. 33, pp. 1112–1113, June 19, 1997.

CHAPTER THREE

Very-Low-Profile Monopoles for Internal Mobile Phone Antennas

3.1 INTRODUCTION

Conventional monopoles for applications in mobile phones are usually protruded from the mobile phone housing. Such a protruded monopole tends to break off easily and also greatly increases the total length of the mobile phone. To overcome this problem, several novel monopole designs with a much reduced height have been developed recently. This kind of very-low-profile monopole is very suitable for integration within the mobile phone housing and operation as a built-in or concealed antenna. To obtain a much lower specific absorption rate (SAR), such a monopole is very promising for application in folder-type mobile phones. In this case, when the mobile phone is in operation the folder cover comprising the speaker is between the monopole and the user's head and can provide a shield for blocking some backward radiation of the monopole to achieve a smaller electromagnetic wave power absorption in the user's head, leading to a low SAR level for the mobile phone.

In this chapter, the design technique of using a branch line planar monopole to achieve GSM/DCS dual-band or GSM/DCS/PCS triple-band operations [1–3] is first discussed in Section 3.2. The design of wrapping the branch line planar monopole into a rectangular boxlike structure to achieve a compact configuration is also presented [2]. Another similar, but simpler, design of wrapping a branch patch planar monopole into a compact structure [4] is described in Section 3.3.

In Section 3.4, the technique of embedding slits in a rectangular or substantially rectangular planar monopole to obtain two separate resonant paths to obtain two wide operating bandwidths in the 900 and 1800 MHz bands is discussed. The design of bending the two-dimensional planar monopole with slits into a three-dimensional structure to effectively lower the monopole height [5] is also demonstrated. Section 3.5 describes the dual-band design of the rectangular spiral planar monopole for applications in mobile phones. Two promising designs of such a rectangular spiral planar monopole are discussed, with the first design of the monopole directly printed on a grounded microwave substrate and the second design of the monopole placed in perpendicular to the grounded microwave substrate.

Section 3.6 introduces the plastic chip antenna for GSM/DCS dual-band operation. The plastic chip antenna is obtained by placing a rectangular spiral metal strip on the top, bottom, and two side surfaces of a rectangular plastic chip, and it has the attractive feature of being surface mountable. Good antenna performance for such a monopole chip antenna has been observed [6]. In Section 3.7, the designs of broadband or dual-band folded planar monopoles [7, 8] are demonstrated. Dual-frequency operations obtained by loading a helix, a meandered wire [9], or a chip inductor to an inverted-L wire are shown in Section 3.8. The PIFA can also be directly printed on a mobile phone circuit board to operate as an integrated or on-board monopole. A design example was recently reported [10] and is described in detail in Section 3.9. Finally, in Section 3.10, monopole designs with reduced backward radiation are shown [11, 12], which have the advantages of enhancing the antenna performance and reducing possible electromagnetic wave power absorption by the user's head.

3.2 BRANCH LINE PLANAR MONOPOLE

3.2.1 Two-Branch Design for Dual-Band Operation

Figures 3.1 and 3.2 show two design examples (designs A and B) of the branch line planar monopole for GSM/DCS dual-band mobile phones. The two monopoles are all printed on an FR4 substrate of thickness 0.8 mm and relative permittivity 4.4 and consist of two branch strips, which are densely meandered to achieve a compact configuration and a very low profile (antenna height on top of the ground is about 12 mm or less, corresponding to less than 4% of the wavelength at 900 MHz). In these designs, two small grounds are placed in both sides of the feed point, through which a coax feed is used to excite the monopole. Both of the two monopoles have a main strip, whose length determines the total monopole height on top of the ground. The total monopole height of design A is 12 mm (about 3.6% of the wavelength at 900 MHz), whereas that of design B is only 9.5 mm (about 2.9% of the wavelength at 900 MHz).

One end of the main strip is connected to the feed point and, at the other end, two branch strips (branch 1 and branch 2) are protruded, with branch 1 having a much longer length. The total length of branch 1 is about 212 and 167 mm for designs A and B, respectively; whereas that of branch 2 is about 48 and 60 mm for designs A and B, respectively. It is observed from experiments that, when only branch 1 is

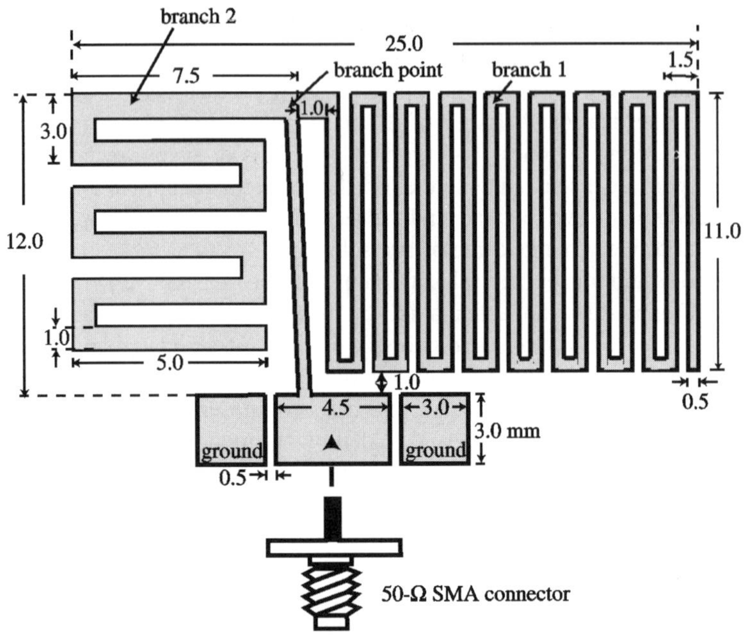

FIGURE 3.1 Geometry of a branch line planar monopole having a size of $12 \times 25\,\text{mm}^2$ (design A) for GSM/DCS dual-band mobile phones.

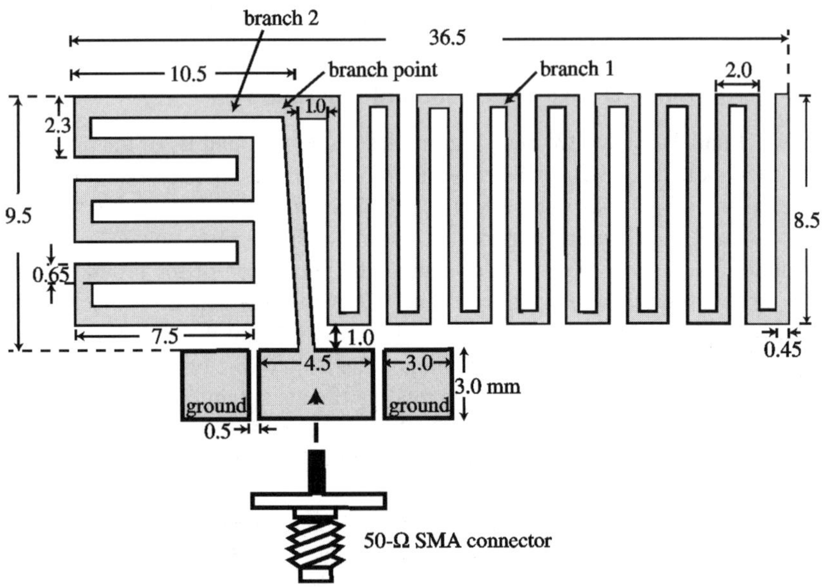

FIGURE 3.2 Geometry of a branch line planar monopole having a size of $9.5 \times 36.5\,\text{mm}^2$ (design B) for GSM/DCS dual-band mobile phones.

present, the lower band at about 900 MHz can be excited with good impedance matching over a wide bandwidth. In this case, branch 1 including the main strip has a resonant length of about 0.68 λ_0 (for design A) or 0.54 λ_0 (for design B) at 900 MHz (λ_0 is the wavelength at resonant frequency). The corresponding resonant lengths are relatively much larger than that (0.25 λ_0) of a simple straight strip monopole. This behavior is largely due to the strong coupling effects between adjacent meandered sections of branch 1, which is densely meandered in the proposed design, especially in design A shown in Figure 3.1.

Although the lower mode at about 900 MHz can well be excited with the presence of branch 1 only, it is difficult to have a second resonant mode excited with good impedance matching at about 1800 MHz. By adding branch 2 of a proper length in the proposed design, it is then found that an upper resonant mode at about 1800 MHz can be excited with good impedance matching over a wide bandwidth, with a very slight effect on the existing lower resonant mode. In designs A and B, the length of branch 2 including the main strip corresponds, respectively, to about 0.36 λ_0 and 0.43 λ_0 at 1800 MHz. Again, probably owing to the coupling effects between adjacent meandered sections, the corresponding resonant lengths of branch 2 in designs A and B are larger than a quarter wavelength.

Figures 3.3 and 3.4 show the measured return loss of the constructed prototypes of designs A and B, respectively. For both designs, two resonant modes at about 900 and 1800 MHz are excited with good impedance matching over a wide bandwidth. For design A, the obtained impedance bandwidths, determined from 10 dB return loss, are 99 MHz (872–971 MHz) and 187 MHz (1703–1890 MHz), covering the required bandwidths of the GSM and DCS systems. For design B, the obtained impedance bandwidths for the lower and upper bands are 114 MHz (867–981 MHz) and 175 MHz (1705–1880 MHz), respectively, which also meet the bandwidth requirements of the GSM and DCS systems.

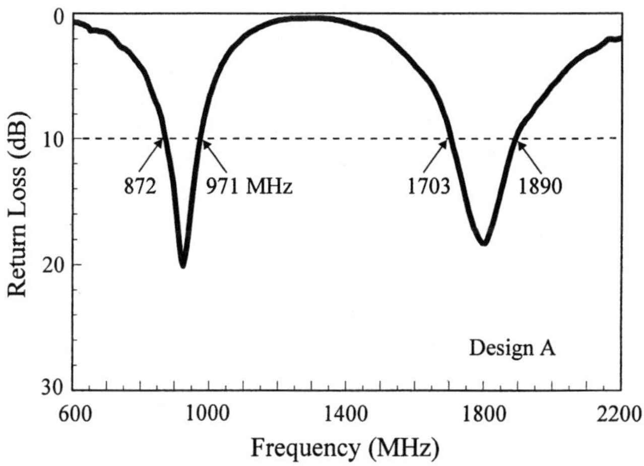

FIGURE 3.3 Measured return loss for the monopole shown in Figure 3.1.

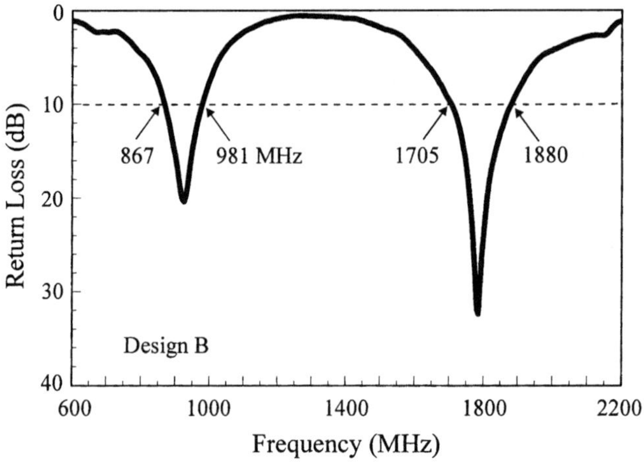

FIGURE 3.4 Measured return loss for the monopole shown in Figure 3.2.

Radiation characteristics of the proposed design were also measured. Figure 3.5 plots the measured radiation patterns at 900 and 1800 MHz for design A. Probably because of the meandering in the proposed design, which causes strong coupling between adjacent meandered sections, and the presence of the supporting FR4 substrate for the monopole, the obtained radiation patterns show some asymmetry. The measured antenna gain is about 0.5 dBi at 900 MHz and about 1.0 dBi at 1800 MHz. Measured radiation characteristics of design B show similar results.

3.2.2 Three-Branch Design for Triple-Band Operation

A planar monopole comprising three branch strips and wrapped into a compact rectangular boxlike structure for very-low-profile GSM/DCS/PCS multiband mobile phone antennas has been implemented [2]. The proposed monopole geometry is shown in Figure 3.6*a*, and Figure 3.6*b* shows the design dimensions, which are obtained with the aid of the simulation software Ansoft HFSS, of the proposed monopole unwrapped into a planar structure. The three branch strips are designed to operate as quarter-wavelength structures at 900, 1800, and 1900 MHz, respectively, and the obtained impedance bandwidths cover the required bandwidths of the GSM, DCS, and PCS bands. The proposed monopole is first printed on a thin FR4 substrate of thickness 0.4 mm and relative permittivity 4.4. Then, by following the dashed lines shown in Figure 3.6*b*, the proposed monopole is wrapped into a three-dimensional compact structure like a rectangular box of dimensions $6 \times 6.5 \times 25\ mm^3$, which makes it suitable for mounting on the top portion of a mobile phone with a small distance of 3 mm to the ground plane, leading to a very low profile of 9.5 mm (less than 3% of the wavelength at 900 MHz) from the ground plane.

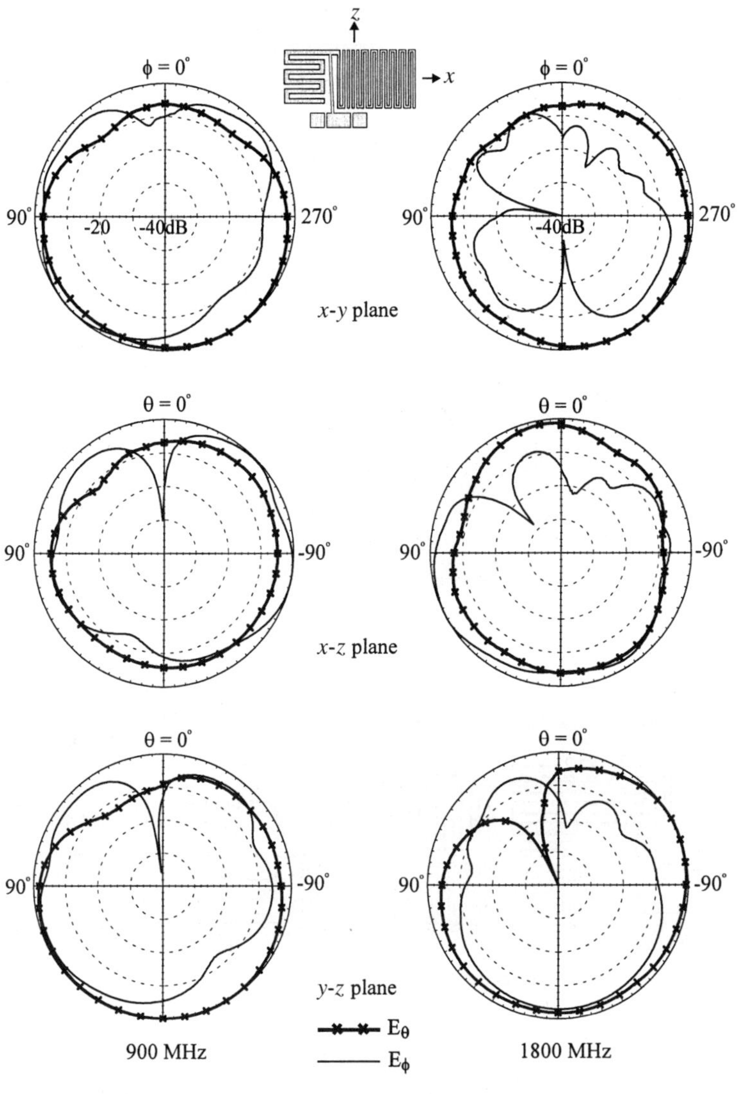

FIGURE 3.5 Measured radiation patterns at 900 and 1800 MHz for the monopole shown in Figure 3.1.

Among the three branch strips in the proposed design, branch 1 is utilized to generate a resonant mode at about 900 MHz for the GSM operation. Note that, starting from point A (the feed point), the strip length (l_1) of branch 1 is 143 mm (about 0.43 wavelength at 900 MHz), which is meandered to achieve a compact structure, and owing to the coupling effect between two adjacent meandered sections the effective electrical length will be less than the physical length. This is largely the

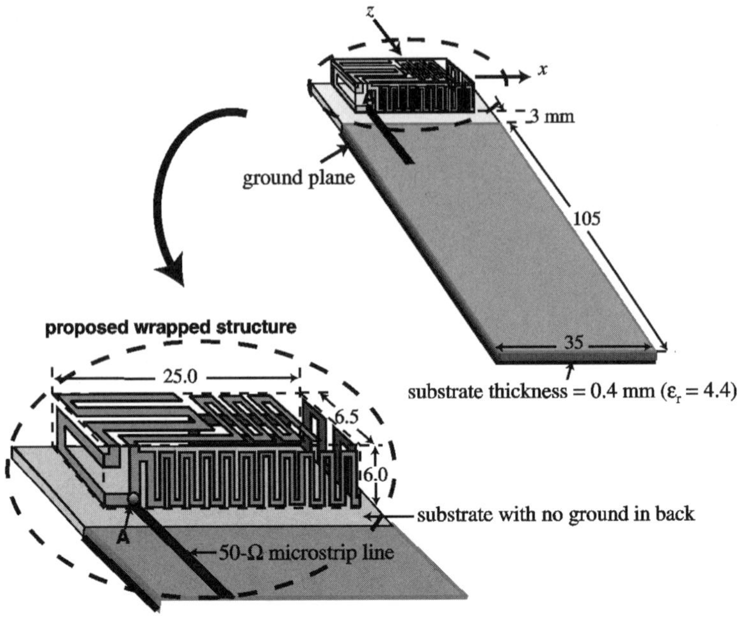

(*a*)

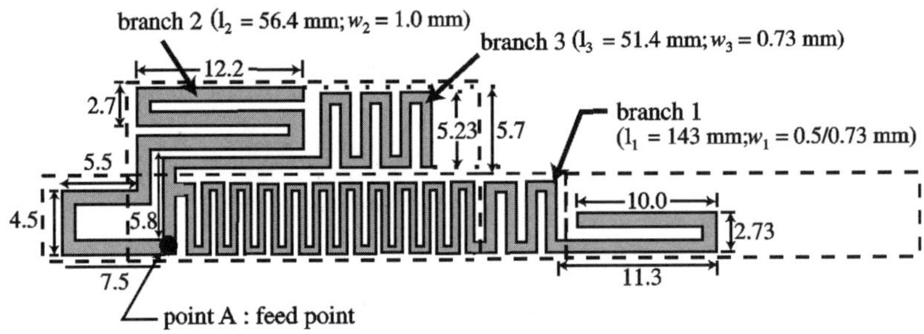

(*b*)

FIGURE 3.6 (*a*) The branch line planar monopole in a wrapped structure for GSM/DCS/PCS multiband mobile phone antenna; (*b*) the monopole unwrapped into a planar structure. (From Ref. 2, © 2002 John Wiley & Sons, Inc.)

reason that the strip length of branch 1 is larger than the required 0.25 wavelength. Also notice that, at the portion close to the end of branch 1, the strip width (0.73 mm) is designed to be larger than that of the other portion (0.5 mm), which is helpful in achieving a wider impedance bandwidth (larger than 70 MHz) for the 900 MHz band.

On the other hand, branch 2 is designed for operating at about 1800 MHz for the DCS band and its strip length (l_2) is 56.4 mm (about 0.34 wavelength at 1800 MHz). Branch 3 is for operating at about 1900 MHz for the PCS band and has a strip length (l_3) of 51.4 mm (about 0.32 wavelength at 1900 MHz). Notice that the strip lengths for both branches 2 and 3 are closer to 0.25 wavelength than that of branch 1, which is largely because branches 2 and 3 have less meandered sections than branch 1, and thus the coupling effect in reducing the effective electrical length is relatively smaller. Through the combined excitation of branches 1 and 2, an upper band centered at about 1850 MHz with a wide bandwidth of about 300 MHz covering the DCS and PCS bands is obtained.

In the experimental study, the proposed monopole is mounted on the top portion of an FR4 substrate (thickness 0.4 mm and relative permittivity 4.4), which can be considered as the circuit board of a practical mobile phone. Note that the top portion of the substrate has no ground plane in back, and the proposed monopole has a distance of 3 mm away from the other portion of the substrate with a ground plane in back, which has dimensions of 35×105 mm^2. The proposed monopole can be easily fed at point A by using a 50-Ω microstrip line. Figure 3.7 shows the measured and simulated return loss for the proposed monopole unwrapped into a planar structure, and the results for the proposed monopole in a wrapped compact structure are shown in Figure 3.8. It is first seen that reasonable agreement between the measurement and the simulation is obtained. In addition, it is seen that there are slight variations in the obtained impedance bandwidths for the proposed monopoles in planar and wrapped structures. From the obtained results in Figure 3.8, the lower band has a bandwidth (2.5:1 VSWR or about 7.3 dB return loss) of 889–960 MHz, covering the required bandwidth of the GSM system. For the upper band, a

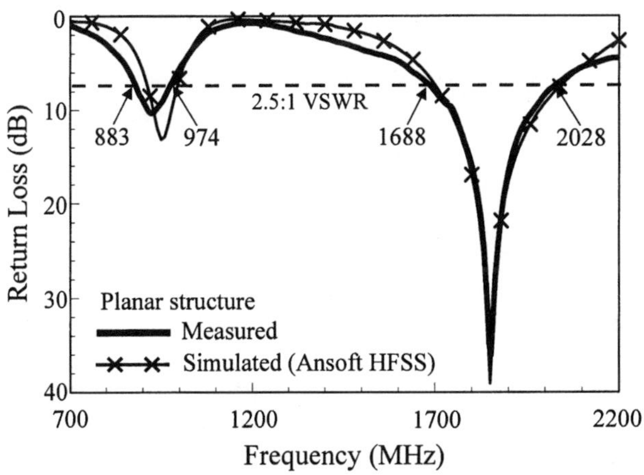

FIGURE 3.7 Measured and simulated return loss for the monopole unwrapped into a planar structure shown in Figure 3.6*b*. (From Ref. 2, © 2002 John Wiley & Sons, Inc.)

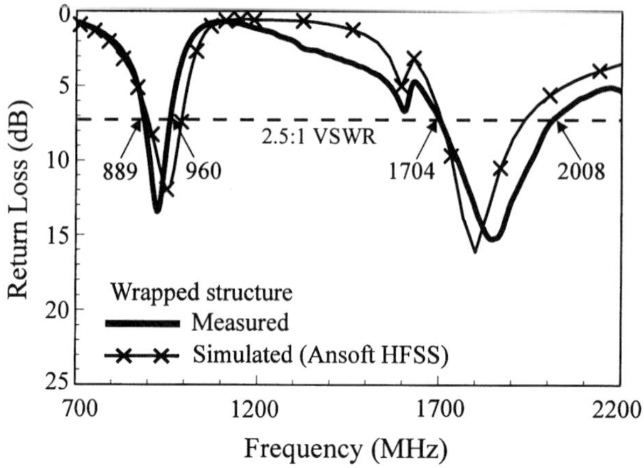

FIGURE 3.8 Measured and simulated return loss for the monopole in the wrapped structure shown in Figure 3.6a. (From Ref. 2, © 2002 John Wiley & Sons, Inc.)

bandwidth of 1704–2008 MHz is obtained, which covers the required bandwidths of the DCS and PCS systems.

Radiation characteristics were also studied. Measured radiation patterns for the proposed monopole at 900, 1800, and 1900 MHz are presented in Figures 3.9, 3.10, and 3.11, respectively. The radiation patterns at 900 MHz are close to those of a simple straight monopole. On the other hand, no prominent differences between the radiation patterns at 1800 and 1900 MHz are seen. Measurements at other operating frequencies show similar radiation patterns. This suggests that the radiation patterns are stable in the upper band for the DCS and PCS operations; so are the radiation patterns in the lower band for the GSM operation. However, in the azimuthal plane (x–y plane), the radiation pattern at 1900 MHz is more like omnidirectional radiation than that at 1800 MHz. Measured antenna gain for the proposed monopole in the lower band for the GSM operation and in the upper band for the DCS and PCS operations is shown in Figure 3.12. The antenna gain is about 1.4–1.7 dBi for the operating frequencies in the GSM band; for the DCS and PCS operations, the antenna gain ranges from about 1.9 to 3.6 dBi.

3.3 BRANCH PATCH PLANAR MONOPOLE IN A WRAPPED STRUCTURE

In addition to using branch metal lines printed on an FR4 substrate as discussed in Section 3.2, the design of a planar monopole comprising two subpatches made of a copper plate and wrapped into a compact structure with a height of 10 mm for the 900 and 1800 MHz operations has also been implemented [4]. The monopole can be mounted on the top portion of a mobile phone housing with a small distance of 4 mm from the ground plane, resulting in a low profile of 14 mm (about 4.2%

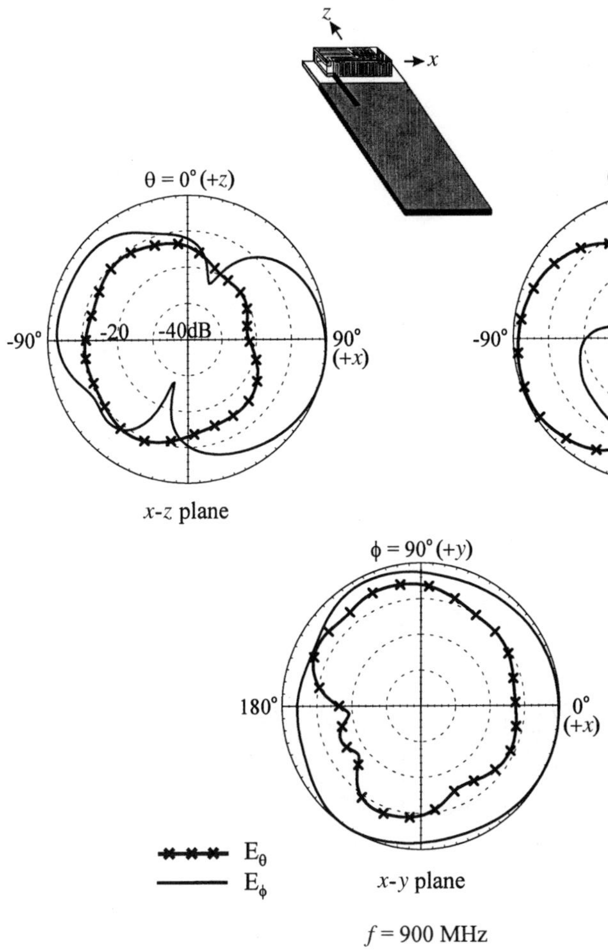

FIGURE 3.9 Measured radiation patterns at 900 MHz for the monopole shown in Figure 3.6*a*. (From Ref. 2, © 2002 John Wiley & Sons, Inc.)

of the wavelength at 900 MHz), and the obtained 10 dB return-loss impedance bandwidths cover the GSM (890–960 MHz) and DCS (1710–1880 MHz) bands.

Figure 3.13*a* shows the geometry of the proposed monopole mounted on top of a grounded FR4 substrate (thickness 0.8 mm and relative permittivity 4.4), and the proposed monopole unwrapped into a planar structure with the design dimensions is shown in Figure 3.13*b*. The design dimensions are also obtained with the aid of the simulation software Ansoft HFSS. The proposed monopole comprises two subpatches: patch 1 is for controlling the DCS operation, whereas patch 2 is for the GSM operation. There are two resonant paths in the proposed monopole: the first path starts from the feed point (point A in the figure) to the end of patch 1

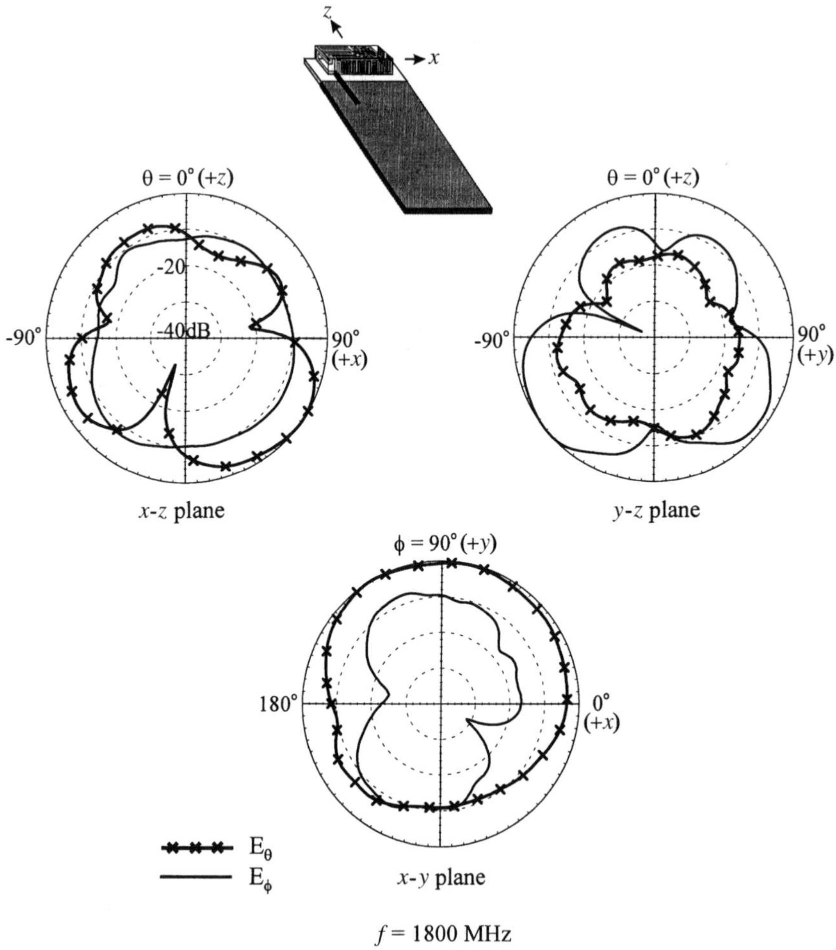

FIGURE 3.10 Measured radiation patterns at 1800 MHz for the monopole shown in Figure 3.6*a*. (From Ref. 2, © 2002 John Wiley & Sons, Inc.)

and has a length 36 mm; the second starts from the feed point, through the connection portion between patches 1 and 2, to the end of patch 2 and has a length about 95 mm. The two resonant paths have their lengths close to one-quarter wavelength at 900 and 1800 MHz, respectively, and generate a lower mode and an upper mode covering the GSM and DCS bands, respectively. Notice that two narrow slits near the connection portion between patches 1 and 2 are inserted to lengthen the length of the resonant path for the lower-mode excitation. Also notice that, by following the dashed lines shown in Figure 3.13*b*, the planar monopole made of copper plate is easily bent and wrapped into the proposed structure shown in Figure 3.13*a*. The proposed wrapped planar monopole is 10 mm in height and

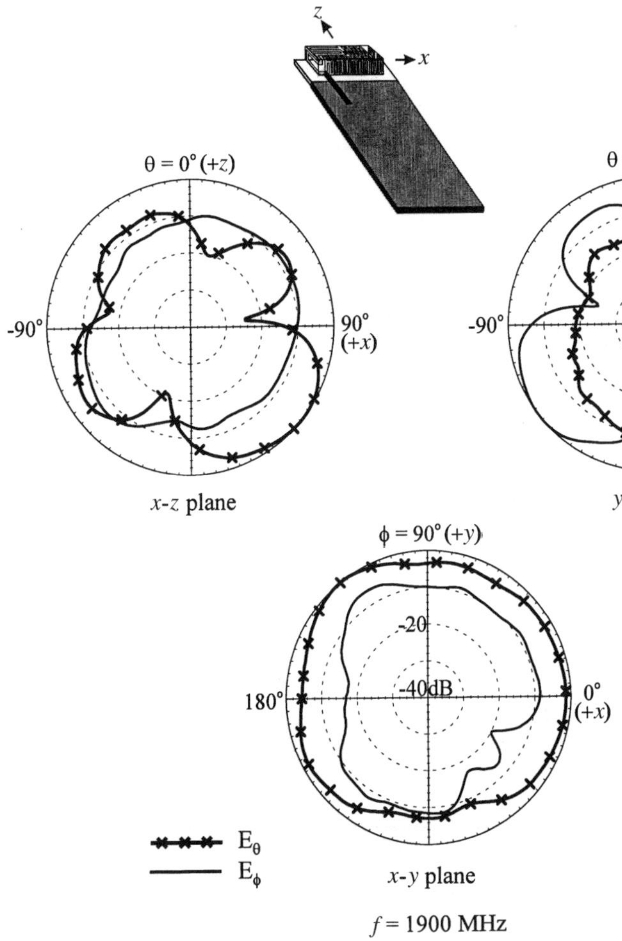

FIGURE 3.11 Measured radiation patterns at 1900 MHz for the monopole shown in Figure 3.6*a*. (From Ref. 2, © 2002 John Wiley & Sons, Inc.)

$9\times32\ \text{mm}^2$ in cross section, and with the inclusion of the 4-mm connection strip to the 50-Ω microstrip feed line, the proposed monopole has a low profile of 14 mm from the ground plane.

Figure 3.14 shows the measured and simulated return loss for the proposed monopole. The measured data agree with the simulated results. Two resonant modes are clearly excited. The 10 dB impedance bandwidth of the lower mode is 89 MHz (873–962 MHz) and covers the GSM band. On the other hand, the bandwidth of the upper mode reaches 181 MHz (1704–1885 MHz) and covers the DCS band. Figures 3.15 and 3.16 plot the measured radiation patterns at 900 and 1800 MHz, respectively. Measured results for other frequencies in the GSM and DCS bands were also studied. The obtained radiation patterns are stable across the GSM and DCS bands and are similar to those shown here. The radiation pattern

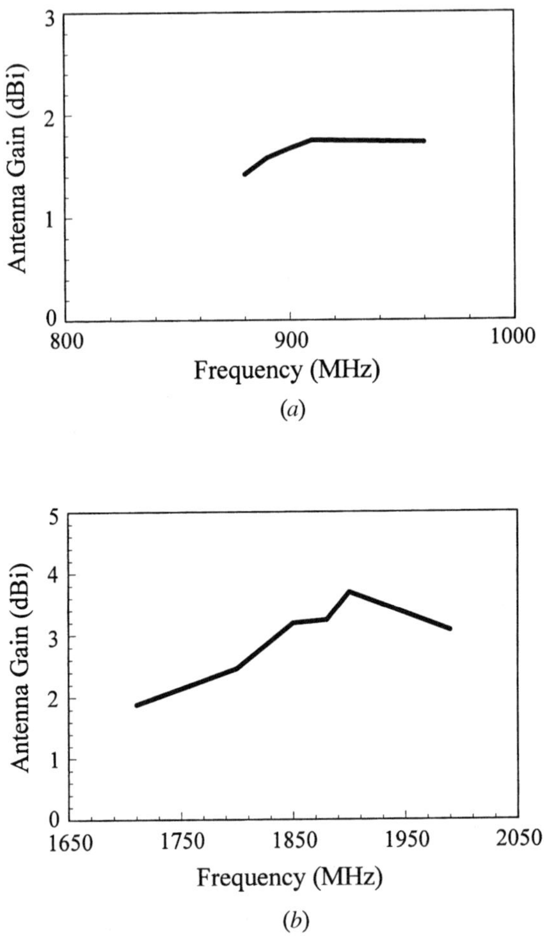

FIGURE 3.12 Measured antenna gain for the monopole shown in Figure 3.6*a*. (*a*) The GSM band; (*b*) the DCS and PCS bands. (From Ref. 2, © 2002 John Wiley & Sons, Inc.)

in the azimuthal plane (x–y plane) is closer to omnidirectional radiation for 900 MHz than for 1800 MHz. The radiation patterns are also more monopole-like for the 900 MHz operation. Figure 3.17 shows the measured antenna gain for frequencies across the GSM and DCS bands. The peak antenna gain is 2.7 and 4.4 dBi across the GSM and DCS bands, respectively.

3.4 PLANAR MONOPOLE WITH SLITS

3.4.1 Two-Dimensional Design

By embedding a folded slit in a rectangular planar monopole, GSM/DCS/PCS triple-band operation can be obtained [13]. A promising example with design dimen-

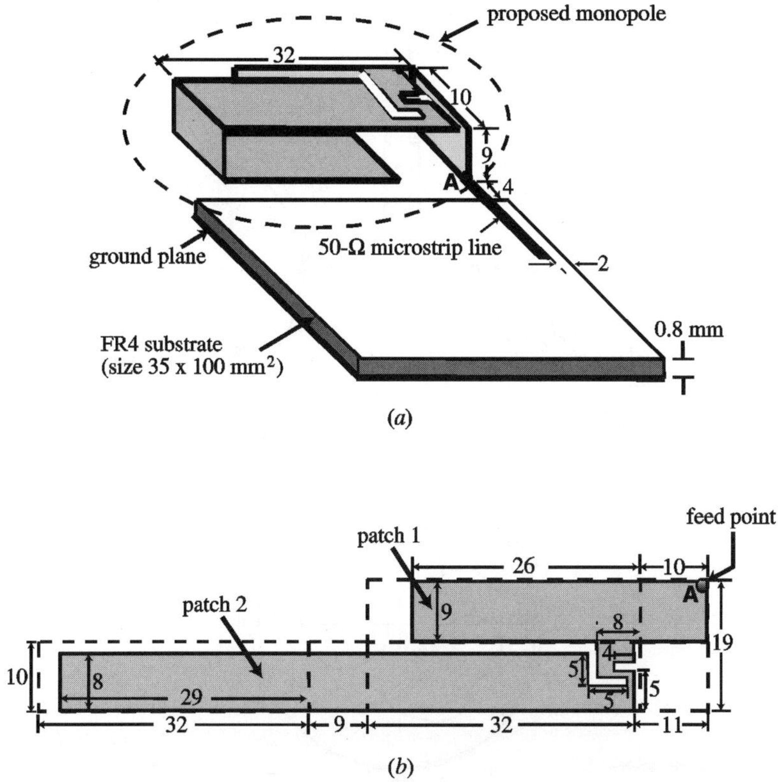

FIGURE 3.13 (*a*) Geometry of the branch patch monopole mounted on top of a grounded substrate as a GSM/DCS mobile phone antenna; (*b*) dimensions of the proposed monopole unwrapped into a planar structure. (From Ref. 4, © 2002 IEE, reprinted with permission)

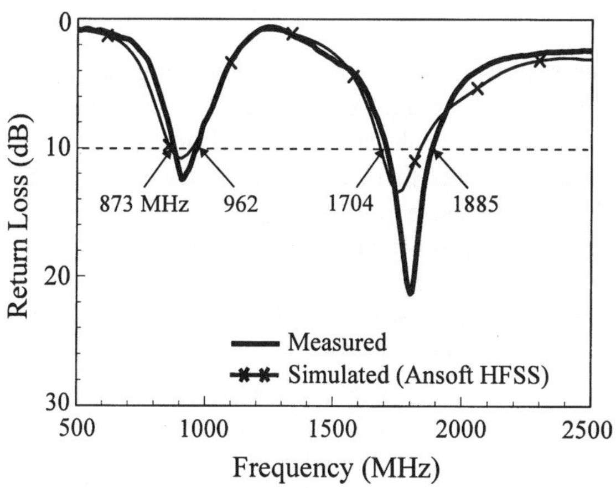

FIGURE 3.14 Measured and simulated return loss for the monopole shown in Figure 3.13. (From Ref. 4, © 2002 IEE, reprinted with permission)

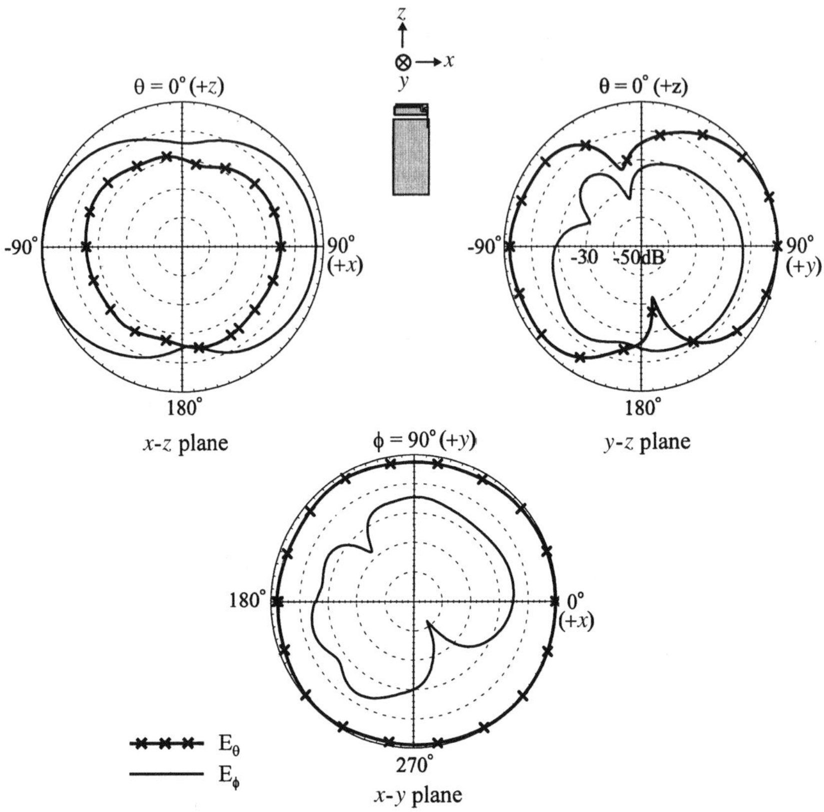

FIGURE 3.15 Measured radiation patterns at 900 MHz for the monopole shown in Figure 3.13. (From Ref. 4, © 2002 IEE, reprinted with permission)

sions is shown in Figure 3.18, in which the radiating patch has dimensions of $10 \times 30\,\text{mm}^2$ and is printed on an inexpensive FR4 substrate. A 50-Ω microstrip line is used to feed the monopole and is printed on the same FR4 substrate. On the other side of the substrate, there is a ground plane below the microstrip feed line. This ground plane has a size of $30 \times 60\,\text{mm}^2$ in the study and can be considered to be the ground plane of a practical mobile phone. Also note that the proposed monopole has a geometry similar to that of the top patch of the PIFA discussed in Section 2.2.2. Similarly, in this design, the folded slit separates the rectangular patch into two subpatches, one smaller inner subpatch encircled by the larger outer subpatch. The proposed monopole is then operated with the inner and outer subpatches resonating as quarter-wavelength structures.

It should be noted that the open end of the folded slit at the patch's bottom edge is placed close to the feed point and the other end inside the patch is also designed to be close to the feed point. In this case, two possible excited surface current paths inside the rectangular patch are obtained. The longer path starts from the feed

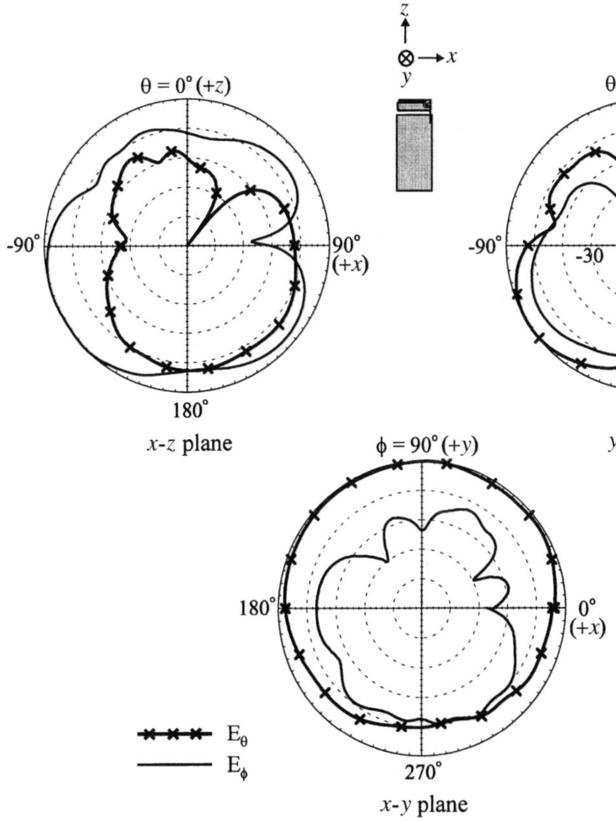

FIGURE 3.16 Measured radiation patterns at 1800 MHz for the monopole shown in Figure 3.13. (From Ref. 4, © 2002 IEE, reprinted with permission)

point and follows the folded slit to the open end of the slit at the patch's bottom edge, and the shorter path is from the feed point to the end of the inner subpatch encircled by the folded slit. It can be seen that the length of the longer path is much greater than the length of the rectangular patch, which makes the fundamental resonant frequency of the proposed monopole greatly lowered. In the proposed design shown in Figure 3.18, this length is about 70 mm, which is slightly less than one-quarter wavelength of the operating frequency at 900 MHz. This difference is largely due to the effect of the supporting FR4 substrate, which reduces the resonant length of the radiating element.

On the other hand, the length of the shorter path in the proposed design is about 30 mm, which makes it possible for the excitation of a quarter-wavelength resonant mode at about 1900 MHz. This resonant mode incorporating the second higher resonant mode of the longer path, which is tuned to about 1800 MHz, forms a

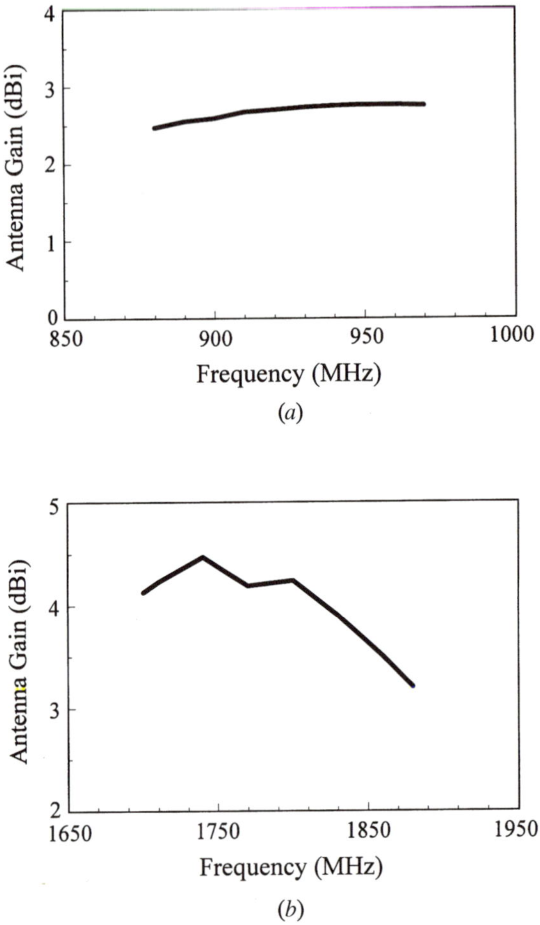

FIGURE 3.17 Measured antenna gain for the monopole shown in Figure 3.13. (*a*) The GSM band; (*b*) the DCS band. (From Ref. 4, © 2002 IEE, reprinted with permission)

wide impedance bandwidth covering the bandwidths of the DCS and PCS bands for the proposed monopole.

A prototype of the proposed design shown in Figure 3.18 has been successfully implemented, and Figure 3.19 shows the measured and simulated return loss of the proposed antenna. Reasonable agreement between the measurement and the simulation is observed, and two separate wide bandwidths are obtained. The bandwidth of the lower band, determined by 2:1 VSWR, reaches 74 MHz and covers the GSM band. On the other hand, the upper band has a bandwidth as large as 454 MHz and covers the DCS and PCS bands. The excited surface current distributions, obtained from the IE3D simulation, on the radiating patch of the proposed antenna at 900, 1800, and 1900 MHz are also presented in Figure 3.20. For the

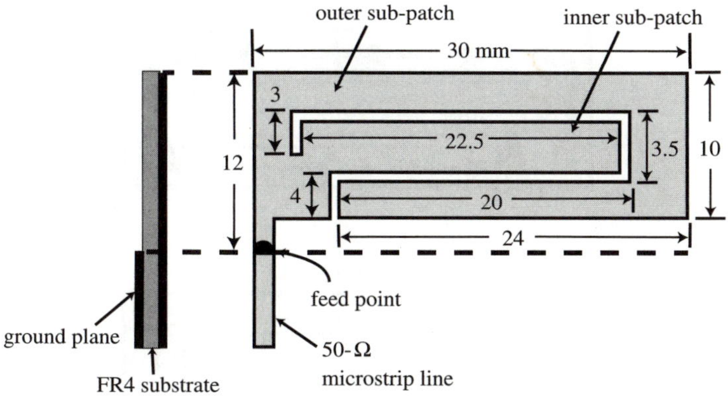

FIGURE 3.18 Geometry of the planar monopole with slits suitable for GSM/DCS/PCS triple-band operation. (From Ref. 13, © 2002 IEEE, reprinted with permission)

900 MHz excitation, a larger surface current distribution is observed for the longer path along the outer subpatch. This suggests that the outer subpatch is the major radiating element for the proposed monopole in the 900 MHz band, and the outer subpatch is operated as a quarter-wavelength structure. For the 1800 and 1900 MHz operations, it is observed that the surface current distribution in the inner subpatch gradually increases. This indicates that the inner subpatch is the major radiating element for the higher operating frequencies of the monopole's upper band. For the lower operating frequencies of the monopole's upper band, the obtained results suggest that both the inner and outer subpatches contribute to the antenna radiation.

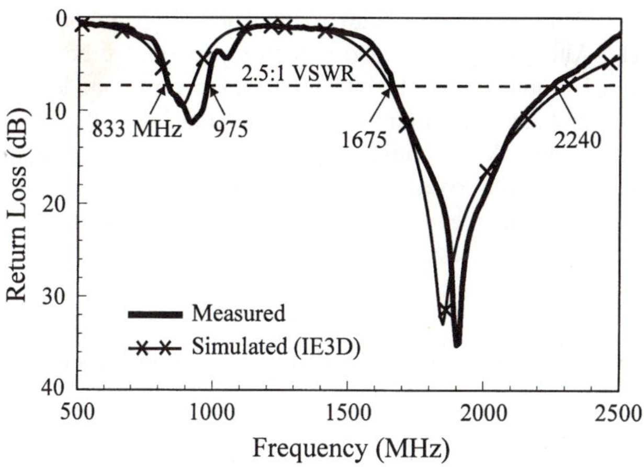

FIGURE 3.19 Measured and simulated return loss for the monopole shown in Figure 3.18. (From Ref. 13, © 2002 IEEE, reprinted with permission)

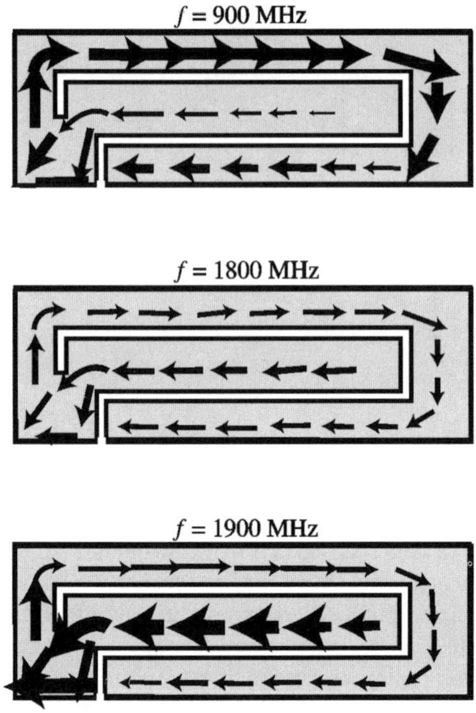

FIGURE 3.20 Simulated IE3D results of the surface current distributions on the radiating patch of the monopole shown in Figure 3.18. (From Ref. 13, © 2002 IEEE, reprinted with permission)

Figure 3.21 plots the measured radiation patterns in the x–y plane (azimuthal plane) and y–z plane (elevation plane) for the proposed monopole at 900, 1800, and 1900 MHz. Although the obtained radiation patterns are not as good as those of a conventional simple monopole antenna having a very good azimuthal omnidirectional pattern and null radiation along the antenna axis ($\theta = 0°$), the proposed antenna shows a monopole-like radiation pattern. For the GSM band, a peak antenna gain of about 2.9 dBi is observed, with gain variations less than 1.5 dBi. For the DCS and PCS bands, the peak antenna gain observed is 3.0 and 3.4 dBi, respectively, and the gain variations are also less than 1.5 dBi.

3.4.2 Three-Dimensional Design with the Monopole in a Bent Structure

By inserting several slits into a rectangular planar monopole, which leads to a meandered monopole, and further bending it into two equal and mutually perpendicular sections, a novel very-low-profile monopole for applications in GSM/DCS dual-band mobile phones has been obtained [5]. Figure 3.22a shows the geometry of the bent planar monopole with slits for GSM/DCS operation, and the side

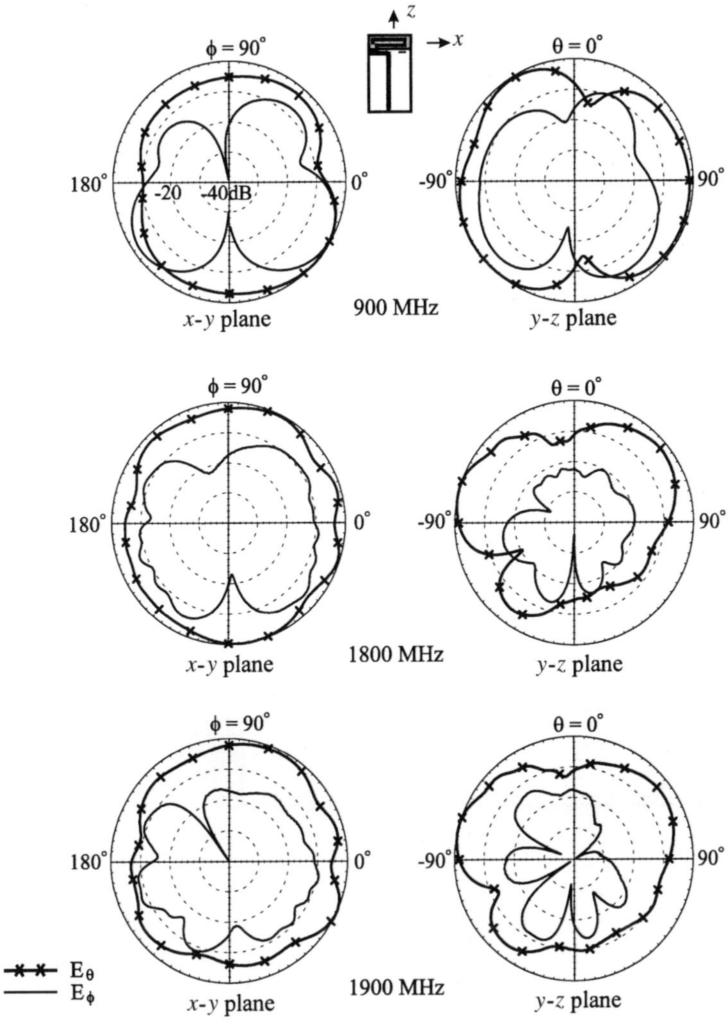

FIGURE 3.21 Measured radiation patterns for the monopole shown in Figure 3.18. (From Ref. 13, © 2002 IEEE, reprinted with permission)

view of the monopole mounted on top of a grounded FR4 substrate is shown in Figure 3.22*b*. The proposed monopole is first printed on a thin (thickness 0.4 mm) FR4 substrate of size $14 \times 30\,\text{mm}^2$. Then, along the centerline (the bending line shown in the planar structure in Fig. 3.22*a*), the proposed monopole is bent into the proposed structure having two mutually perpendicular sections of dimensions $7 \times 30\,\text{mm}^2$. Between the two sections, a foam material of dimensions $6.6 \times 6.6 \times 30\,\text{mm}^3$ is inserted to support the monopole (see Fig. 3.22*b*). With the bending design, the proposed bent monopole has dimensions $7 \times 7 \times 30\,\text{mm}^3$ and,

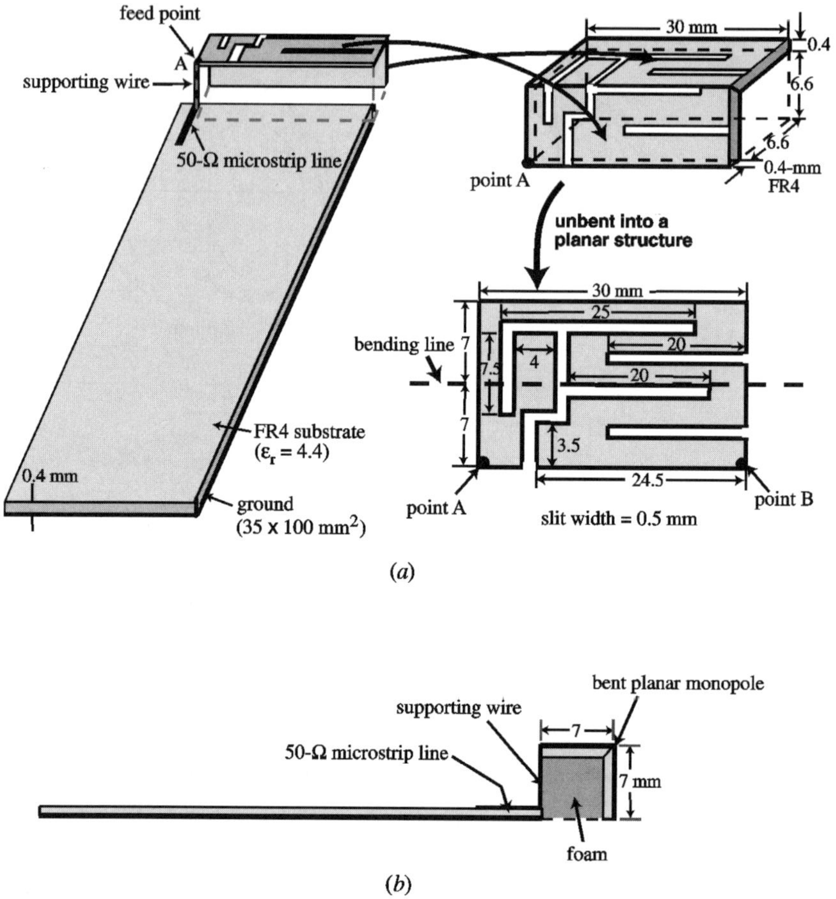

FIGURE 3.22 (*a*) Geometry and dimensions of the bent planar monopole with slits for GSM/DCS operation; (*b*) side view of the monopole mounted on top of a grounded FR4 substrate. (From Ref. 5, © 2002 John Wiley & Sons, Inc.)

when mounted on a mobile phone, the monopole's profile is only about 7 mm (about 2.1% of the wavelength at 900 MHz) from the ground plane. For feeding the proposed monopole, a conducting wire in perpendicular to the grounded substrate is connected between the proposed monopole and a 50-Ω microstrip line printed on the grounded substrate. The microstrip line is further connected to a 50-Ω SMA connector placed in the ground plane side of the substrate, through which a coax is used to excite the monopole in the experiment.

Several slits are inserted in the proposed monopole to achieve a much-lengthened resonant path. All the widths of the inserted slits are fixed to 0.5 mm, and the optimal slit dimensions are obtained with the aid of the simulation software Ansoft HFSS. By following the slits' boundaries, a meandered resonant path starting from point

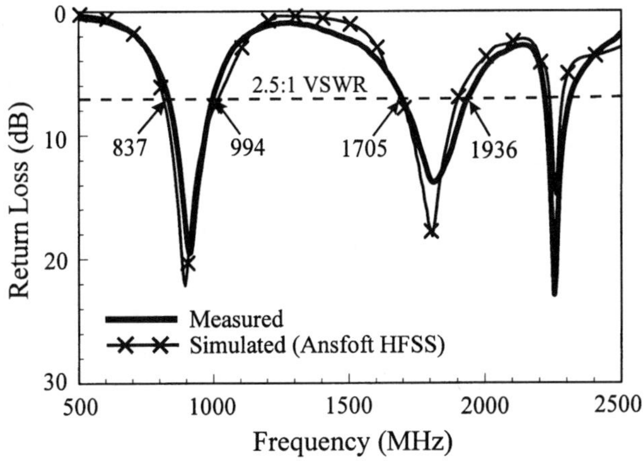

FIGURE 3.23 Measured and simulated return loss for the monopole shown in Figure 3.22. (From Ref. 5, © 2002 John Wiley & Sons, Inc.)

A to point B (see the planar structure in Fig. 3.22*a*) has a length of about 135 mm, which corresponds to about 0.4 times the wavelength at 900 MHz. This resonant length is larger than the required 0.25 wavelength for a monopole operation, which is largely owing to the coupling effect between adjacent meandered subpaths. In addition to obtaining a much-lengthened resonant path, there is a small rectangular section of dimensions $7.5 \times 4\,\text{mm}^2$ near point A whose dimensions are tuned to achieve good impedance matching for the proposed monopole. Also, probably owing to the meandering, the first two resonant modes of the proposed monopole have a frequency ratio of about 2.0 and cover the GSM and DCS bands, which makes the proposed monopole suitable for applications in GSM/DCS dual-band mobile phones.

Figure 3.23 shows the measured and simulated return loss for the constructed prototype of the proposed monopole. The measured data agree well with the simulated results. It is clearly seen that the lower and upper modes occur near 900 and 1800 MHz, respectively. The lower mode has an impedance bandwidth (2.5:1 VSWR) of about 157 MHz (837–994 MHz), which covers the required bandwidth for the GSM operation. On the other hand, the upper band has a bandwidth of 231 MHz (1705–1936 MHz), also covering the DCS band.

Figures 3.24 and 3.25 plot the measured radiation patterns at 900 and 1800 MHz, respectively. Radiation patterns at other operating frequencies in the GSM and DCS bands were also measured and showed similar radiation patterns; that is, the radiation patterns are stable across the respective operating bandwidth. It is also observed that the measured radiation patterns at 900 MHz show monopole-like radiation patterns. The measured antenna gain for frequencies across the GSM and DCS bands is presented in Figure 3.26. For the GSM band, the peak antenna gain is about 2.9 dBi and gain variations are less than 1.6 dBi. As for the DCS

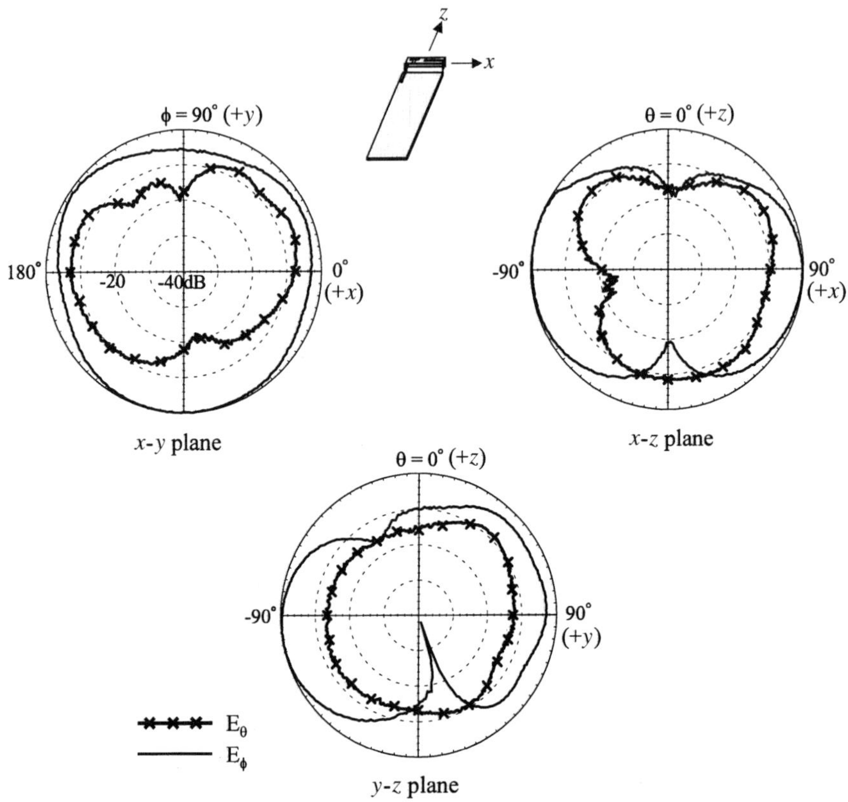

FIGURE 3.24 Measured radiation patterns at 900 MHz for the monopole shown in Figure 3.22. (From Ref. 5, © 2002 John Wiley & Sons, Inc.)

band, the peak antenna gain is about 2.8 dBi but with smaller gain variations (less than 1 dBi).

3.5 RECTANGULAR SPIRAL PLANAR MONOPOLE

In the monopole designs described in Sections 3.2–3.4, the technique of obtaining dual-frequency operation is based on utilizing two separate resonant paths of different lengths. However, one can also use a single resonant path and fine-tune its first two resonant frequencies to the desired frequencies, such as 900 and 1800 MHz for GSM and DCS dual-band operations. In this case, because only one resonant path is required, a more compact size for the dual-frequency monopole can be obtained. A very promising design of such a dual-frequency monopole is the rectangular spiral planar monopole, and two possible arrangements are shown in Figures 3.27 and 3.28. The rectangular spiral planar monopole can be directly

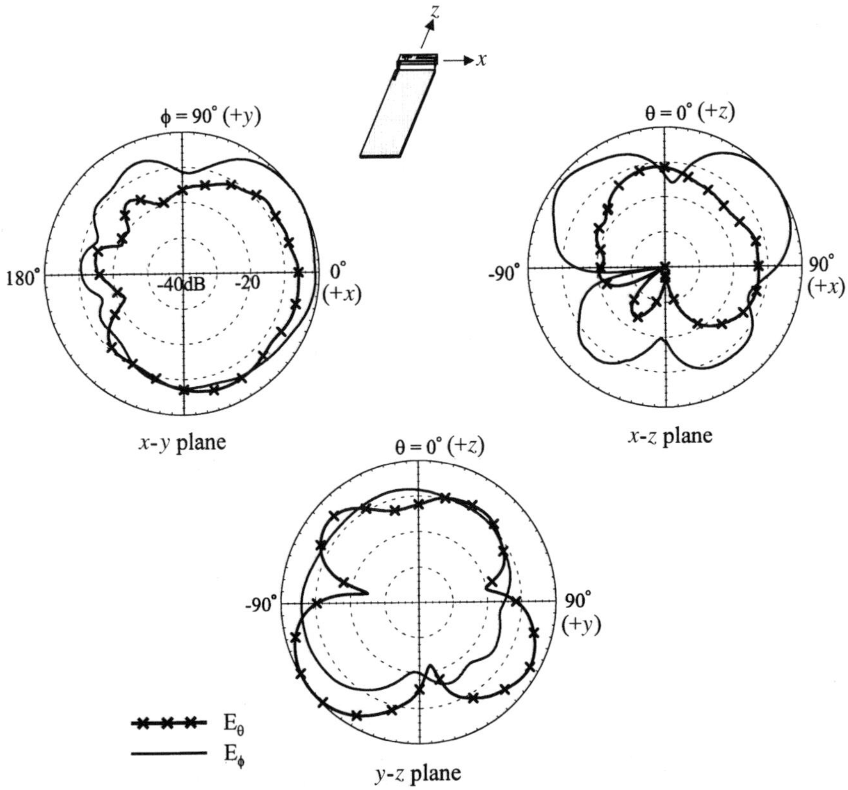

FIGURE 3.25 Measured radiation patterns at 1800 MHz for the monopole shown in Figure 3.22. (From Ref. 5, © 2002 John Wiley & Sons, Inc.)

printed on a grounded microwave substrate, as shown in Figure 3.27, or placed in perpendicular to the grounded microwave substrate, as shown in Figure 3.28. The former design has the advantage of easy integration with the associated circuitry. The latter design can greatly reduce the total monopole height from the ground plane; that is, a very-low-profile dual-frequency monopole can be obtained [14]. This characteristic makes the latter design very attractive for applications in mobile phones.

A prototype of the design shown in Figure 3.28 has also been successfully implemented. The rectangular spiral planar monopole is first printed on an FR4 substrate of thickness 0.4 mm and area $5 \times 27\,\text{mm}^2$, which serves as a support for the monopole, and then placed in perpendicular to the grounded microwave substrate, which can be considered as the mobile phone circuit board, with a distance of 5 mm (*d* in Fig. 3.28) from the ground plane. In this case, the total monopole height is only 5.4 mm, which is less than 2% of the wavelength at 900 MHz. In addition, the

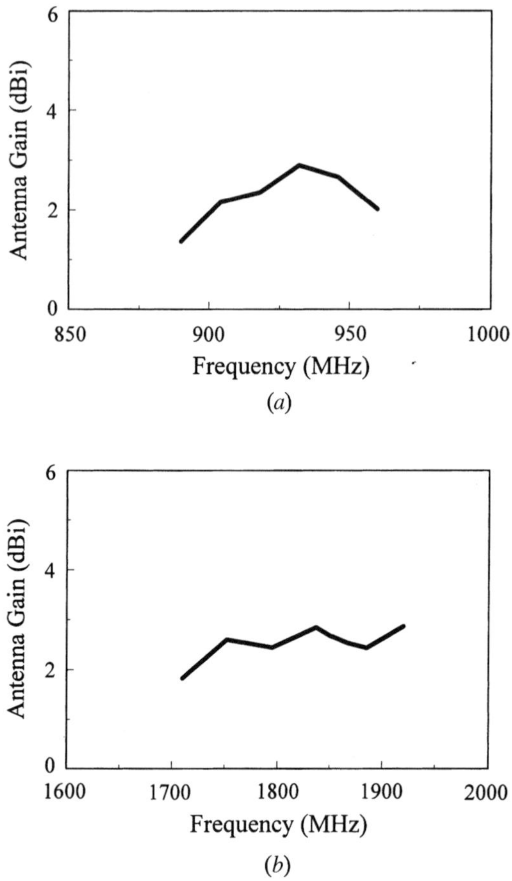

FIGURE 3.26 Measured antenna gain for the monopole shown in Figure 3.22. (*a*) The lower band for GSM operation; (*b*) the upper band for DCS operation. (From Ref. 5, © 2002 John Wiley & Sons, Inc.)

width of the planar monopole is only 5 mm, which makes it very promising for placement within the housing of most present-day mobile phones.

Note that a 50-Ω microstrip line printed on the grounded microwave substrate (thickness 0.8 mm, relative permittivity 4.4, and size $35 \times 100\,\text{mm}^2$) is used to feed the monopole at the feed point. On the other side of the microwave substrate, the ground plane has an area of about $35 \times 95\,\text{mm}^2$; that is, there is a region of about $35 \times 5\,\text{mm}^2$ on the top portion of the microwave substrate without ground plane. Also note that, to fine-tune the monopole's first two resonant frequencies to be about 900 and 1800 MHz, the resonant path of the rectangular spiral planar monopole starting from the feed point at the outer open end to the inner encircled open end is first selected to have a length of about one-quarter wavelength at 900 MHz, and then the widths in each linear section of the rectangular spiral planar monopole are varied

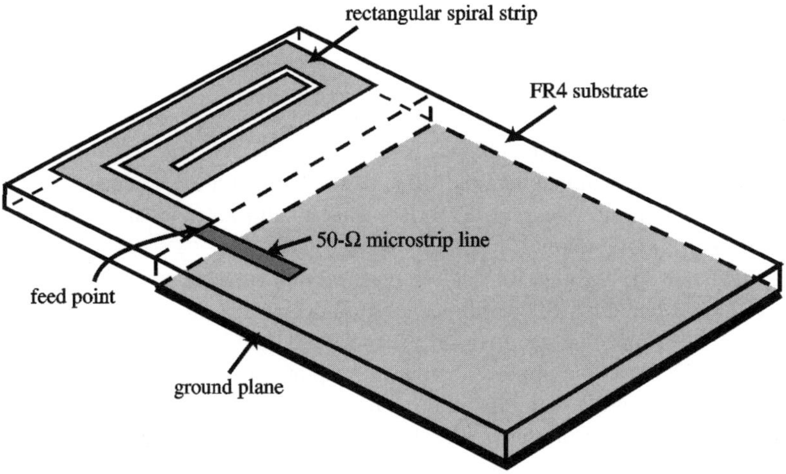

FIGURE 3.27 Geometry of the rectangular spiral planar monopole printed on a grounded microwave substrate.

to fine-tune the frequency ratio of the first two resonant frequencies to be about 2.0. Experimental results of the constructed prototype of the proposed design showed that two separate wide impedance bandwidths at about 900 and 1800 MHz can be obtained and can cover both the required bandwidths of the GSM and DCS bands. Good radiation characteristics have also been observed.

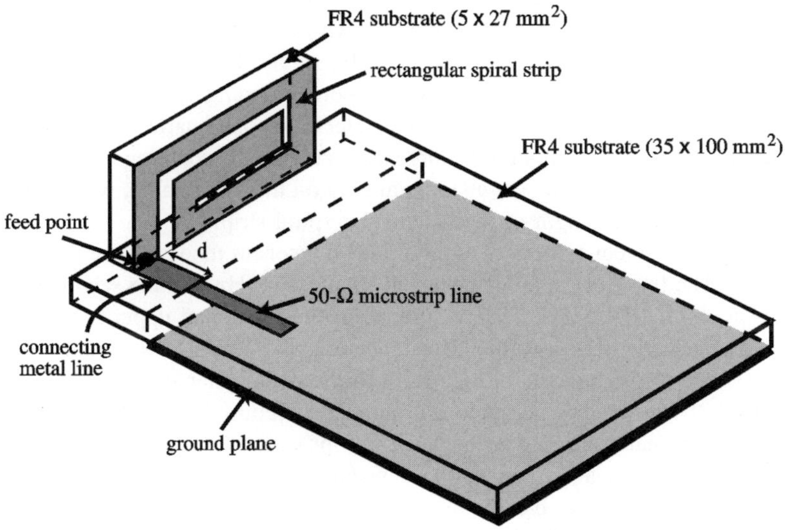

FIGURE 3.28 Geometry of the rectangular spiral planar monopole placed in perpendicular to a grounded microwave substrate.

3.6 DUAL-BAND MONOPOLE CHIP ANTENNA

Conventional chip antennas are usually constructed with ceramic chips, which are fragile in nature and tend to break easily. In this section, we present a new design of chip antennas using a plastic chip. This kind of plastic chip antenna is also surface mountable, like conventional ceramic chip antennas. In addition, because the plastic chip is lightweight and does not break, the plastic chip antennas are attractive for applications in mobile phones.

To show the applicability of such a plastic chip antenna, a dual-band design capable of GSM and DCS operations for mobile phones has been implemented [6]. Figure 3.29 shows the geometry of the proposed dual-band plastic chip antenna for the GSM/DCS mobile phone. The plastic chip antenna has compact dimensions of $5 \times 10 \times 16\,\text{mm}^3$ and is mounted on the top portion of an FR4 substrate (thickness 0.8 mm and relative permittivity 4.4), which can be treated as the circuit board of a practical mobile phone. Notice that the top portion has no ground plane in back, and a distance of 5 mm is required for the proposed antenna to be away from the grounded portion of the FR4 substrate, which has dimensions $35 \times 105\,\text{mm}^2$. For exciting the antenna, a conducting strip having a length 5 mm and the same width as the 50-Ω microstrip line printed on the FR4 substrate is connected to point A (the feed point) at the lower right corner of the plastic chip. This feed design allows the proposed antenna to be surface mountable. Also note that, in the experiment, the 50-Ω microstrip line was further connected to a 50-Ω SMA connector placed in the ground-plane side of the FR4 substrate. For practical applications, the 50-Ω microstrip line can be directly connected to the transmitter/receiver of the mobile phone.

The radiating element on the surfaces (the top, bottom, and two side surfaces) of the rectangular plastic chip has a simple structure of a rectangular spiral metal strip (see the metal strip in the planar structure shown in Fig. 3.29, in which the design dimensions are also given). With the inclusion of the connecting strip, the total length of the radiating element is 109 mm, corresponding to about 33% of the wavelength at 900 MHz. This length is greater than one-quarter wavelength for the operation of a straight monopole, largely because of the coupling effect between two nearby sections of the rectangular spiral strip on the plastic chip, which reduces the effective electrical length of the rectangular spiral strip.

The proposed antenna operates at the first two resonant modes of the rectangular spiral strip. The first (lower) mode occurs at about 900 MHz, and the second (upper) mode is at about 1800 MHz. The occurrence of the lower and upper modes at the desired frequencies is mainly tuned by adjusting the length and position of the innermost section (section $\overline{BC}$ in Fig. 3.29) of the rectangular spiral strip.

Figure 3.30 shows the measured and simulated return loss of the constructed prototype. Reasonable agreement between the measurement and the simulation is observed, and it is clearly seen that two resonant modes are excited. The lower mode has an impedance bandwidth, determined from 2.5:1 VSWR or about 7.3 dB return loss, of 75 MHz (890–965 MHz) covering the GSM band; the upper band has an impedance bandwidth of 174 MHz (1706–1880 MHz), which meets the bandwidth requirement of the DCS band.

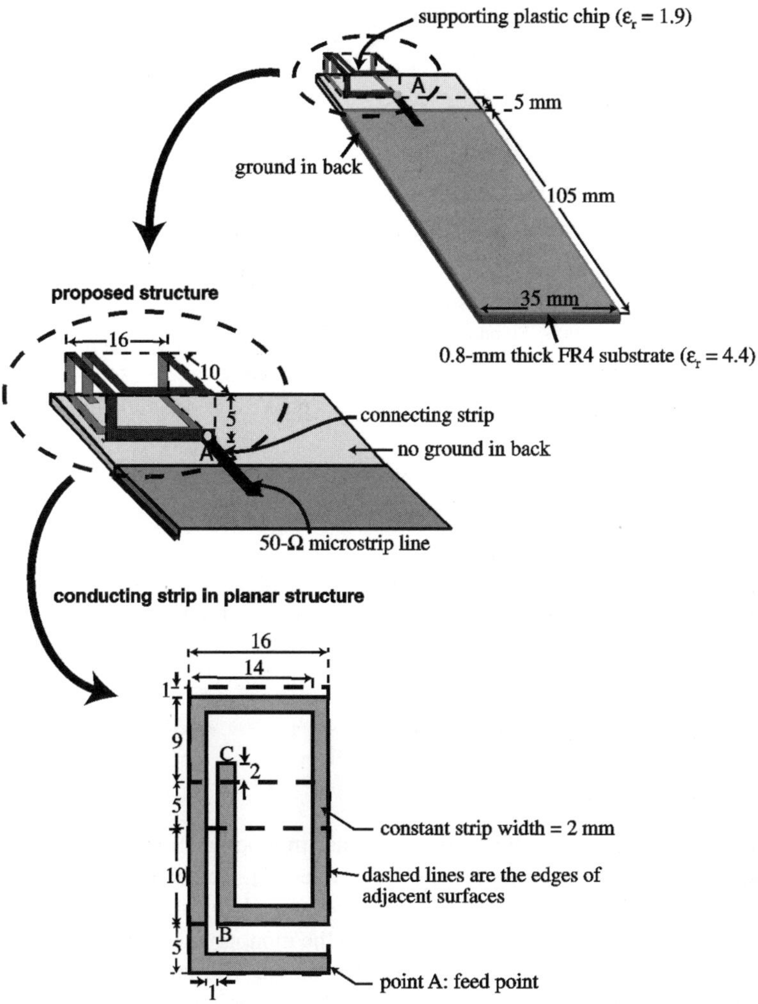

FIGURE 3.29 Geometry of the dual-band plastic chip antenna for a GSM/DCS mobile phone. (From Ref. 6, © 2002 John Wiley & Sons, Inc.)

The radiation characteristics were also studied. Figures 3.31 and 3.32 present the measured radiation patterns at 900 and 1800 MHz, respectively. For the 900 MHz case, monopole-like radiation patterns are observed, and good omnidirectional radiation pattern in the azimuthal plane (x–y plane) is seen. For the 1800 MHz case, the radiation in the azimuthal plane is also close to omnidirectional. Radiation patterns for other operating frequencies were also measured, and the radiation patterns are stable for frequencies across their respective GSM and DCS bands. Measured antenna gain for the GSM and DCS bands is presented in Figure 3.33.

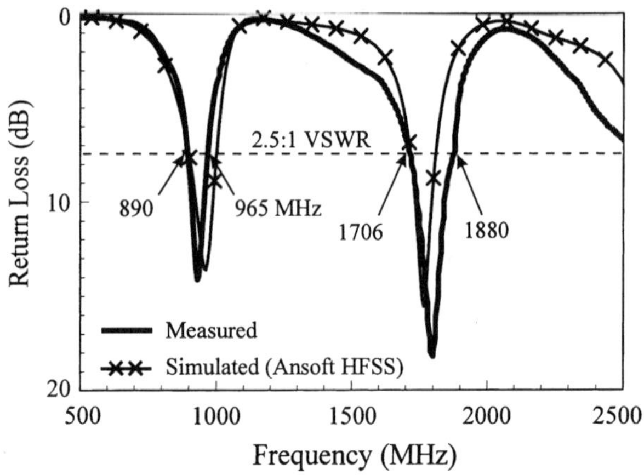

FIGURE 3.30 Measured and simulated return loss of the antenna shown in Figure 3.29. (From Ref. 6, © 2002 John Wiley & Sons, Inc.)

For the GSM or the lower band, the antenna gain is about 1.0–1.5 dBi; for the DCS or the upper band, the antenna gain has a range of about 1.6–2.5 dBi.

3.7 FOLDED PLANAR MONOPOLE

3.7.1 A Simple Folded Patch

This section presents a broadband folded planar monopole suitable for DCS, PCS, UMTS, and WLAN operations for mobile phones [7]. The monopole geometry is shown in Figure 3.34. The proposed monopole is obtained by folding a planar monopole of a rectangular shape, which makes the monopole length greatly reduced. In addition, a wide operating bandwidth is obtained. The proposed folded planar monopole has compact dimensions of $10 \times 17.5 \times 5\,\text{mm}^3$ and shows a very wide operating bandwidth of about 900 MHz, from about 1.7 to 2.6 GHz. For applications in a mobile phone, the total monopole height from the ground plane is only 13 mm.

As shown in Figure 3.34, the folded monopole is mounted on top of a grounded FR4 substrate (thickness 0.4 mm, relative permittivity 4.4, and size $35 \times 100\,\text{mm}^2$), which again can be considered to be the circuit board of a practical mobile phone. The folded monopole comprises a center section of dimensions $10 \times 5\,\text{mm}^2$ and two opposite sections of dimensions $10 \times 17.5\,\text{mm}^2$, which are in parallel to each other. At one end of the folded monopole, a conducting wire, which is also used to support the monopole, connects a 50-Ω microstrip feed line printed on the grounded substrate to the monopole.

It is also noted that there is a gap of 3 mm between the folded planar monopole and the ground plane. This gap is usually required for a planar monopole placed above a ground plane to achieve good impedance matching over a wide bandwidth.

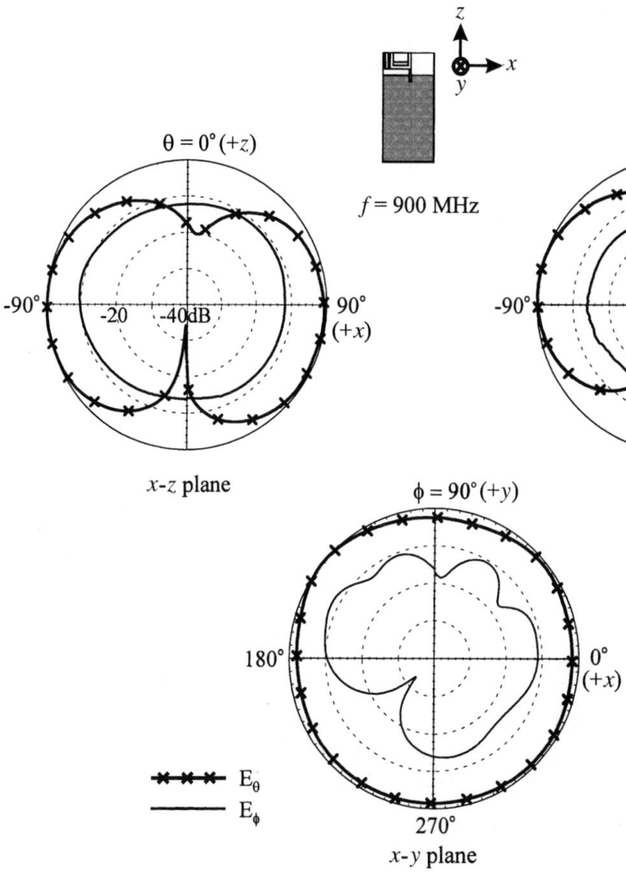

FIGURE 3.31 Measured radiation patterns at 900 MHz for the antenna shown in Figure 3.29. (From Ref. 6, © 2002 John Wiley & Sons, Inc.)

As for the width (5 mm in this design) of the monopole's center section, when it is increased, the length (17.5 mm in this design) of the monopole can be further decreased. However, the increase in this width should be limited to the thickness of practical mobile phones. On the other hand, when the width of the center section decreases, there will be some coupling between the two opposite sections, and impedance matching of the proposed antenna will be greatly affected.

Figure 3.35 presents the measured and simulated return loss of the constructed prototype. The measurement in general agrees with the simulation, and a wide impedance bandwidth (2.5:1 VSWR) of about 1676–2582 MHz is obtained, which covers the DCS, PCS, UMTS, and WLAN bands. The radiation characteristics have also been studied. Figure 3.36 shows the measured radiation patterns at 1800 and 1900 MHz. The obtained patterns are in general close to the monopole-like patterns (a null or a dip in $\theta = 0°$ and 180° directions in the elevation planes

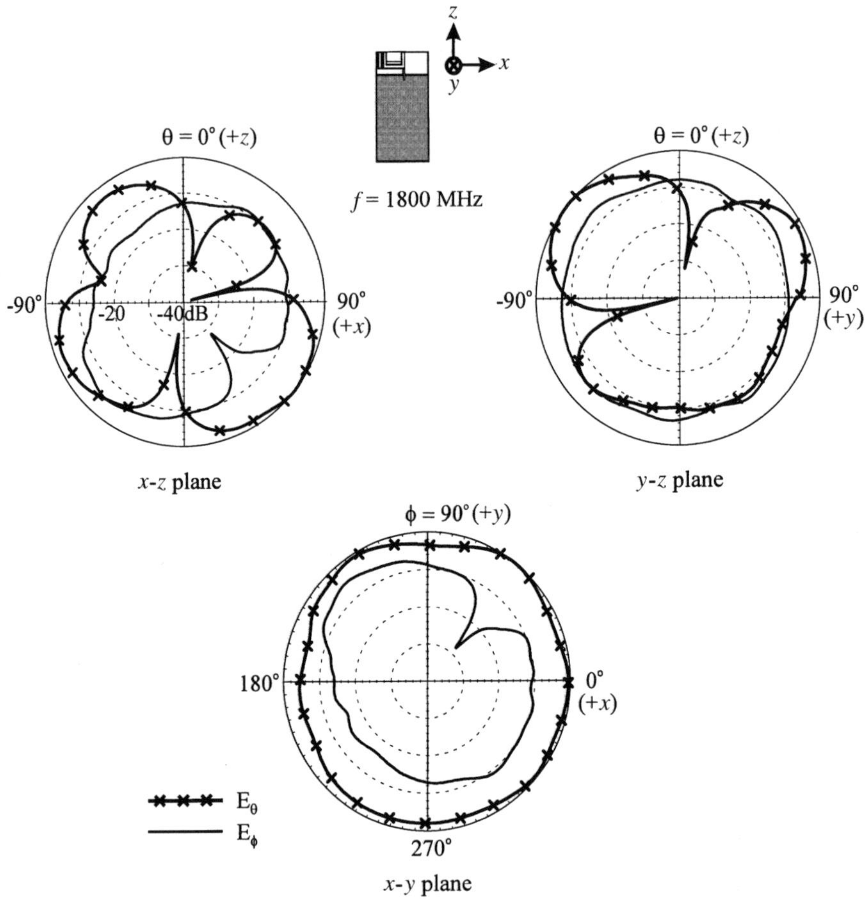

FIGURE 3.32 Measured radiation patterns at 1800 MHz for the antenna shown in Figure 3.29. (From Ref. 6, © 2002 John Wiley & Sons, Inc.)

and omnidirectional or near-omnidirectional radiation in the azimuthal plane), except that there are dips near the x and y directions in the elevation planes (x–z and y–z planes). This characteristic is probably due to the cancellation of some radiated fields contributed from two opposite sections in the proposed folded monopole.

Figure 3.37 shows the measured radiation patterns at 2050 and 2450 MHz. For the patterns at 2050 MHz, characteristics similar to those observed in Figure 3.36 are seen. For 2450 MHz, the radiation in the elevation planes is also similar to that at other frequencies. However, in the azimuthal plane, relatively stronger radiation in the right-hand half-plane ($\varphi = 90°$–$0°$–$270°$ half-plane) is seen. Figure 3.38 shows the measured antenna gain across the operating bandwidth. A peak antenna gain of about 5.2 dBi at about 2.1 GHz is seen, and the gain variations

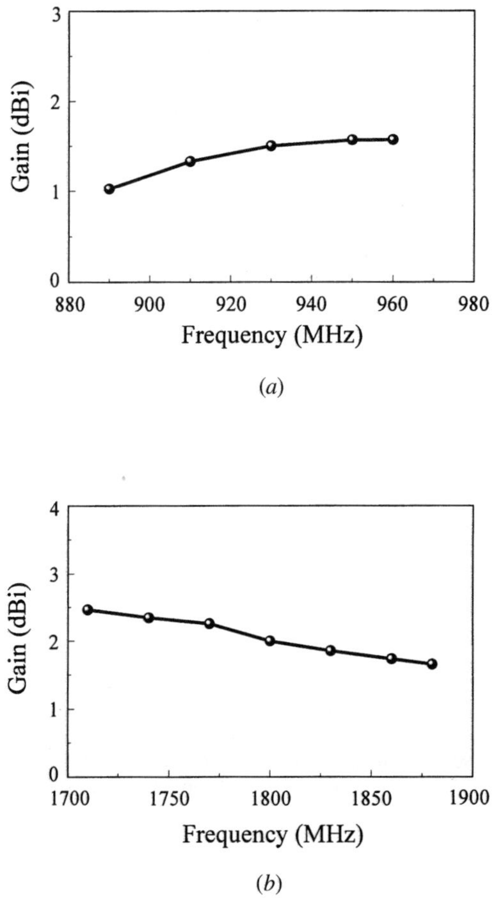

FIGURE 3.33 Measured antenna gain for the antenna shown in Figure 3.29. (*a*) The GSM band; (*b*) the DCS band. (From Ref. 6, © 2002 John Wiley & Sons, Inc.)

are less than 2.6 dBi for frequencies from 1.7 to 2.5 GHz. Finally, it should be noted that further reduction in the monopole length (17.5 mm in this design) is possible, if a larger width of the center section of the folded monopole is chosen. In addition, one can also choose the center section of the folded monopole to face upward (see the geometry shown in Fig. 3.39). This arrangement is also a very promising design in achieving a broadband folded monopole for applications in a mobile phone.

3.7.2 A Meandered, Folded Patch

Another folded monopole design for applications in a mobile phone has been reported [8]. The monopole geometry is shown in Figure 3.40. In this design, the

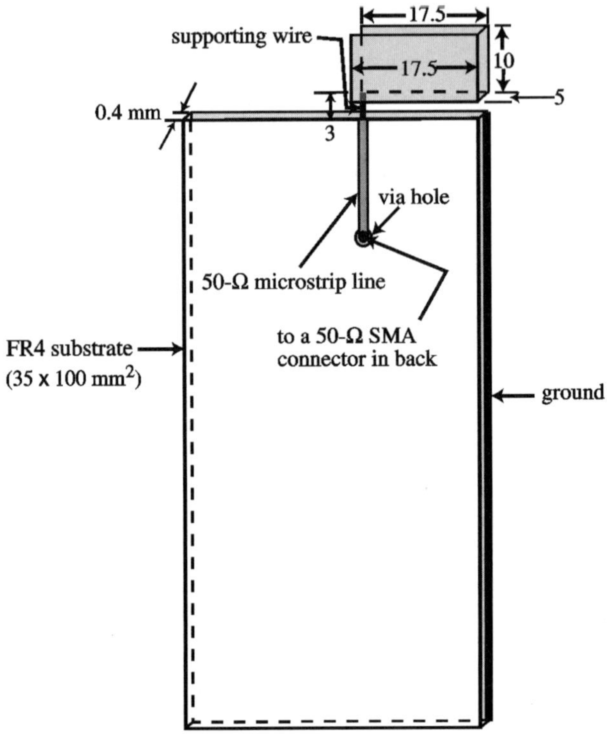

FIGURE 3.34 Geometry of the broadband folded monopole for DCS/PCS/UMTS/WLAN operations of a mobile phone. (From Ref. 7, © 2002 John Wiley & Sons, Inc.)

monopole was easily constructed by folding a planar meandered patch (the dashed lines shown in the figure are the folding lines) into a rectangular tubelike structure and had a compact size of $9 \times 9 \times 34\,\text{mm}^3$ to achieve 900 and 1800 MHz operations. The width (34 mm) of the constructed monopole was chosen to be less than that (35 mm) of the grounded substrate such that it is possible for the proposed monopole to be placed within the mobile phone housing. Thus, when the width of the mobile phone housing is larger, for example 40 mm, it is possible for the height of the proposed monopole to be further lowered to be less than the 9 mm studied here. For the thickness of the proposed monopole, a reasonable thickness of 9 mm was chosen, which is less than that of most present-day mobile phones. Also notice that the planar meandered patch shown in Figure 3.40*b* was made of a copper plate of thickness 0.2 mm in the experiment. In this case, good rigidity of the constructed prototype is obtained. For practical applications, one can also use a thicker copper plate to achieve a more rigid structure of the proposed monopole.

For feeding the proposed monopole, a copper line of length 5 mm was connected between the feed point (point A) of the proposed monopole to the 50-Ω microstrip feed line printed on the grounded substrate. The copper line also serves as a supporting post of the proposed monopole on top of the grounded substrate. In

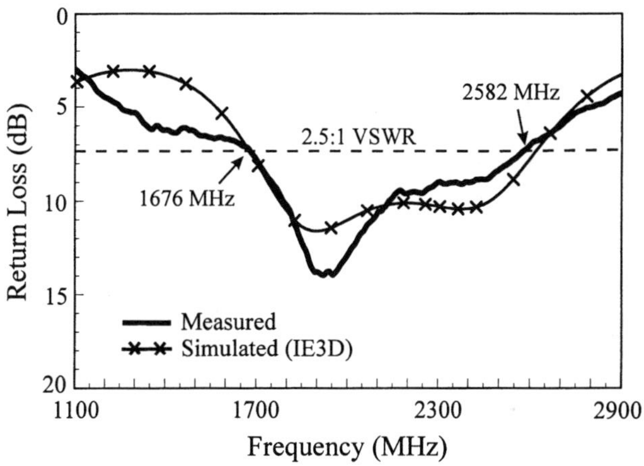

FIGURE 3.35 Measured and simulated return loss for the monopole shown in Figure 3.34. (From Ref. 7, © 2002 John Wiley & Sons, Inc.)

this case, the total height of the proposed monopole from the ground is 14 mm or only about 4.2% of the wavelength at 900 MHz.

As seen in Figure 3.40*b*, there are two resonant paths (paths 1 and 2) in the proposed monopole. Through the meandering caused by the three identical linear slits of width 1 mm and length 29 mm, path 1 has a length of about 110 mm, which is mainly used for generating the 900 MHz resonant mode. With the inclusion of the 5-mm connecting copper line, the length of patch 1 corresponds to about 34% of the wavelength at 900 MHz. Note that this resonant length is larger than one-quarter wavelength of the resonant frequency, and this behavior is largely due to the coupling effect between two meandered sections, which reduces the effective length of the resonant path. On the other hand, path 2 is for generating the 1800 MHz resonant mode and has a length of about 34 mm, which corresponds to about 23% of the wavelength at 1800 MHz by including the length of the connecting copper line. Note that this length is close to one-quarter wavelength at 1800 MHz, which is probably because there is no meandering in path 2.

A prototype of the proposed monopole was constructed and tested. Figure 3.41 shows the measured and simulated return loss of the constructed prototype. Good agreement between the simulation and the measurement is obtained. From the measured data, two resonant modes are clearly excited. The lower mode is at about 900 MHz, as designed, and has a bandwidth (10 dB return loss) of 94 MHz (879–973 MHz), which satisfies the operating bandwidth of the GSM system. For the upper resonant mode, a much wider bandwidth of 270 MHz (1615–1885 MHz) is obtained, which also covers the required bandwidth of the DCS system.

The radiation characteristics were also studied. Figures 3.42 and 3.43 plot the measured radiation patterns at 900 and 1800 MHz, respectively, for the constructed

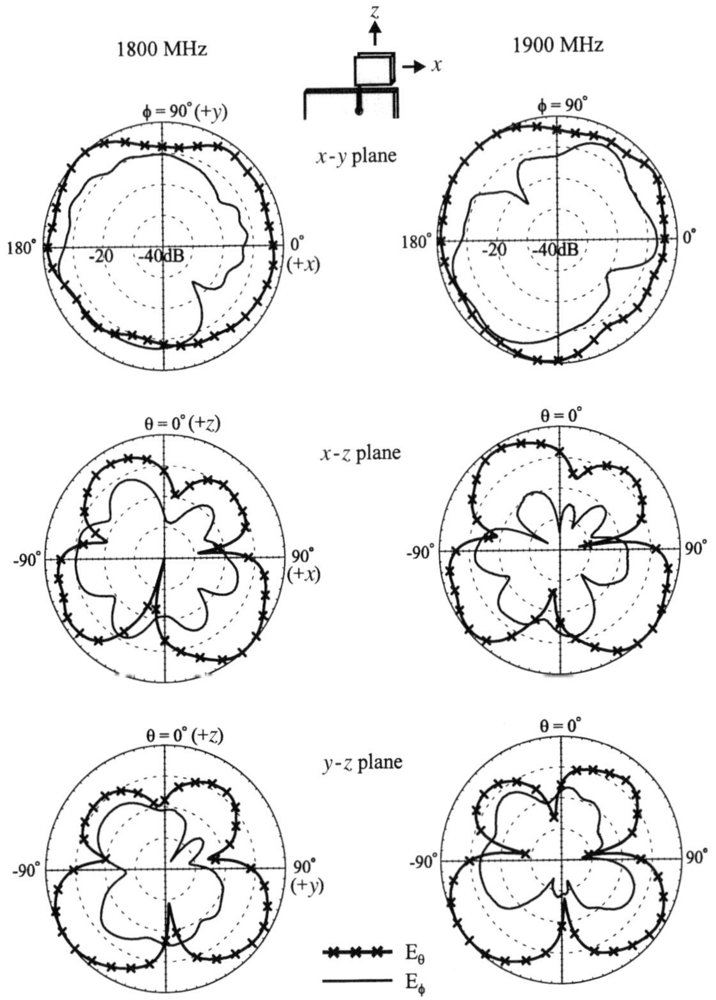

FIGURE 3.36 Measured radiation patterns at 1800 and 1900 MHz for the monopole shown in Figure 3.34. (From Ref. 7, © 2002 John Wiley & Sons, Inc.)

prototype. It is seen that the radiation pattern is more omnidirectional in the elevation plane (x–y plane) at 900 MHz than at 1800 MHz, and the radiation patterns in the principal planes are also more monopole-like for the 900 MHz operation. This behavior is partly because the ground plane effect for the 900 MHz operation is relatively much smaller, because of its relatively larger wavelength compared with the 1800 MHz operation. Radiation patterns for other operating frequencies across the GSM and DCS bands were also measured, and the obtained patterns are stable across their respective operating band and similar to those shown in Figures 3.42 and 3.43. Figure 3.44 shows the measured antenna gain for frequencies in the

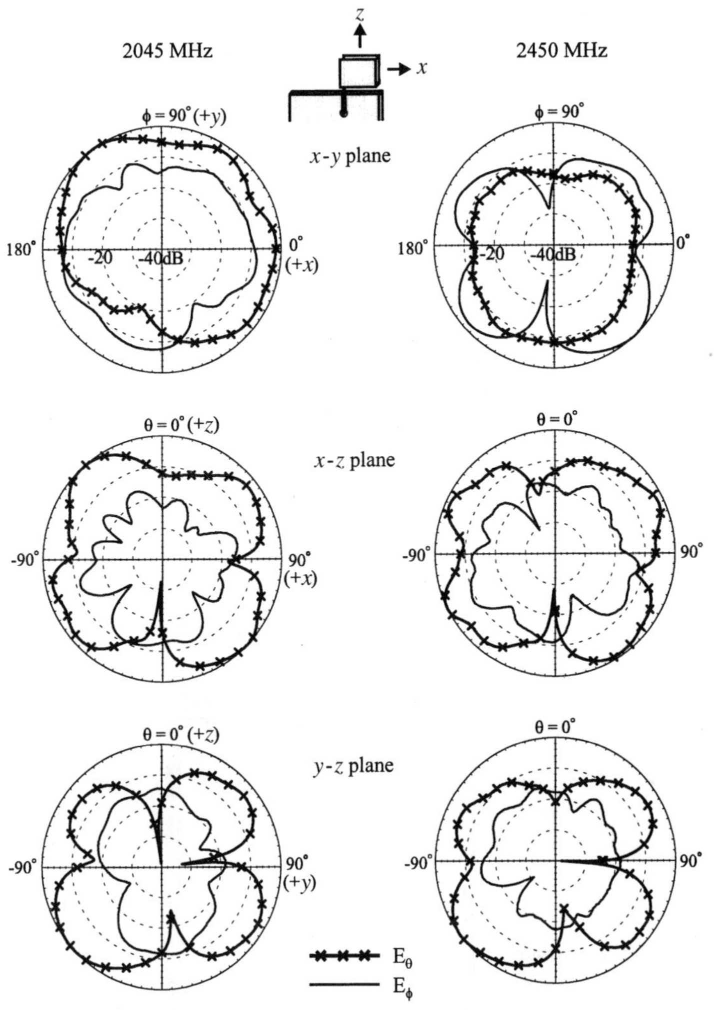

FIGURE 3.37 Measured radiation patterns at 2050 and 2450 MHz for the monopole shown in Figure 3.34. (From Ref. 7, © 2002 John Wiley & Sons, Inc.)

GSM and DCS bands. The antenna gain across the GSM and DCS bands are about 2.2–2.8 dBi and 3.0–4.5 dBi, respectively.

3.8 INVERTED-L WIRE MONOPOLE

3.8.1 With a Helix Loading

Inverted-L wire monopoles are generally operated as quarter-wavelength structures and have a relatively much lower profile compared with a conventional straight

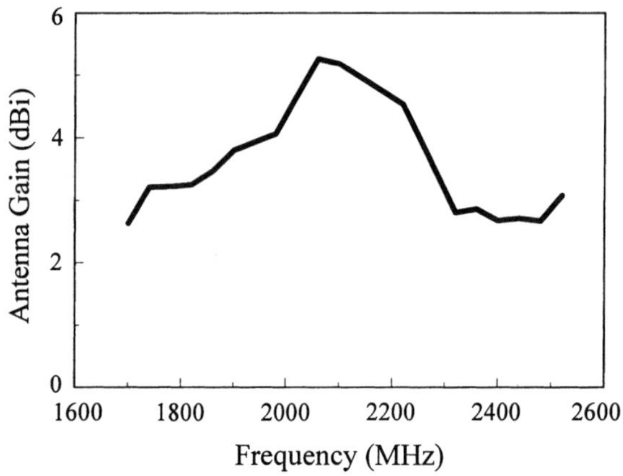

FIGURE 3.38 Measured antenna gain for the monopole shown in Figure 3.34. (From Ref. 7, © 2002 John Wiley & Sons, Inc.)

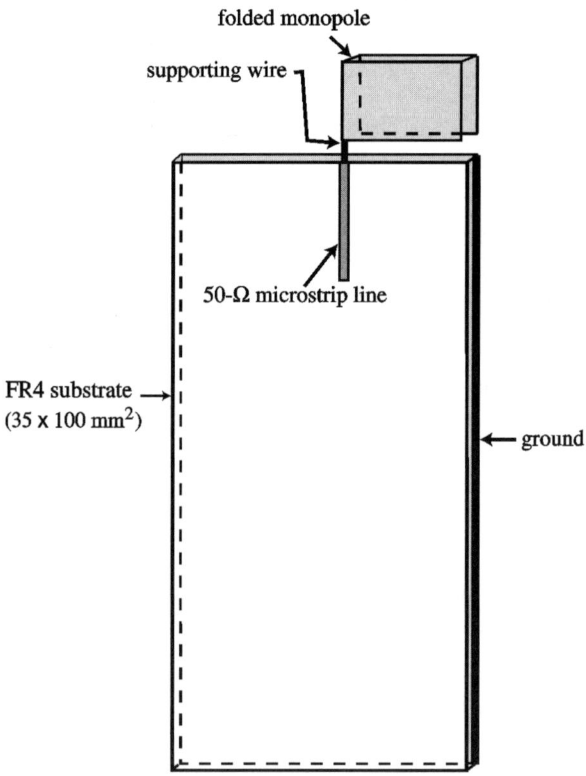

FIGURE 3.39 Geometry of the broadband folded monopole for applications in a mobile phone.

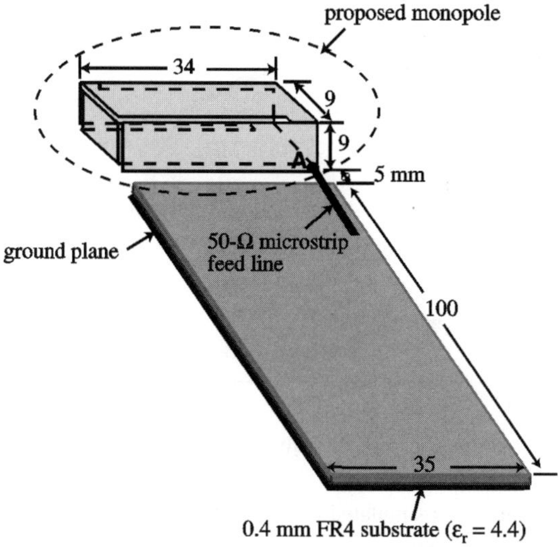

(*a*)

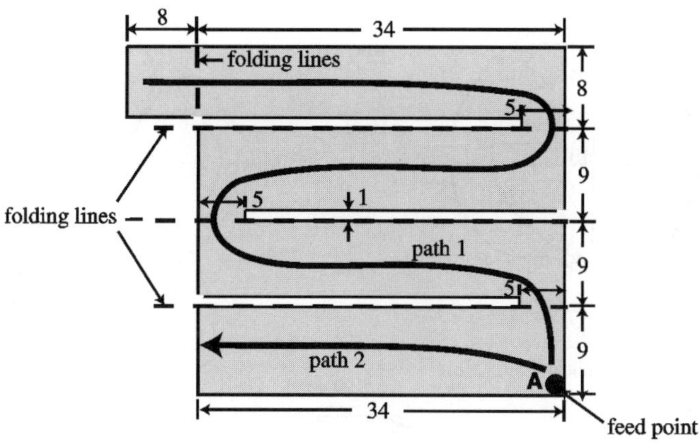

(*b*)

FIGURE 3.40 (*a*) Geometry of the folded meandered-patch monopole mounted on top of a grounded substrate; (*b*) dimensions of the proposed monopole unfolded into a planar structure. (From Ref. 8, © 2002 John Wiley & Sons, Inc.)

monopole. By taking advantage of their low profile (usually less than 0.05 wavelength of the lowest operating frequency), a new design of an inverted-L monopole for applications in GSM/DCS/PCS multiband mobile phone has been studied. The monopole geometry is shown in Figure 3.45. The monopole comprises an inverted-L wire and a loading helix, and the inverted-L wire mainly determines the size of the proposed monopole. Thus, to achieve a minimum monopole size,

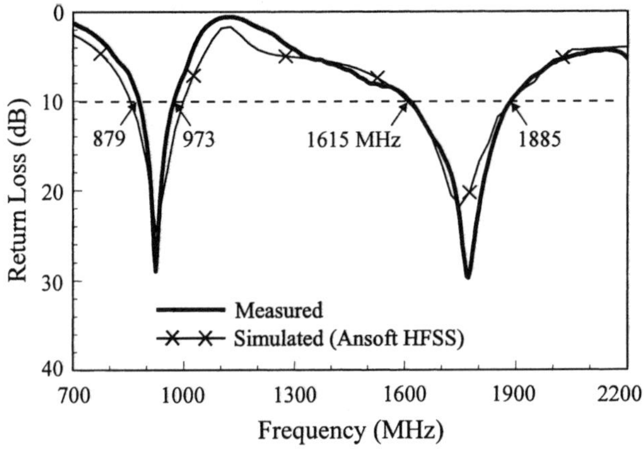

FIGURE 3.41 Measured and simulated return loss for the monopole shown in Figure 3.40. (From Ref. 8, © 2002 John Wiley & Sons, Inc.)

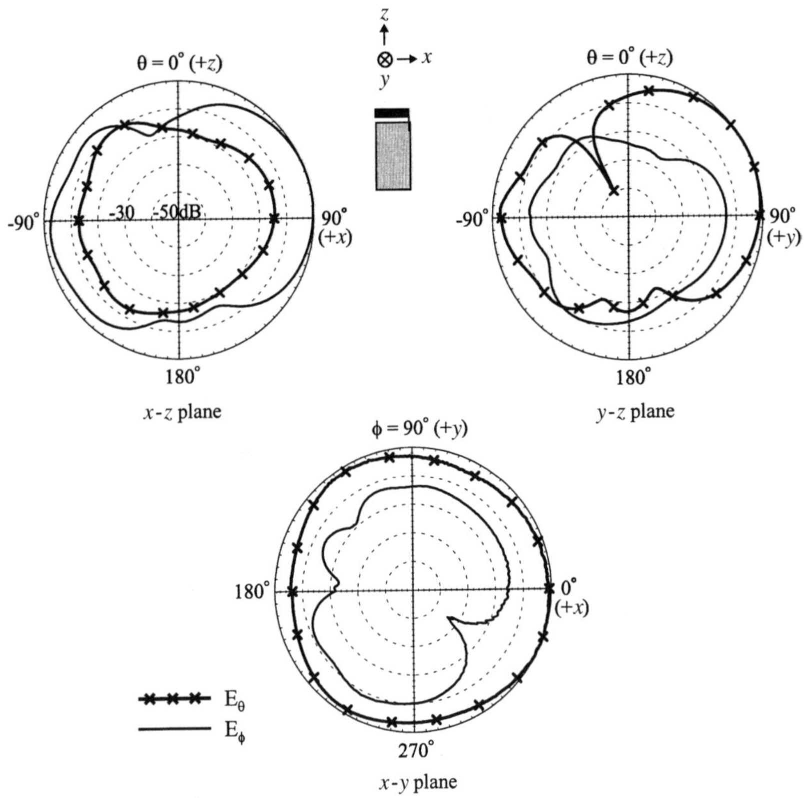

FIGURE 3.42 Measured radiation patterns at 900 MHz for the monopole shown in Figure 3.40. (From Ref. 8, © 2002 John Wiley & Sons, Inc.)

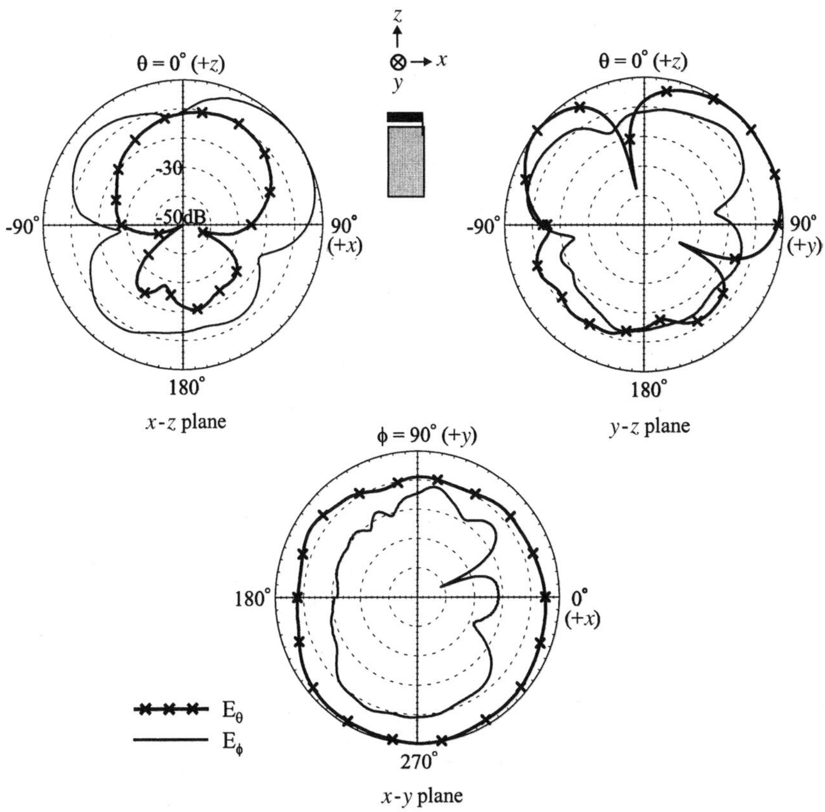

FIGURE 3.43 Measured radiation patterns at 1800 MHz for the monopole shown in Figure 3.40. (From Ref. 8, © 2002 John Wiley & Sons, Inc.)

the inverted-L wire is designed to operate at about 1850 MHz for the upper band covering the DCS and PCS bands, which can limit the monopole dimensions (30 mm in width and 11.5 mm in height in this design) to be suitable for mounting on the top portion of a standard mobile phone. With comparison to the conventional dual-frequency designs using a straight wire and a helix [15–17], the proposed design shows a relatively much lower profile.

For operating at about 900 MHz for the lower band covering the GSM band, a helix, a conductor wound into a right-hand or left-hand coil that provides a much shorter linear dimension than the straight monopole structure when the operating frequency is the same, is loaded on the inverted-L monopole (see Fig. 3.45). Moreover, without the need of additional matching networks, the proposed monopole can be easily excited, with good impedance matching for frequencies across the GSM, DCS, and PCS bands.

With the design dimensions shown in Figure 3.45, prototypes with and without the loading helix have been constructed and tested. It has been observed that,

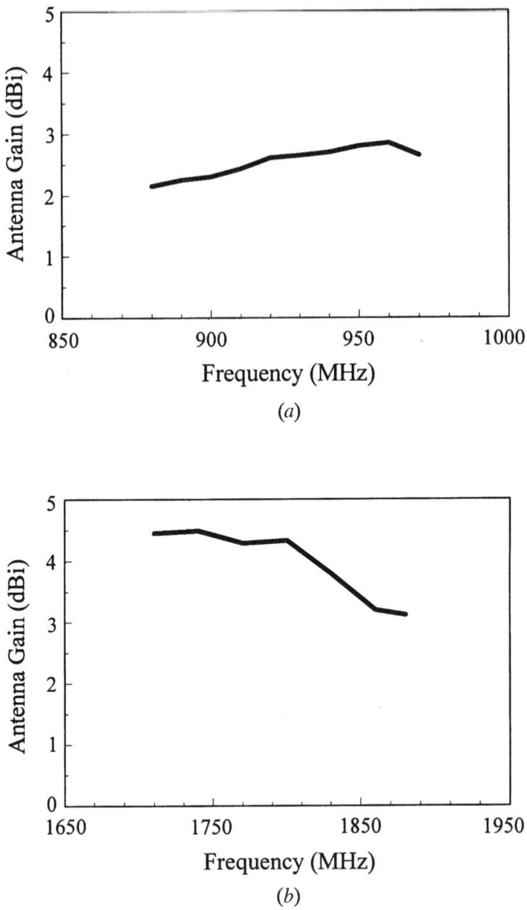

FIGURE 3.44 Measured antenna gain for the monopole shown in Figure 3.40. (*a*) The GSM band; (*b*) the DCS band. (From Ref. 8, © 2002 John Wiley & Sons, Inc.)

when the loading helix is present, an additional resonant mode (the lower mode) is excited at about 900 MHz, with the resonant mode at about 1850 MHz (the upper mode) almost unaffected. The obtained impedance bandwidths for the lower and upper modes also cover the GSM, DCS, and PCS bands. Good radiation characteristics have also been observed, and the measured gain level is about 1.6 dBi for the GSM operation and about 3.5 dBi across the DCS and PCS bands.

3.8.2 With a Meandered Wire

In addition to using a loading helix, a design incorporating the use of an inverted-L wire and a meandered wire for achieving the 900 and 1800 MHz operations has been

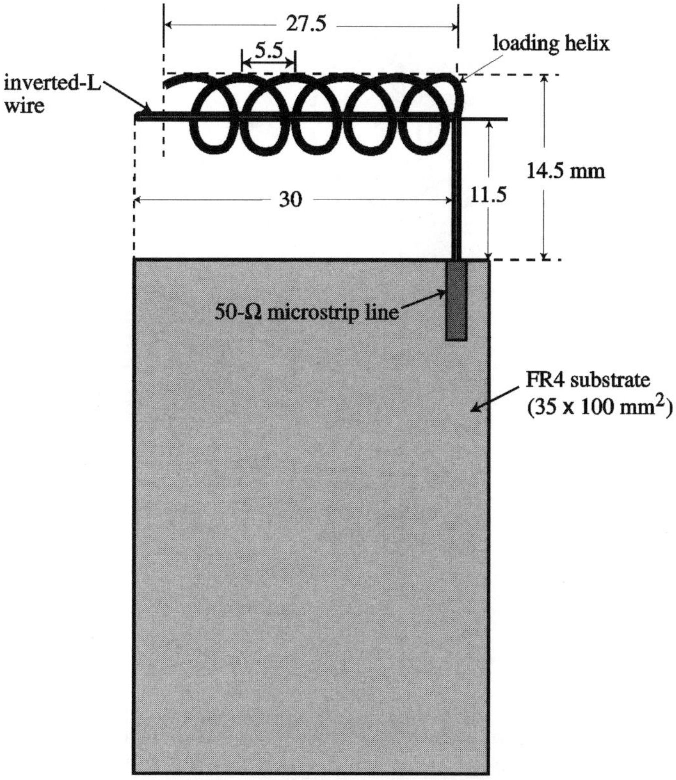

FIGURE 3.45 Geometry of the low-profile inverted-L monopole loaded with a helix for a GSM/DCS/PCS multi-band mobile phone.

demonstrated [9]. Figure 3.46 shows the geometry of such a dual-frequency monopole suitable for applications in a mobile phone. The proposed monopole comprises an inverted-L wire and a meandered wire, both operated as quarter-wavelength structures. The design dimensions are optimized with the aid of the simulation software Ansoft HFSS. The inverted-L wire is for DCS operation at 1800 MHz and consists of a vertical section (between points A and B, length 12 mm) and a horizontal section (between points B and C, length 32 mm). The total length of the inverted-L wire is 44 mm, which corresponds to about 0.25 wavelength at 1800 MHz. On the other hand, the meandered wire having a length 108 mm (starting from points B to D) is for the GSM operation at 900 MHz and is loaded at point B, the joint point of the vertical and horizontal sections of the inverted-L wire. Note that, with the inclusion of the length of the vertical section of the inverted-L wire, the total length is 120 mm, which is about 0.36 wavelength at 900 MHz, much larger than the required 0.25 wavelength for operating as a quarter-wavelength structure. This is largely because there is some coupling between

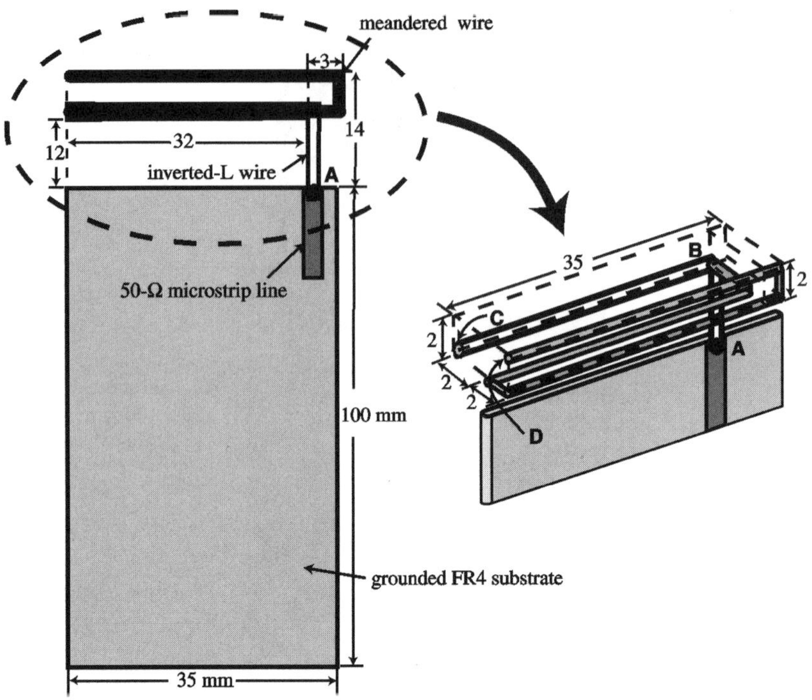

FIGURE 3.46 Geometry of the inverted-L monopole loaded with a meandered wire for a GSM/DCS dual-band mobile phone. (From Ref. 9, © 2002 John Wiley & Sons, Inc.)

adjacent meandered sections, which reduces the effective electrical length of the meandered wire.

It should be noted that the horizontal section of the inverted-L wire and the meandered wire are configured to be on two adjacent surfaces (the bottom and the front side surfaces) of a long rectangular box with dimensions $2 \times 4 \times 35\,\text{mm}^3$. This configuration makes the proposed monopole easy to construct. Also note that the proposed monopole requires a spacing (or the length of the vertical section of the inverted-L wire) of 12 mm away from the ground plane to achieve two large enough bandwidths covering the GSM and DCS bands. This makes the total monopole height from the top of the ground plane to be 14 mm, which is only about 4.2% of the wavelength at 900 MHz.

In the experiment, the proposed monopole is fed by a coax feed (placed in the ground plane side of the substrate), which is connected to the 50-Ω microstrip line through a via-hole in the substrate. Figure 3.47 shows the measured and simulated return loss. Good agreement between the measurement and the simulation is observed. Two resonant modes at about 900 and 1800 MHz are excited with good impedance matching. The impedance bandwidth, determined from 2.5:1 VSWR or about 7.3 dB return loss, for the lower band is 140 MHz (872–1012 MHz) and covers the required bandwidth of the GSM operation. For the upper band, an impedance

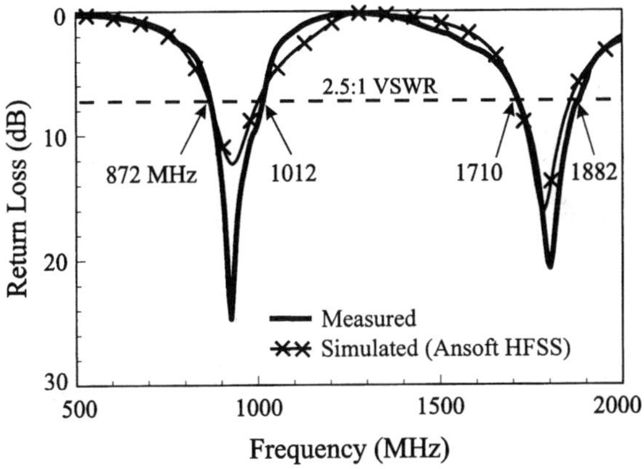

FIGURE 3.47 Measured and simulated return loss for the monopole shown in Figure 3.46. (From Ref. 9, © 2002 John Wiley & Sons, Inc.)

bandwidth of 172 MHz (1710–1882 MHz) is obtained, which also satisfies the bandwidth requirement of the DCS operation.

Radiation characteristics for both lower and upper bands were also studied. Measured radiation patterns at 900 MHz are plotted in Figure 3.48, and results for other operating frequencies are similar to those shown here. From the measurement monopole-like radiation patterns are seen, and nearly omnidirectional radiation is observed in the azimuthal plane (x–y plane). The measured radiation patterns at 1800 MHz are presented in Figure 3.46. Again, the measured radiation patterns for other frequencies across the upper band are stable and similar to those presented here. The measured antenna gain is shown in Figure 3.50. For the lower (GSM) band, the antenna gain is about 1.5–2.1 dBi. For the upper (DCS) band, the antenna gain reaches about 4.2–5.4 dBi.

3.8.3 With a Chip Inductor

With the loading of a chip inductor, the inverted-L monopole can also perform a dual-frequency operation. Figure 3.51 shows a promising design of the inverted-L strip monopole loaded with a chip inductor for a GSM/DCS dual-band mobile phone. This design technique is very similar to that applied in Section 2.2.9 for the PIFA with a rectangular spiral strip loaded with a chip inductor. When a chip inductor of suitable inductance is loaded across a gap cut in the inverted-L strip, it will perform like a short circuit for lower operating frequencies and, conversely, like an open circuit for higher operating frequencies.

For the short-circuit condition, the inverted-L strip is operated as a quarter-wavelength structure, and its total length will roughly correspond to 0.25 wavelength of the center frequency in the lower band. On the other hand, for the open-circuit condition, the portion of the inverted-L strip from the gap to the open end can be

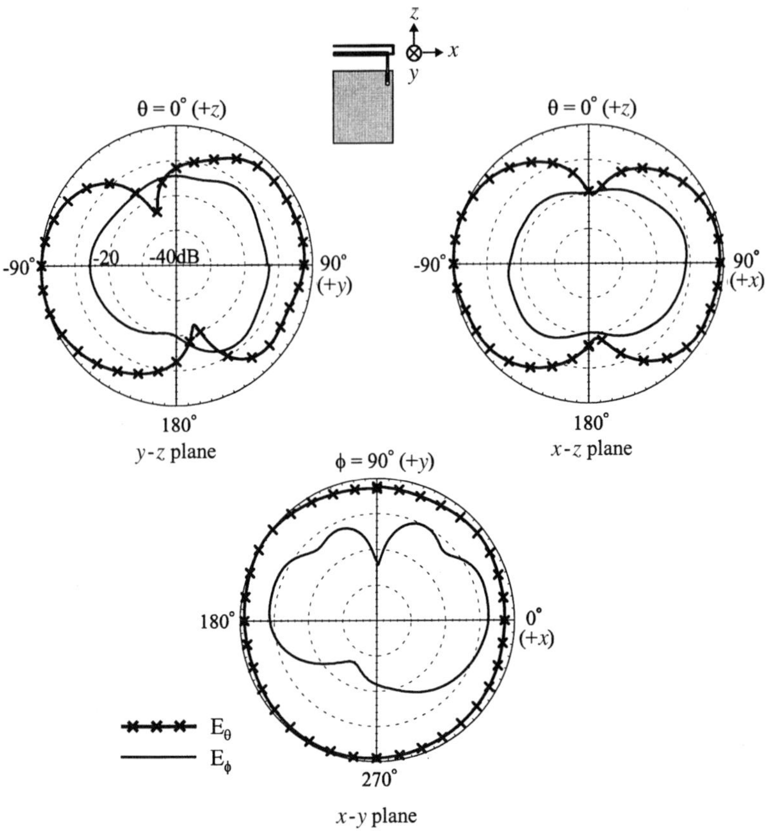

FIGURE 3.48 Measured radiation patterns at 900 MHz for the monopole shown in Figure 3.46. (From Ref. 9, © 2002 John Wiley & Sons, Inc.)

treated to be disconnected, and the portion of the inverted-L strip from the feed point to the gap can be operated as a quarter-wavelength structure. In this case, by controlling the position of the loading chip inductor, the desired upper operating frequency can be obtained, thereby leading to a dual-frequency operation for an inverted-L strip monopole.

3.9 PRINTED INVERTED-F MONOPOLE

An inverted-F strip monopole can also be directly printed on a mobile phone circuit board to operate as an integrated monopole [10]. The geometry of such a monopole design is shown in Figure 3.52, in which a folded inverted-F strip monopole is printed on a dielectric substrate and fed by a coplanar waveguide. In this case, no via-hole is required in shorting the inverted-F strip to the ground plane. In addition,

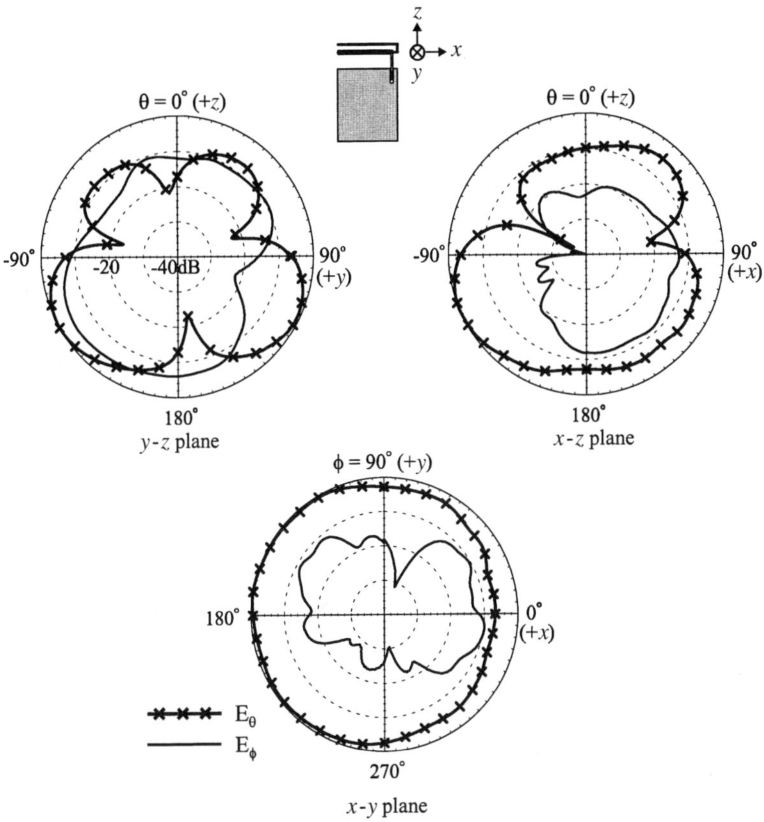

FIGURE 3.49 Measured radiation patterns at 1800 MHz for the monopole shown in Figure 3.46. (From Ref. 9, © 2002 John Wiley & Sons, Inc.)

the open end of the shorted horizontal strip is folded toward the center strip of the inverted-F strip monopole, which decreases the required realty space for the proposed design.

In this study, an inexpensive FR4 substrate of thickness 1.6 mm and relative permittivity 4.4 is used for the dielectric substrate. The folded inverted-F strip has a uniform width of w and comprises a vertical shorting strip of length h, a horizontal strip of length l, a center strip of length $h - w$, and a folded portion of length $h_1 + l_1$. The center strip of the folded inverted-F strip is connected to the signal strip (width w_c) of the CPW feed line, which is designed to have a characteristic impedance of 50 Ω. The spacing between the signal strip and the coplanar grounds is denoted as g.

The design parameters of the proposed monopole can easily be determined. With the desired center operating frequency known, the total length, $h + l + h_1 + l_1$, of the folded inverted-F strip can be first determined to be approximately one-quarter wavelength of the center operating frequency in free space (the relative permittivity

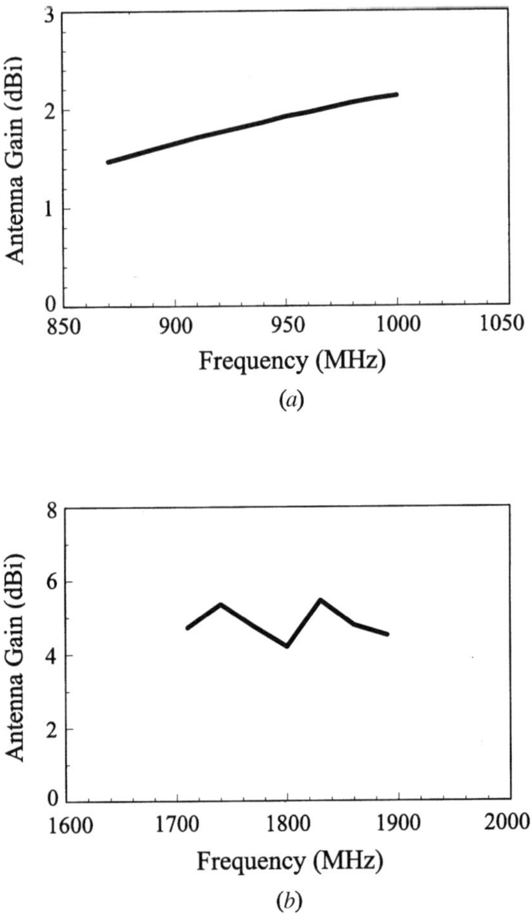

FIGURE 3.50 Measured antenna gain for the monopole shown in Figure 3.46. (*a*) The lower band; (*b*) the upper band. (From Ref. 9, © 2002 John Wiley & Sons, Inc.)

of the dielectric substrate used here has small effects on the resonant length of the inverted-F strip). Then, by fine-tuning the length l_1 of the folded portion of the inverted-F strip and adjusting the distance d between the center strip and the shorting strip, good impedance matching over a wide frequency range can be achieved for the proposed design.

A prototype of the proposed design for the application in the UMTS band has been constructed and studied. The designed center frequency was at 2045 MHz, which is the center frequency of the UMTS band (1920–2170 MHz). For this reason, the total length of the folded inverted-F strip was adjusted to be 43 mm, which is about 29% of the wavelength at 2045 MHz. Also note that, in this design, the length of the shorting strip was selected to be only 8 mm (about 5% of the wavelength at

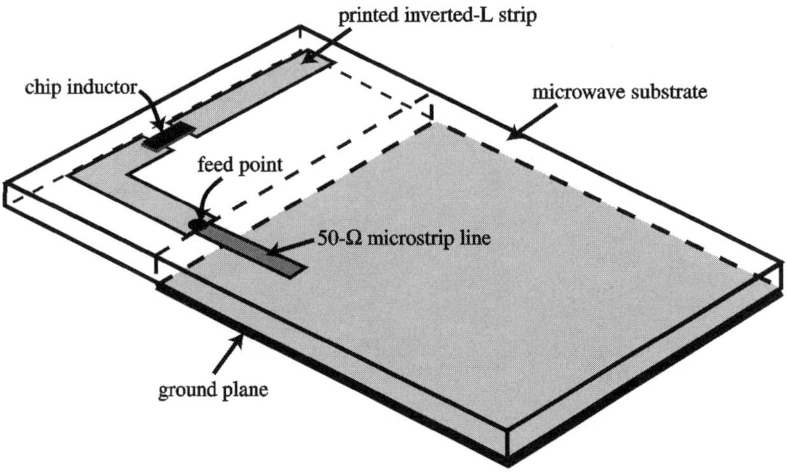

FIGURE 3.51 Geometry of the inverted-L strip monopole loaded with a chip inductor for a GSM/DCS dual-band mobile phone.

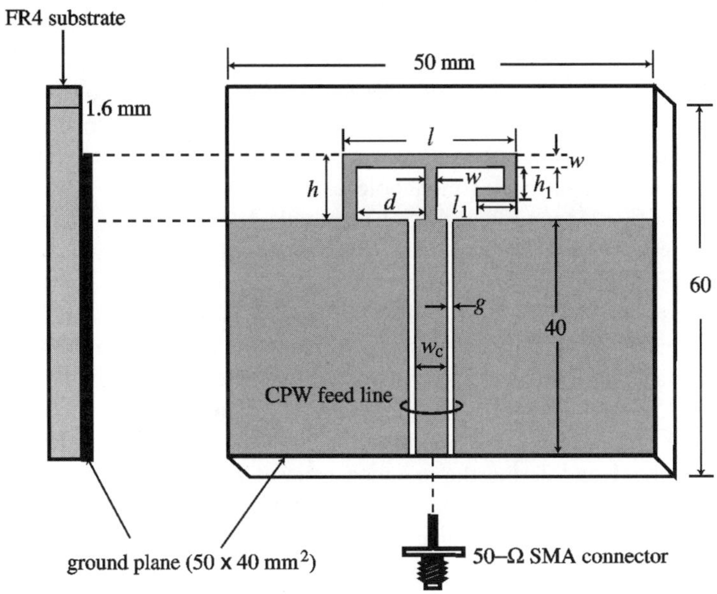

FIGURE 3.52 Geometry of the CPW-fed inverted-F monopole. (From Ref. 10, © 2002 John Wiley & Sons, Inc.)

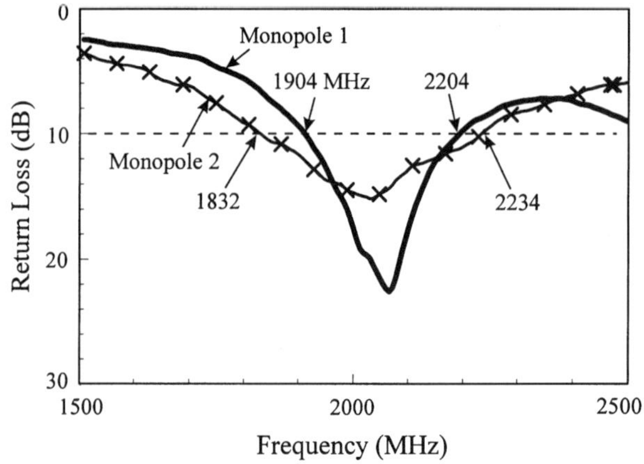

FIGURE 3.53 Measured and simulated return loss for the monopole shown in Figure 3.52 with $h = 8$ mm, $d = 12$ mm, $w = 2.6$ mm, $g = 0.5$ mm, $w_c = 6.4$ mm. Monopole 1: $l = 33$ mm, $l_1 = h_1 = 0$. Monopole 2: $l = 25$ mm, $l_1 = 6$ mm, $h_1 = 4$ mm. (From Ref. 10, © 2002 John Wiley & Sons, Inc.)

2045 MHz) to achieve a lower height for the proposed monopole. In addition, it is found that the total strip length excluding the shorting strip length is 35 mm, very close to one-quarter wavelength at 2045 MHz.

Figure 3.53 shows the measured return loss for the constructed prototype (denoted as monopole 2) with a center frequency at about 2045 MHz. A corresponding inverted-F strip antenna without the folded portion (i.e., $l_1 = h_1 = 0$) was also constructed, and the measured results (denoted as monopole 1) are also shown in the figure. To operate at about 2045 GHz, the required total strip length for monopole 1 was adjusted to 41 mm (about 0.28 resonant wavelength), which is slightly smaller than that of monopole 2. However, monopole 2 has a smaller occupied area (8×25 mm^2) than that of monopole 1 (8×33 mm^2), mainly because of the presence of the folded portion of the proposed design. From the obtained results, it is also seen that the impedance bandwidth, determined by 10 dB return loss, reaches 402 MHz (1832–2234 MHz), which is about 20% referenced to 2045 GHz and is much larger than that of monopole 1 (300 MHz). The obtained impedance bandwidth also covers the required bandwidth of the UMTS band.

Figures 3.54 and 3.55 plot the measured radiation patterns at 2045 GHz for monopoles 1 and 2, respectively. Similar radiation characteristics are observed for both monopoles 1 and 2, and the radiation patterns are also observed to be stable for operating frequencies across the UMTS band. It is also noted that, for monopole 2, the patterns in the azimuthal plane (x–y plane) and the elevation plane (y–z plane) are nearly omnidirectional, which gives the proposed design large radiation coverage in practical applications. The measured antenna gain for monopoles 1 and 2 for frequencies within the UMTS band is also presented in Figure 3.56. Monopole 2 shows a peak antenna gain of about 2.3 dBi, whereas that of monopole 1 is about

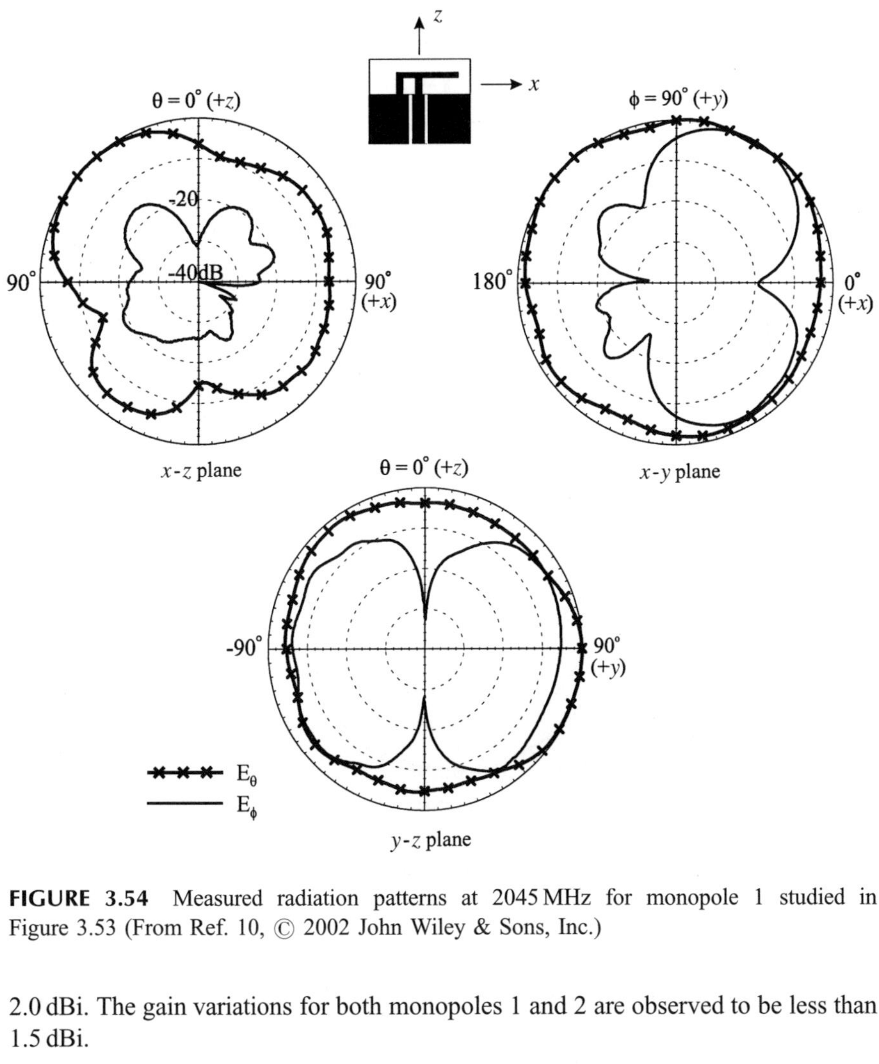

FIGURE 3.54 Measured radiation patterns at 2045 MHz for monopole 1 studied in Figure 3.53 (From Ref. 10, © 2002 John Wiley & Sons, Inc.)

2.0 dBi. The gain variations for both monopoles 1 and 2 are observed to be less than 1.5 dBi.

3.10 SHORTED MONOPOLE WITH REDUCED BACKWARD RADIATION

It was recently shown that, by bending a planar monopole into an inverted-U shape and short-circuiting one of its ends to the ground, a low-profile monopole with reduced backward radiation can be obtained [11]. The proposed antenna is easily fed by using a capacitively coupled feed and is suitable for applications in mobile phones for providing reduced backward radiation toward the user's head. The monopole design is shown in Figure 3.57, and the design dimensions given in the

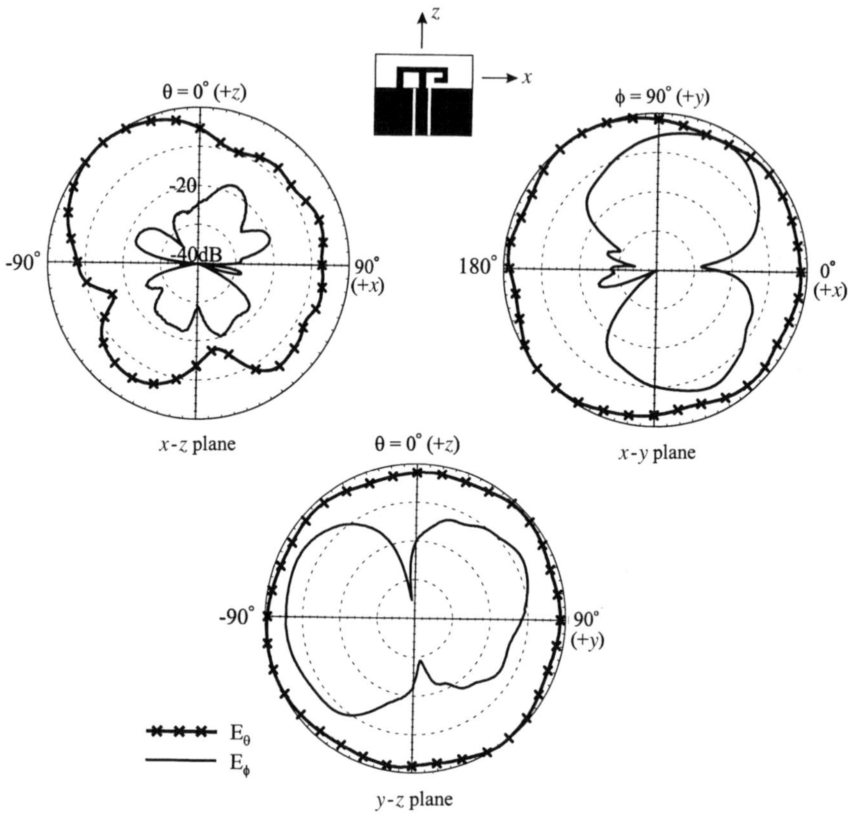

FIGURE 3.55 Measured radiation patterns at 2045 MHz for monopole 2 studied in Figure 3.53 (From Ref. 10, © 2002 John Wiley & Sons, Inc.)

figure are for the DCS operation. The proposed monopole is mounted on an L-shaped ground plane with a vertical ground (ground A in the figure) of dimensions $70 \times 40\,mm^2$ and a horizontal ground (ground B) of dimensions $40 \times 14\,mm^2$. Ground A can be considered as the ground plane of the circuit board of a practical mobile phone, and ground B is an additional ground plane placed on the top portion of the mobile phone to accommodate the proposed monopole. The proposed monopole has a simple structure and is formed by bending a planar monopole of width 10 mm and length 40 mm, which is about one-quarter wavelength of the frequency at 1800 MHz, into an inverted-U shape and short-circuiting one of its ends to ground B.

The dimensions of the proposed monopole shown in the figure are for the DCS application. In this case, the proposed monopole comprises three sections (sections A, B, and C). Section A ($13 \times 10\,mm^2$) has a distance of 2 mm to ground B and is 3 mm from the feed patch ($8 \times 8\,mm^2$), through which the monopole is capacitively

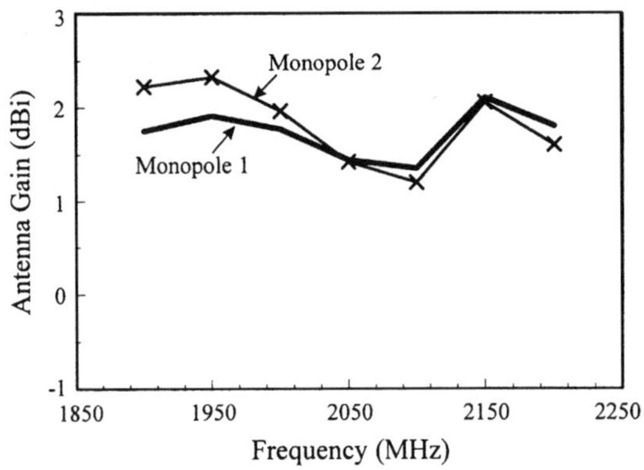

FIGURE 3.56 Measured antenna gain for frequencies across the UMTS band for the monopoles studied in Figure 3.53. (From Ref. 10, © 2002 John Wiley & Sons, Inc.)

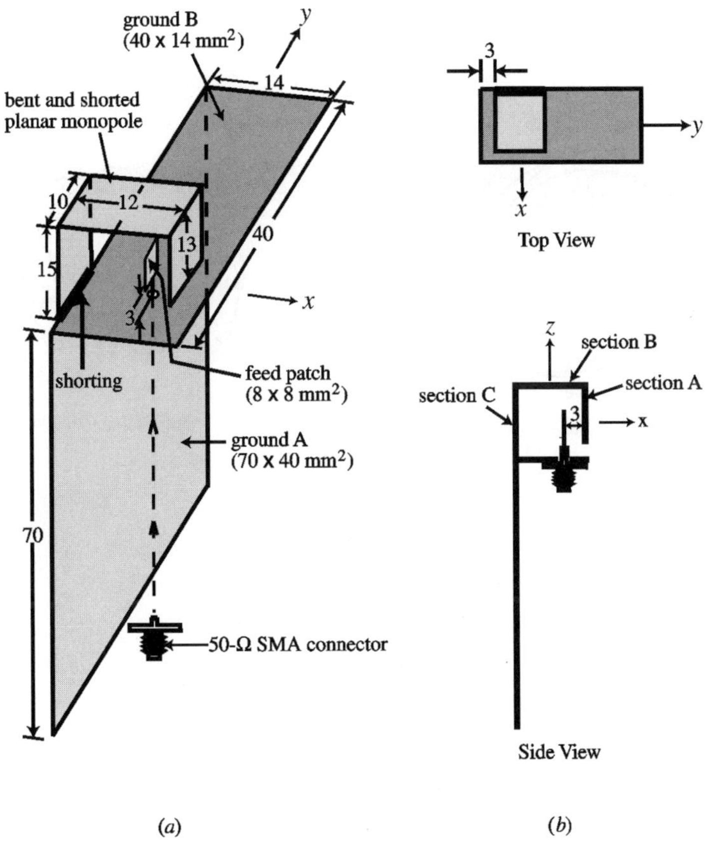

FIGURE 3.57 (*a*) Geometry of the proposed antenna on an L-shaped ground plane; (*b*) monopole's top view and side view. (From Ref. 11, © 2002 John Wiley & Sons, Inc.)

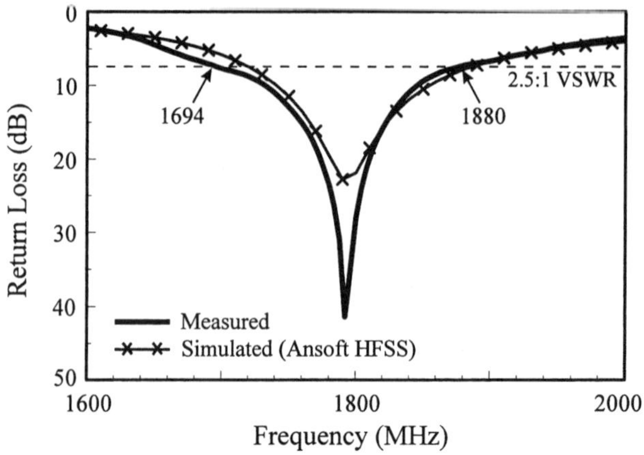

FIGURE 3.58 Measured and simulated return loss for the monopole shown in Figure 3.57. (From Ref. 11, © 2002 John Wiley & Sons, Inc.)

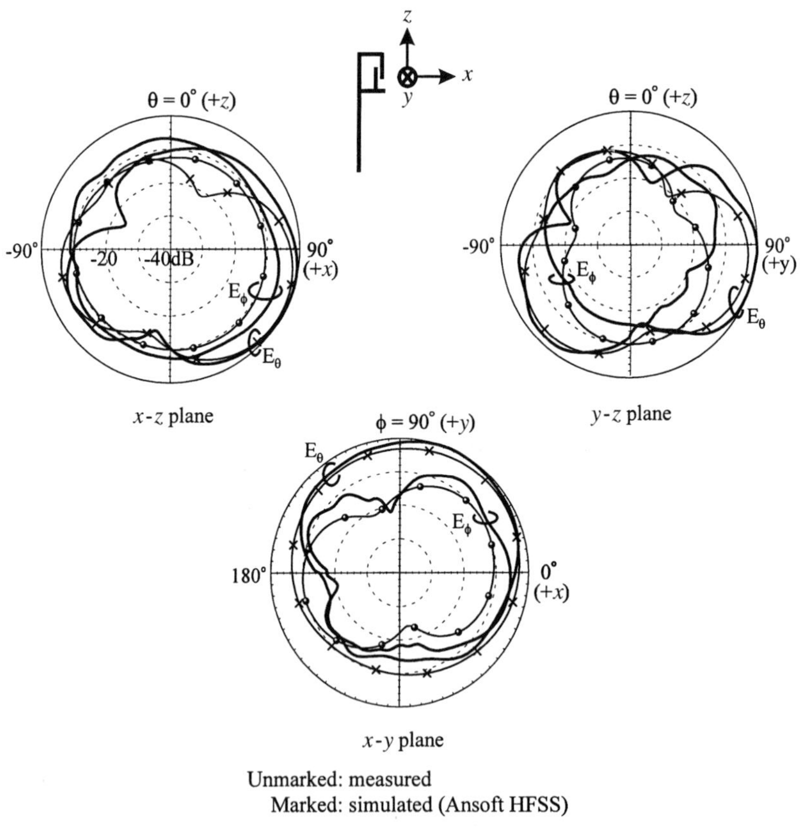

FIGURE 3.59 Measured radiation patterns at 1800 MHz for the monopole shown in Figure 3.57. (From Ref. 11, © 2002 John Wiley & Sons, Inc.)

excited. Section B ($12 \times 10\,mm^2$) is in parallel to ground B, and section C ($15 \times 10\,mm^2$) is shorted to the ground. Note that section C is in the same plane to ground A (see Fig. 3.57*b*), and in the proposed monopole, section C can also be considered as an extension of the ground plane such that it can shield a certain portion of the radiated power toward the monopole's back or the $-x$ direction (see Fig. 3.57). This is probably the reason that reduced backward radiation is achieved for the proposed monopole. In addition, the proposed monopole has a compact size of dimensions $15 \times 10 \times 12\,mm^3$, with a low profile of only 15 mm (9% of the wavelength at 1800 MHz).

With the design dimensions shown in Figure 3.57, a prototype of the proposed monopole has been constructed. The design dimensions are obtained with the aid of the simulation software Ansoft HFSS. Figure 3.58 shows the measured and simulated return loss of the constructed prototype. Good agreement between the measurement and the simulation is seen. The obtained impedance bandwidth (2.5:1 VSWR) reaches 186 MHz (1694–1880 MHz), which covers the required bandwidth of the DCS system. Measured and simulated radiation patterns at 1800 MHz are presented in Figure 3.59. The measured data in general agree with the simulated results. It is also clearly seen that, in both the x–z plane (the elevation plane) and the x–y plane (the azimuthal plane), stronger radiation intensity in the forward direction ($+x$ direction) than in the backward direction ($-x$ direction) is observed. This behavior probably occurs because section C in the proposed design also serves as a shield for blocking some of the monopole's backward radiation. Radiation patterns for other frequencies across the DCS band were also measured, and radiation patterns similar to those shown here were observed. Figure 3.60 shows the measured antenna gain for frequencies across the DCS band. The antenna gain is stable across the DCS band and has a gain level of about 4 dBi.

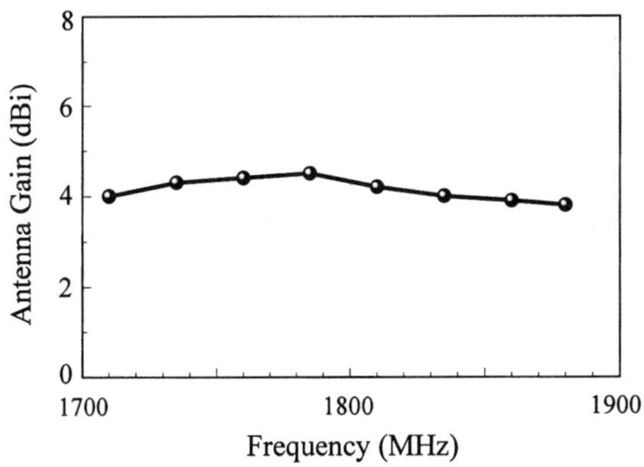

FIGURE 3.60 Measured antenna gain for the monopole shown in Figure 3.57. (From Ref. 11, © 2002 John Wiley & Sons, Inc.)

REFERENCES

1. Z. Ying, "Multiple band, multiple branch antenna for mobile phone," World Intellectual Property Organization, International Publication No. WO 99/22420, May 6, 1999.
2. P. L. Teng and K. L. Wong, "Planar monopole folded into a compact structure for very-low-profile multi-band mobile phone antenna," Microwave Opt. Technol. Lett., vol. 33, pp. 22–25, April 5, 2002.
3. K. Rutkowski and G. J. Hayes, "Multiple frequency band branch antennas for wireless communications," U.S. Patent No. 6198442, March 6, 2001.
4. F. S. Chang, S. H. Yeh, and K. L. Wong, "Planar monopole in wrapped structure for low-profile GSM/DCS mobile phone antenna," Electron. Lett., vol. 38, pp. 499–500, May 23, 2002.
5. G. Y. Lee and K. L. Wong, "Very-low-profile bent planar monopole antenna for GSM/DCS dual-band mobile phones," Microwave Opt. Technol. Lett., vol. 35, Sept. 20, 2002.
6. K. L. Wong, S. W. Su, T. W. Chiou, and Y. C. Lin, "Dual-band plastic chip antenna for GSM/DCS mobile phones," Microwave Opt. Technol. Lett., vol. 33, pp. 330–332, June 5, 2002.
7. G. Y. Lee, S. H. Yeh, and K. L. Wong, "A broadband folded planar monopole antenna for mobile phones," Microwave Opt. Technol. Lett., vol. 33, pp. 165–167, May 5, 2002.
8. F. S. Chang and K. L. Wong, "Folded meandered patch monopole antenna for low-profile GSM/DCS dual-band mobile phone," Microwave Opt. Technol. Lett., vol. 34, pp. 84–86, July 20, 2002.
9. H. C. Tung, C. Y. Fang, and K. L. Wong, "An inverted-L monopole antenna loaded with a meandered wire for GSM/DCS dual-band mobile phones," Microwave Opt. Technol. Lett., vol. 33, pp. 212–214, May 5, 2002.
10. Y. L. Kuo and K. L. Wong, "Coplanar waveguide-fed folded inverted-F antenna for UMTS application," Microwave Opt. Technol. Lett., vol. 32, pp. 364–366, March 5, 2002.
11. Y. H. Yeh, Y. Y. Chen, and K. L. Wong, "A low-profile, bent and shorted monopole antenna with reduced backward radiation for mobile handsets," Microwave Opt. Technol. Lett., vol. 33, pp. 146–147, April 20, 2002.
12. S. Eggleston, "Antenna assembly, and associated method, having parasitic element for altering antenna pattern characteristics," U.S. Patent No. 6249255, June 19, 2001.
13. G. Y. Lee, T. W. Chiou, K. L. Wong, and C. Wang, "Low-profile planar monopole antenna for GSM/DCS/PCS triple-band mobile phone," in 2002 IEEE Antennas Propagat. Soc. Int. Symp. Dig., vol. 3, pp. 26–29.
14. C. M. Chiang, "Built-in dual frequency antenna," U.S. Patent No. 6337663, Jan. 8, 2002.
15. P. Haapala, P. Vainikainen, and P. Eratuuli, "Dual frequency helical antennas for handsets," in Proc. 1996 IEEE Vehicular Technol. Conf., pp. 336–338.
16. P. Haapala, "Antenna operating in two frequency ranges," U.S. Patent No. 6054966, April 25, 2000.
17. D. Maldonado, "2band antenna," U.S. Patent No. 5812097, Sept. 22, 1998.

CHAPTER FOUR

Base Station Antennas for Cellular Communication Systems

4.1 INTRODUCTION

Patch antennas with a thick air-layer substrate can have a wide impedance bandwidth and are very promising for applications in cellular system base stations. A variety of broadband patch antenna designs have been described in Reference 1. Applying these broadband designs and some recently reported design techniques [2–17], the patch antennas satisfying the operating bandwidths of present-day cellular communication systems such as the GSM, DCS, PCS, and UMTS systems are presented in this chapter. In Section 4.2, recent advanced patch antenna designs for single-band cellular system base stations are introduced. These designs include the use of slot-coupled feeds [5], probe feeds [6–10], coplanar probe feeds [11], microstrip line feeds, capacitively coupled feeds [12], and so on.

Section 4.3 addresses some novel dual-band (GSM/DCS) and multiband (GSM/DCS/PCS) designs that have recently been developed. These recent designs include the use of a radiating patch incorporating a parasitic shorted patch for GSM/DCS operations and a radiating patch incorporating a U-shaped ground plane for GSM/DCS/PCS operations [13]. In comparison to the conventional design of using two separate radiating patches, these designs have a more compact antenna size. Some recent advanced patch antenna designs capable of broadband dual-polarized operations for cellular system base stations are discussed in Sections 4.4 and 4.5, in which designs utilizing two same feeds and two hybrid feeds are presented, respectively. Finally, Section 4.6 shows a compact patch antenna design for achieving GSM/DCS dual-band dual-polarized radiation. Details of the design considerations are given, and experimental results of the constructed prototype are presented.

4.2 ANTENNA FOR SINGLE-BAND OPERATION

4.2.1 Slot-Coupled Feed Design

A slot-coupled patch antenna with a cross-slot for compact dual-polarized operation in the DCS band has been implemented [5]. The antenna geometry is shown in Figure 4.1. Note that, to obtain a wide impedance bandwidth for the DCS operation, a thick air-layer substrate of thickness 12.8 mm (about 7.7% of the wavelength at 1800 MHz) is used. In the square patch, whose dimensions are chosen to be $50 \times 50\ \text{mm}^2$, a cross-slot of equal lengths is embedded. Note that, when there is no cross-slot presence, the required patch size for operating in the 1800 MHz band is about $60 \times 60\ \text{mm}^2$; that is, the cross-slot-loaded patch antenna studied here has a patch size only about 0.69 times that of the design using a simple square patch. With the antenna size reduction achieved, good dual-polarized performance has also been obtained.

In this design, the two arms of the embedded cross-slot have a length of 44 mm and a width of 1 mm. The two linearly polarized waves are excited by two slot-coupled feeds with H-shaped coupling slots. For port 1 excitation, the central arm and two side arms of the H-shaped coupling slot are, respectively, $1\ \text{mm} \times H_{12}$ and $1\ \text{mm} \times 2H_{11}$, and the coupling slot is centered below the square patch. As for port 2 excitation, the coupling slot's central arm and two side arms are $0.25\ \text{mm} \times H_{22}$ and $1\ \text{mm} \times H_{21}$, respectively. The center-to-center distance of the two H-shaped coupling slots is 20 mm in this design. Also note that, to achieve

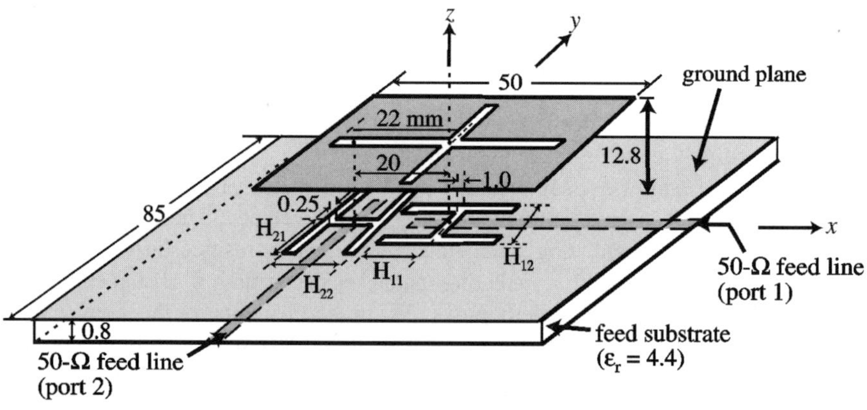

FIGURE 4.1 Geometry of the compact dual-polarized slot-coupled patch antenna for DCS operation. (From Ref. 5, © 2001 John Wiley & Sons, Inc.)

good isolation between the two feeding ports, the central arm of the coupling slot for port 2 is aligned in the direction of the microstrip feed line for port 1. By selecting suitable values of H_{11}, H_{12}, H_{21}, H_{22}, and tuning stub lengths of the two 50-Ω microstrip feed lines for ports 1 and 2, good impedance matching for the excitation of both linear polarizations can be obtained.

With the design parameters selected, a prototype was constructed and tested. The ground-plane size was chosen to be 85 mm×85 mm, which is only about $0.5\,\lambda \times 0.5\,\lambda$ (λ is the free-space wavelength at 1800 MHz), and an inexpensive FR4 substrate of thickness 0.8 mm and relative permittivity 4.4 was used for the feed substrate. Figure 4.2 shows the measured return loss for port 1 and port 2 excitation. The obtained impedance bandwidth (10 dB return loss) reaches about 10% and covers the 1800 MHz band. Measured isolation between the two feeding ports is presented in Figure 4.3. The isolation is seen to be below −37.3 dB across the entire operating bandwidth. For the lower frequency range of the operating bandwidth, the isolation obtained is even less than −45 dB. Figure 4.4 plots the radiation patterns measured at two principal planes for port 1 and port 2 excitation at 1800 MHz. Good cross-polarization levels (about 20 dB) are observed for the two orthogonal linearly polarized waves, and the measured antenna gain is about 7.0 and 7.1 dBi for port 1 and port 2 excitation, respectively. The obtained antenna performance is very suitable for applications in a DCS system.

4.2.2 Probe-Fed Design

In this section, promising broadband probe-fed patch antenna designs developed very recently for cellular system base stations are presented. Two major design techniques have been used in achieving a wide impedance bandwidth of the patch antenna. The first technique utilizes a probe feed with a short probe pin to excite a patch antenna with a thick air-layer substrate. Because of the short probe pin,

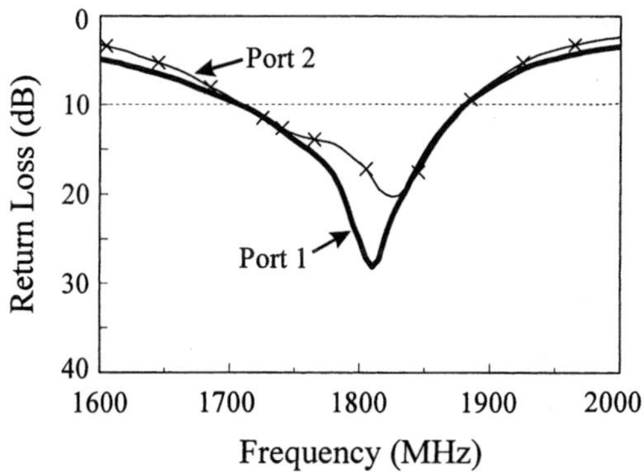

FIGURE 4.2 Measured return loss of ports 1 and 2 for the antenna shown in Figure 4.1 with $H_{11} = 10$ mm, $H_{12} = 20$ mm, $H_{21} = 12.5$ mm, $H_{22} = 11$ mm. (From Ref. 5, © 2001 John Wiley & Sons, Inc.)

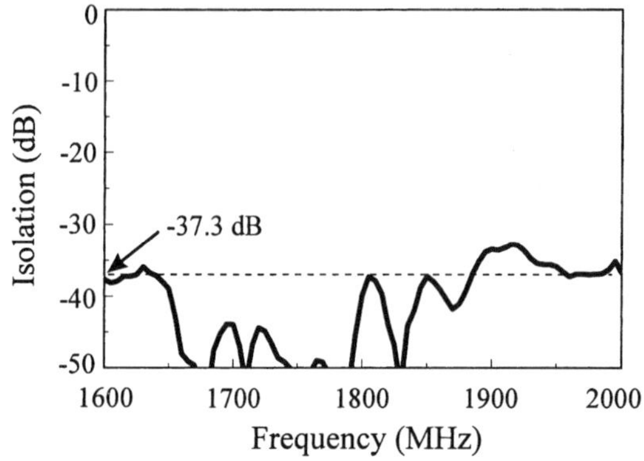

FIGURE 4.3 Measured isolation (S_{21}) between ports 1 and 2 for the antenna shown in Figure 4.1. (From Ref. 5, © 2001 John Wiley & Sons, Inc.)

the probe inductance contributed to the antenna's input impedance is small and can be neglected, which makes possible a wide impedance bandwidth for the patch antenna. To achieve this goal, designs using a three-dimensional microstrip transition feed, an L-shaped probe feed, a capacitively coupled top-loaded probe feed, a V-shaped radiating patch, and so on, have been reported and discussed in Reference 1. Very recently, designs using a conducting cylinder transition [6], a

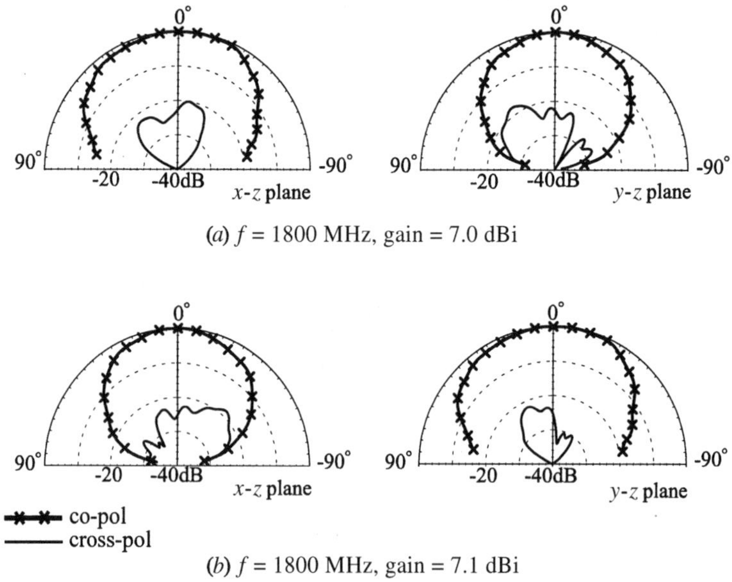

FIGURE 4.4 Measured radiation patterns at 1800 MHz for the antenna shown in Figure 4.1. (*a*) Port 1 excitation; (*b*) port 2 excitation. (From Ref. 5, © 2001 John Wiley & Sons, Inc.)

triangular transition patch [7], a ground plane with a small elevated portion [8], and a W-shaped ground plane [9] have been shown to be very promising in obtaining broadband patch antennas for cellular system base stations. These recent designs are studied in this section.

The second major design technique utilizes the embedding of suitable slots or slits in the radiating patch, such as a small annular slot around the feed point in the patch, a U-shaped slot inside the patch (U-slotted patch), or a pair of wide slits inserted at the patch boundary (E-patch) [1]. In these cases, although a long probe pin is used, the embedded slots or slits can compensate the effect of the inductance of the long probe pin in the thick substrate layer, making possible an enhanced impedance bandwidth for the patch antenna. With the use of the E-patch design, it was recently shown that, by incorporating a U-shaped ground plane or using two E-patches fed out of phase, reduced cross-polarization radiation over a wide frequency range can also be obtained [10]. Detailed results of these two designs are included in this section.

4.2.2.1 With a Conducting Cylinder Transition Figure 4.5 shows the geometry of the probe-fed patch antenna with a short probe pin and a conducting cylinder transition. In this design, a simple rectangular patch of dimensions $L \times W$ is used. The thickness of the air-layer substrate is h, and in the lower region of the substrate layer is the probe pin of the 50-Ω SMA connector, which is chosen to be only 3 mm in the proposed design. This short probe pin will introduce small or negligible inductance to the antenna's input impedance, and its effect on the antenna's impedance matching can be minimized.

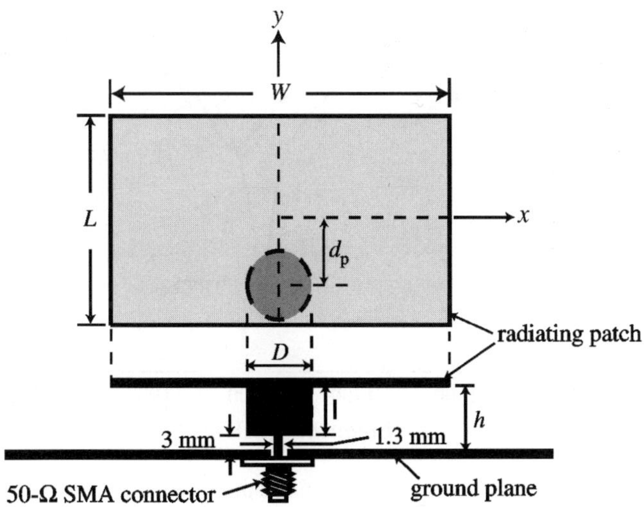

FIGURE 4.5 Geometry of the broadband probe-fed patch antenna with a short probe pin and a conducting cylinder transition. (From Ref. 6, © 2001 John Wiley & Sons, Inc.)

In the upper region of the substrate layer, there is a conducting cylinder transition with a relatively larger diameter D, compared to that (about 1.3 mm) of the probe pin of a conventional SMA connector. This conducting cylinder has a length of ℓ $(=h-3\,\text{mm})$ and is hollow inside, but with conducting plates on its upper and bottom sides, which are connected to the radiating patch and the short probe pin, respectively. In the proposed design, this conducting cylinder mainly serves as a smooth transition region from the source to the radiating patch, which improves the impedance matching of the proposed antenna with a thick air-layer substrate. By using the short probe pin and the conducting cylinder transition, an impedance bandwidth (1.5:1 VSWR) larger than 10% can easily be obtained, which makes the proposed antenna suitable for practical applications in a DCS cellular system base station. Also note that the feed position d_p is defined here to be the distance from the center of the conducting cylinder to the patch center.

In this study, conventional probe-fed patch antennas with a regular probe pin (reference antenna in this study) were first constructed and studied. The center operating frequency was designed to be about 1.8 GHz. The dimensions of the rectangular patch were thus selected to be $72\times115\,\text{mm}^2$. Figure 4.6 shows the measured return loss for the reference antenna in which a probe feed with a long probe pin is used. Cases with different air-substrate thicknesses are studied. The measured return loss is obtained for the antennas with their optimal feed positions. For the case with $h=6$ mm, the impedance bandwidth obtained is about 60 MHz or only 3.3%. When a larger air-substrate thickness is chosen, the antenna's resonant frequency is decreased and it also becomes difficult to achieve good impedance matching better than about 14 dB (1.5:1 VSWR) for desired operating frequencies.

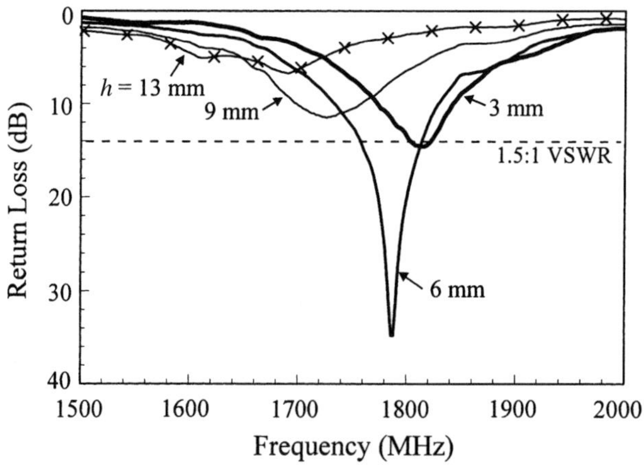

FIGURE 4.6 Measured return loss of a conventional probe-fed patch antenna [the geometry shown in Figure 4.5 with $D=1.3$ mm (that is, a regular probe feed), $L=72$ mm, $W=115$ mm, ground plane size $=150\times225\,\text{mm}^2$]. (From Ref. 6, © 2001 John Wiley & Sons, Inc.)

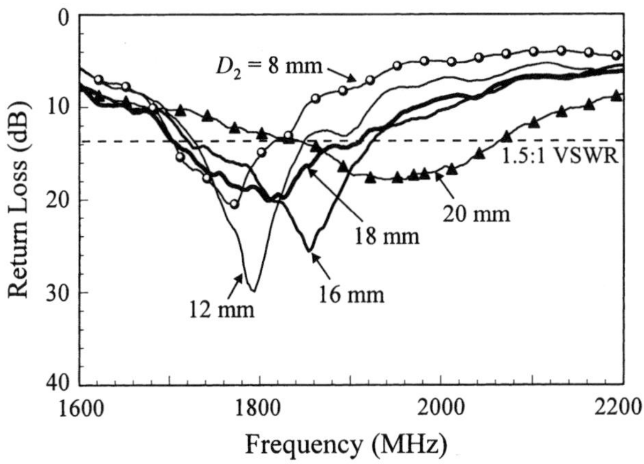

FIGURE 4.7 Measured return loss of the antenna shown in Figure 4.5 with $L = 72$ mm, $W = 115$ mm, ground plane size = 150×225 mm^2. (From Ref. 6, © 2001 John Wiley & Sons, Inc.)

Figure 4.7 presents the measured return loss for the proposed antenna. It is seen that greatly enhanced impedance matching is obtained for the proposed antenna compared to the reference antenna. The corresponding measured data are also listed in Table 4.1 for comparison. For the cases with $D = 16$, 18, and 20 mm, the impedance bandwidths obtained are all larger than 10%, which is suitable for applications in a DCS cellular system base station.

The radiation characteristics of the proposed antenna have also been studied. Figure 4.8 plots the measured radiation patterns for the antenna with $D = 18$ mm at 1700, 1800, and 1880 MHz. Good broadside radiation patterns are obtained. However, large cross-polarization radiation in the H-plane pattern is also observed. This characteristic is similar to that reported for broadband probe-fed patch antennas with a thick air-layer substrate [1], and by using two out-of-phase feeds for a single antenna element, using two proposed antenna elements fed out of phase in a two-element array, or using a U-shaped ground plane, the large cross-polarization

TABLE 4.1 Performance of the Antennas Studied in Figure 4.7 [6]. Impedance Bandwidth (BW) is Determined From 1.5:1 VSWR; f_c is the Center Frequency of the Impedance Bandwidth.

D (mm)	d_p (mm)	f_c (MHz)	BW (MHz, %)
8	3.2	1757	105, 6.0
12	3.0	1787	117, 6.5
16	2.8	1822	199, 10.9
18	2.7	1800	205, 11.4
20	2.6	1950	205, 10.5

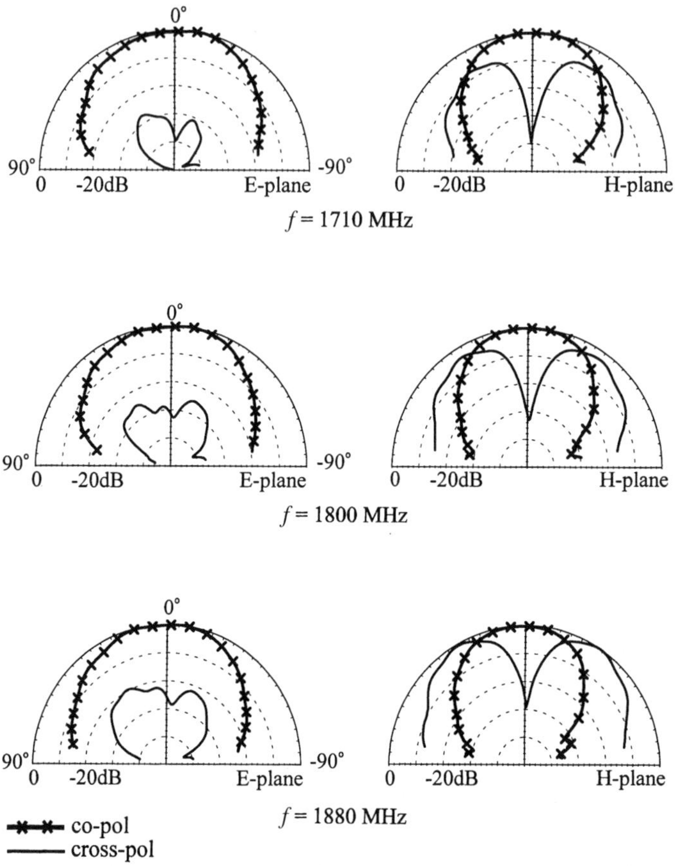

FIGURE 4.8 Measured radiation patterns of the antenna studied in Figure 4.7 with $D = 18$ mm. (*a*) $f = 1710$ MHz; (*b*) $f = 1800$ MHz; (*c*) $f = 1880$ MHz. (From Ref. 6, © 2001 John Wiley & Sons, Inc.)

radiation in the H-plane can be effectively reduced. The related designs are discussed in Sections 4.2.2.5 and 4.2.2.6. The measured antenna gain is also presented in Figure 4.9. For the operating bandwidth within 1710–1880 MHz (the DCS band), the obtained antenna gain reaches about 9.8 dBi and the gain variations are less than 3 dBi.

4.2.2.2 With a Triangular Transition Patch Figure 4.10 shows the geometry of the probe-fed patch antenna with a short probe pin and a triangular transition patch. The radiating patch has dimensions of $L \times W$ and is supported by plastic posts above a ground plane. The distance between the radiating patch and the ground plane is h. At one of the patch's radiating edge, a triangular conducting patch with a flare angle of α and a height of $h - 3$ mm is added and oriented straight toward the ground

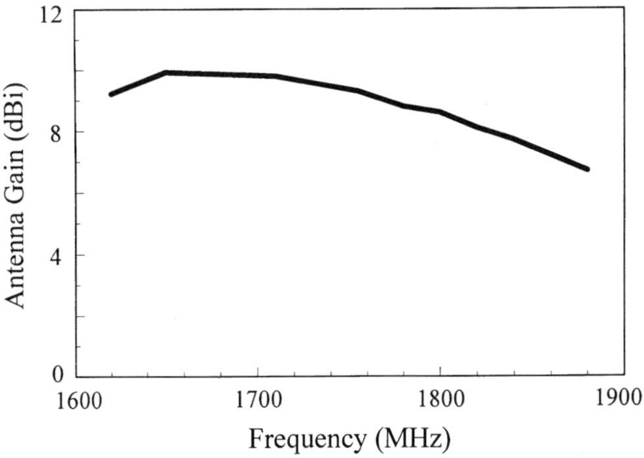

FIGURE 4.9 Measured antenna gain of the antenna studied in Figure 4.7 with $D = 18\,\text{mm}$. (From Ref. 6, © 2001 John Wiley & Sons, Inc.)

plane. This triangular patch is also centered with respect to the centerline (y-axis in the figure) of the radiating patch. A conventional probe feed with a short probe pin (3 mm long in this study) is then connected to the tip of the triangular patch. By further selecting a suitable flare angle of the triangular patch, the proposed antenna can be excited with good impedance matching over a wide frequency range. This behavior is partly because the length of the probe pin is kept small, which introduces negligible inductance to the antenna's input impedance, and partly because the triangular patch provides a smooth transition region between the source and the radiating patch, which improves the impedance matching of the proposed antenna with a thick air-layer substrate.

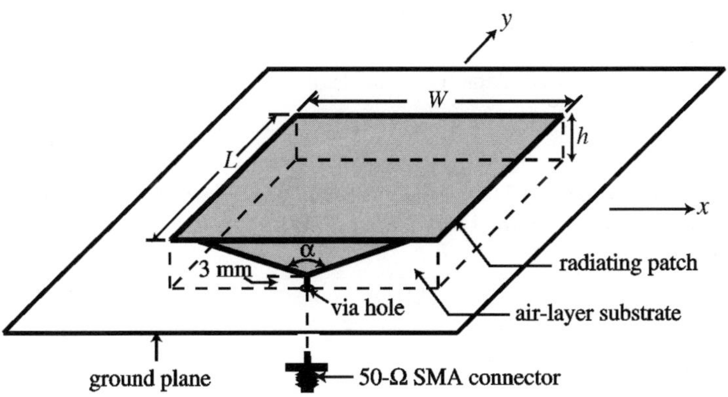

FIGURE 4.10 Geometry of the broadband probe-fed patch antenna with a short probe pin and a triangular transition patch. (From Ref. 7, © 2001 John Wiley & Sons, Inc.)

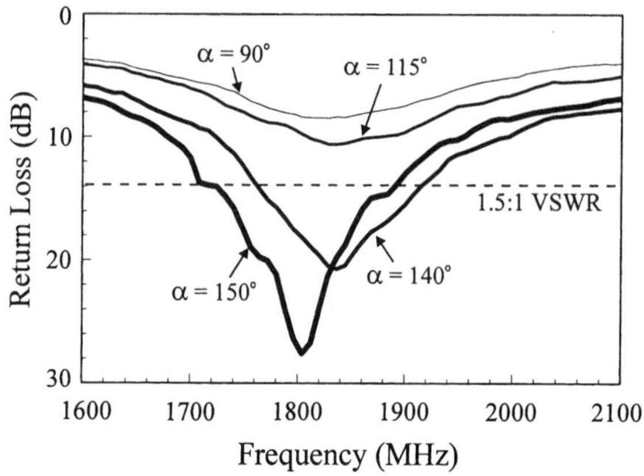

FIGURE 4.11 Measured return loss for the antenna shown in Figure 4.10; $L = 65$ mm, $W = 107$ mm, $h = 16$ mm, ground-plane size $= 220 \times 150$ mm^2. (From Ref. 7, © 2001 John Wiley & Sons, Inc.)

In the experiment, the center operating frequency was designed to be about 1800 MHz. For this purpose, the radiating patch with dimensions of 65×107 mm^2 ($L \times W$) was chosen, and the air-layer substrate with a thickness (h) of 16 mm was also selected, which is about 9.6% of the wavelength at 1800 MHz. By varying the flare angle (α) of the triangular patch between the radiating patch and the probe feed with a short probe pin of 3-mm length, several prototypes of the proposed antenna were constructed. Also note that, in the proposed design, the probe feed is at the position just below the center of one of the patch's radiating edges, which greatly simplifies the design considerations for the proposed antenna.

Figure 4.11 shows the measured return loss for the proposed antenna with various angles of α. It is clearly seen that, with an increase of the flare angle α, the impedance matching is greatly improved. For the case of $\alpha = 150°$, the impedance bandwidth, determined by 1.5:1 VSWR (about 14 dB return loss), reaches 180 MHz or about 10% with respect to the measured center frequency at 1803 MHz (see measured data listed in Table 4.2). Also note that, for the case

TABLE 4.2 Performance of the Antennas Studied in Figure 4.11 [7]. Impedance Bandwidth (BW) is Determined From 1.5:1 VSWR; f_c is the Center Frequency of the Impedance Bandwidth.

α (°)	f_c (MHz)	BW (MHz, %)
90	—	—
115	—	—
140	1840	152, 8.3
150	1803	180, 10.0

with $\alpha = 150°$ in this study, the edge of the triangular patch facing the flare angle α has a length almost the same as that of the patch's radiating edge. It should also be noted that the obtained impedance bandwidth for the case with $\alpha = 150°$ covers the required bandwidth of the DCS system.

The measured radiation patterns for the antenna with $\alpha = 150°$ are plotted in Figure 4.12. For the radiation patterns measured at 1710, 1800, and 1880 MHz, good broadside radiation patterns are observed. Large cross-polarization radiation in the H-plane (x–z plane) patterns is also seen. On the other hand, it is interesting to see that the cross-polarization radiation in the E-plane (y–z plane) patterns is always lower than -35 dB. Figure 4.13 presents the measured antenna gain for the antenna with $\alpha = 150°$. Within the antenna's impedance bandwidth, a peak antenna gain of about 8.1 dBi is seen and the gain variations are less than 1.6 dBi.

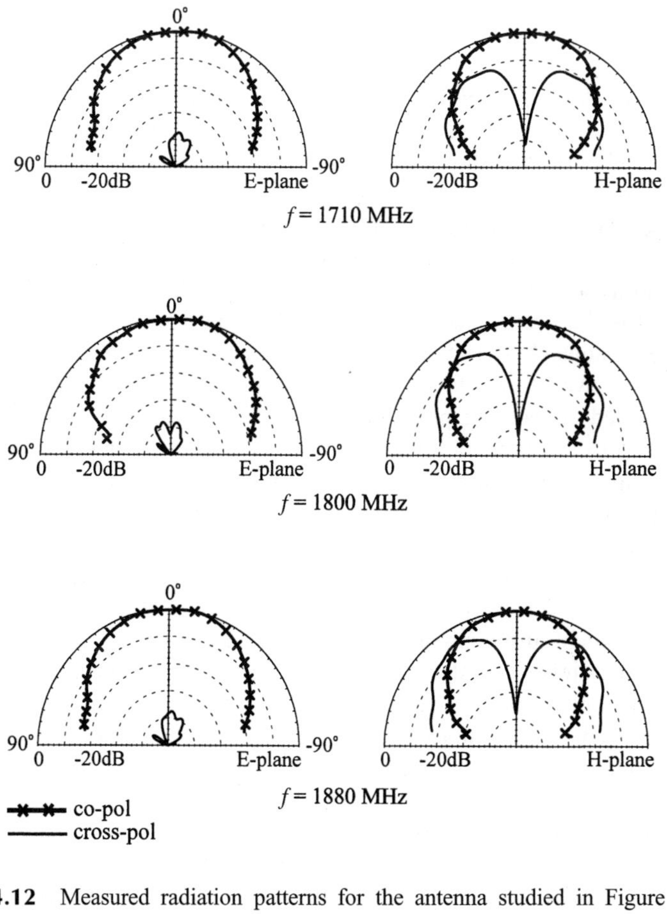

FIGURE 4.12 Measured radiation patterns for the antenna studied in Figure 4.11 with $\alpha = 150°$. (From Ref. [7], © 2001 John Wiley & Sons, Inc.)

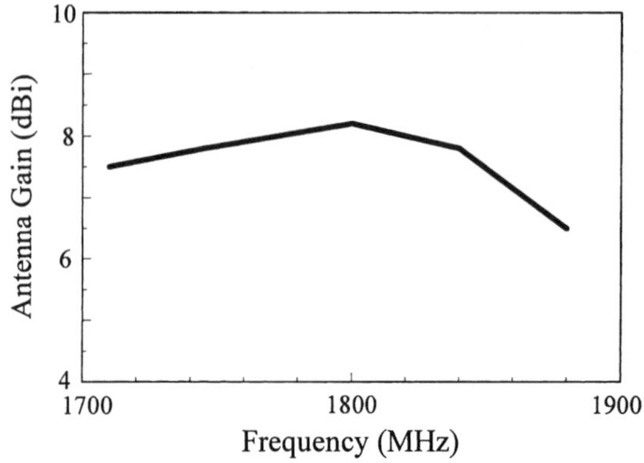

FIGURE 4.13 Measured antenna gain for the antenna studied in Figure 4.11 with $\alpha = 150°$. (From Ref. 7, © 2001 John Wiley & Sons, Inc.)

4.2.2.3 With the Ground Plane Having a Small Elevated Portion The geometry of the broadband probe-fed patch antenna with its ground plane having a small elevated portion is shown in Figure 4.14. The radiating patch has a length of L and a width of W and is supported by plastic posts (not shown in the figure) above the ground plane. Along the antenna's resonant direction (y–z plane in the figure), a small square portion of the ground plane with a side length of S is elevated to be close to the radiating patch with a distance of t. This square elevated ground plane portion is then connected to the original ground plane with two conducting plates of dimensions $(h - t) \times S$ facing the antenna's resonant direction. Note that, in this case, the two other sides of the elevated ground plane portion can be with or without added conducting plates. However, when there are conducting plates only at the two sides of the elevated ground plane portion facing in parallel to the x-axis in the figure, good impedance matching for the proposed antenna is difficult to achieve in the experiment. The feed position for the coax feed is placed at the center of the elevated ground plane and away from the patch center with a distance of d_p. For the proposed antenna, good impedance matching over a wide frequency range can be easily achieved with a regular coax feed with a short probe pin of length t, independent of the air-substrate thickness h chosen.

To begin the study, several conventional probe-fed patch antennas without an elevated ground plane portion (denoted as the reference antenna in this study) were constructed and studied. A radiating patch with dimensions of $70\,\text{mm} \times 105\,\text{mm}$ ($L \times W$) was chosen in the study. The obtained impedance bandwidths and the optimal feed positions for the cases with $h = 3\,\text{mm}$ (reference 1) and $6\,\text{mm}$ (reference 2) are listed in Table 4.3 for comparison. It is observed that the optimal feed position is quickly shifted toward the patch edge when the air-substrate thickness increases from 3 to 6 mm, and the obtained impedance bandwidth is only about 2.0% or less. In addition, when the air-substrate thickness

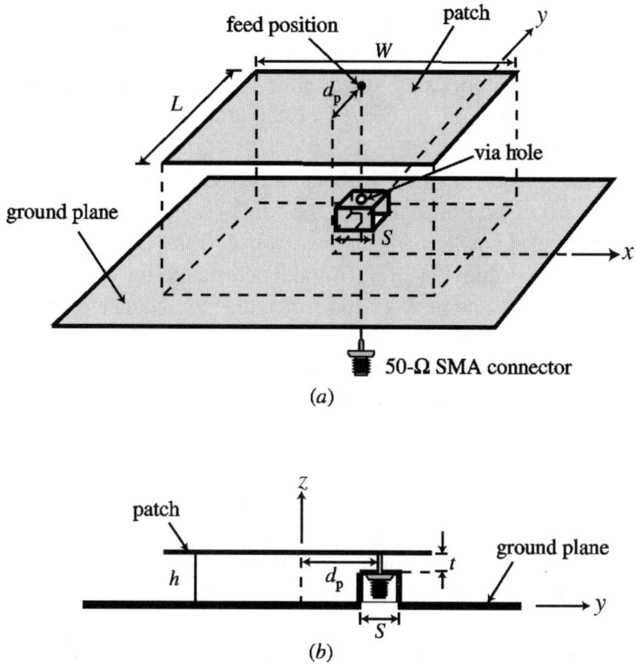

FIGURE 4.14 Geometry of the broadband probe-fed patch antenna with its ground plane having a small elevated portion. (From Ref. 8, © 2001 IEEE, reprinted with permission)

increases further, measured return loss less than about 14 dB (1.5:1 VSWR) cannot be achieved for the reference antennas.

By using the proposed antenna shown in Figure 4.14 with the length of t fixed at 3 mm and the side length (S) of the elevated ground plane portion chosen to be 25 mm, several prototypes with different air-substrate thicknesses of 9 mm (denoted as antenna 1 in the study), 10.5 mm (antenna 2), 12 mm (antenna 3), and 15 mm (antenna 4) were constructed and tested. Note that, with $S = 25$ mm, the elevated ground plane portion occupies a small part (less than 10%) of the antenna volume

TABLE 4.3 Performance of the Antennas Studied in Figure 4.14 [8]. Impedance Bandwidth (BW) is Determined From 1.5:1 VSWR.

	h (mm)	t (mm)	d_p (MHz)	BW (MHz, %)
Reference 1	3.0	—	18	38, 2.0
Reference 2	6.0	—	34	30, 1.6
Antenna 1	9.0	3	20	123, 7.1
Antenna 2	10.5	3	22	200, 11.1
Antenna 3	12.0	3	22	264, 14.5
Antenna 4	15.0	3	31	420, 22.3

between the radiating patch and the ground plane ($70 \times 105 \times h\,\text{mm}^3$), which could minimize the decrease in the effective volume of the proposed antenna and maximize the possible achievable impedance bandwidth. Results of the measured return loss are presented in Figure 4.15, and the corresponding optimal feed positions and impedance bandwidths are given in Table 4.3. It is clearly seen that an increasing impedance bandwidth is obtained with an increase in the air-substrate thickness. This behavior is expected and confirms the use of a coax feed with a short probe pin, which introduces a small, negligible reactance to the antenna's input impedance and is very useful in achieving good impedance matching over a wide frequency range. In the case of antenna 4, the obtained impedance bandwidth reaches 420 MHz (1670–2090 MHz) or 22.3% referenced to the center frequency at 1880 MHz. Also note that, in comparison to the conventional probe-fed patch antenna, the optimal feed position for the proposed antenna varies much more slowly with increasing air-substrate thickness, especially for h increasing from 9 to 12 mm where d_p varies only from 20 to 22 mm. This behavior makes it relatively easier to locate the optimal feed position for the proposed antenna.

It is also observed that the impedance bandwidth (1706–1906 MHz) of antenna 2 with $h = 10.5$ mm meets the required bandwidth of the DCS system. Figure 4.16 plots the measured radiation patterns in two principal planes of antenna 2. Results for three operating frequencies at 1710, 1800, and 1880 MHz are shown. It is seen that good broadside radiation patterns are obtained. However, relatively larger cross-polarization in the H-plane radiation is also seen. Measured antenna gain for antenna 2 is shown in Figure 4.17. Results indicate that, across the DCS band (1710–1880 MHz), the peak antenna gain reaches about 9.8 dBi and the gain variations are less than 2.5 dBi. This antenna is suitable for applications in a DCS cellular system base station.

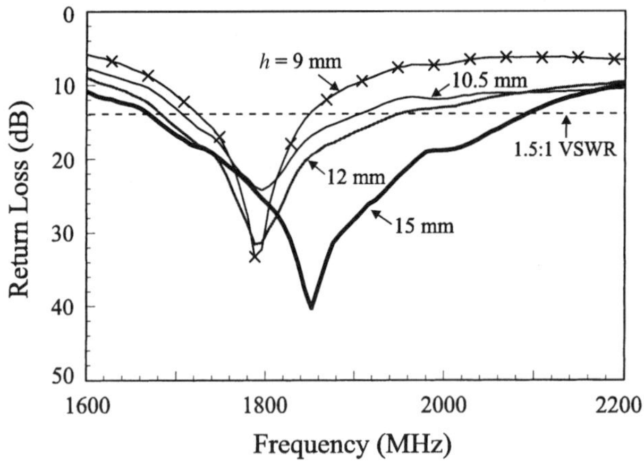

FIGURE 4.15 Measured return loss for the antenna shown in Figure 4.14 with $L = 70$ mm, $W = 105$ mm, $t = 3$ mm, $S = 25$ mm, ground plane size $= 165 \times 120\,\text{mm}^2$. (From Ref. 8, © 2001 IEEE, reprinted with permission)

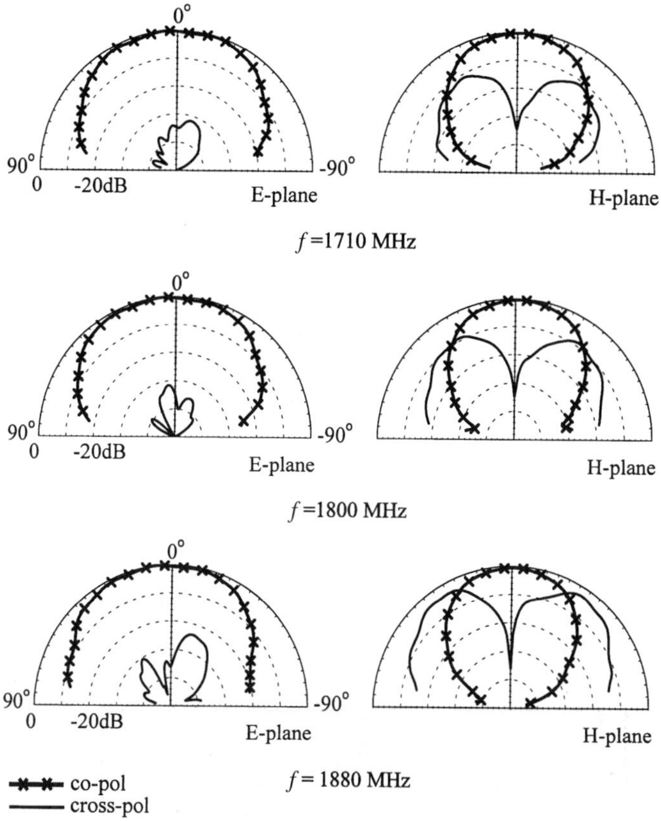

FIGURE 4.16 Measured radiation patterns for the antenna studied in Figure 4.15 with $h = 10.5$ mm (antenna 2 in Table 4.3). (From Ref. 8, © 2001 IEEE, reprinted with permission)

4.2.2.4 *With a W-Shaped Ground Plane* Figure 4.18 shows the geometry of the broadband probe-fed patch antenna with a W-shaped ground plane. The rectangular patch of dimensions $L \times W$ (L is along the resonant direction) is supported by plastic posts above the centerline (y-axis) of the proposed W-shaped ground plane. The W-shaped ground plane has a length of L_g in the antenna's resonant direction and a bent angle of α. There is also a small flat section of 9 mm in width and L_g in length for accommodating the 50-Ω SMA connector in this study. The distance between the rectangular patch and the small flat section is h, which is chosen to be small (about 2% of the center operating wavelength) to reduce the required probe pin length for feeding the patch antenna. In this case, good impedance matching across a wide operating bandwidth can easily be obtained by placing the probe feed along the patch's centerline (y-axis). The distance between the optimal feed position and the patch center is denoted as d_p here.

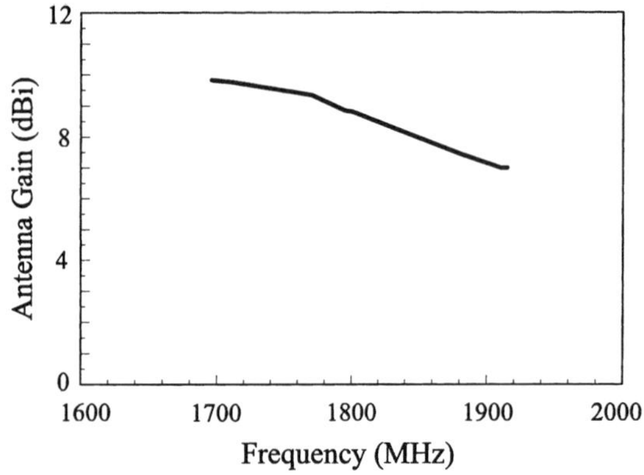

FIGURE 4.17 Measured antenna gain for the antenna studied in Figure 4.15 with $h = 10.5$ mm (antenna 2 in Table 4.3). (From Ref. 8, © 2001 IEEE, reprinted with permission)

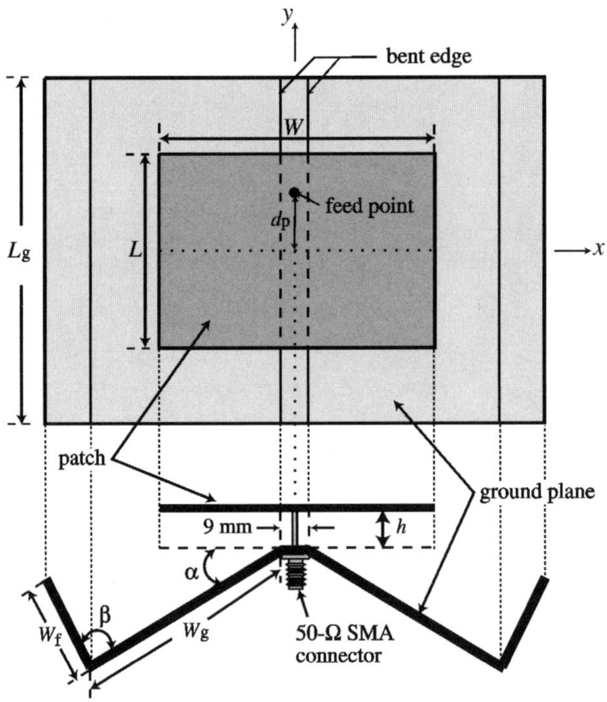

FIGURE 4.18 Geometry of the broadband probe-fed patch antenna with a W-shaped ground plane. (From Ref. 9, © 2002 IEEE, reprinted with permission)

In this design, it can be seen that the effective average substrate thickness increases with an increase of the angle α; that is, the antenna's operating bandwidth can be increased with an increase in α. The width of the bent section of the W-shaped ground plane is W_g, and, at the edge of the bent section, a flange of height W_f is added, with an angle β from the bent section. By choosing a proper angle of β and a larger height of W_f, the antenna gain and cross-polarization level (XPL) for the proposed antenna can be enhanced. Many prototypes of the proposed antenna with various bent angles (α) and flange parameters (β, W_f) have been constructed and tested.

The case for a simple bent ground plane (i.e., without the flange section in the proposed design) was first studied. Figure 4.19 shows the measured return loss for the constructed prototypes with various angles of α. The rectangular patch has dimensions of 60 mm×90 mm ($L \times W$). The distance h from the top of the bent ground plane to the rectangular patch was fixed to be 3 mm (about 2% of the wavelength at the center operating frequency). [Also note that, when larger values of h are chosen, good impedance matching becomes difficult or impossible to achieve for the conventional patch antenna with a flat ground plane ($\alpha = 0°$ in this study).] The two bent sections of the ground plane were chosen to have a width of 90 mm (W_g), which is the same as the width (W) of the rectangular patch. [Also note that the cases of $W_g = 1.5W$ and $2.0W$ were also studied, and almost the same results are obtained for the case of $W_g = W$ presented here.] The optimal feed position was at $d_p = 22$ mm, which was found to be insensitive to the variation in the bent angle α.

From the results obtained in Figure 4.19, it is seen that the impedance bandwidth is larger when the bent angle α is larger. In the figure, the case with $\alpha = 0°$ represents the result for a conventional probe-fed patch antenna with a thin air substrate. The

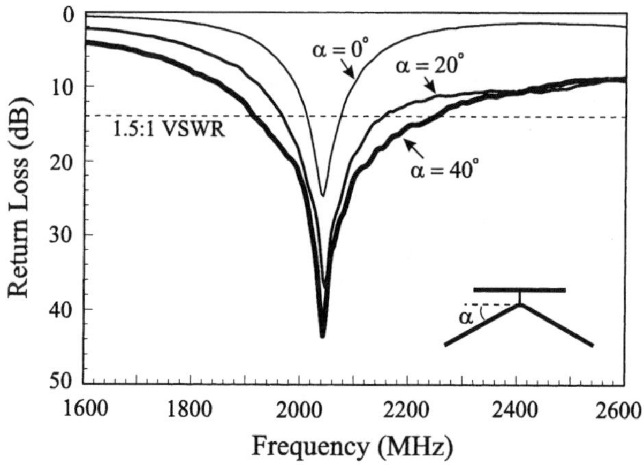

FIGURE 4.19 Measured return loss for the antenna shown in Figure 4.18 with $W_f = 0$; $L = 60$ mm, $W = 90$ mm, $h = 3$ mm, $L_g = 120$ mm, $W_g = 90$ mm, $d_p = 22$ mm. (From Ref. 9, © 2002 IEEE, reprinted with permission)

TABLE 4.4 Performance of the Antennas Shown in Figure 4.18 with $W_f = 0$ [9]; Antenna Parameters are Given in Figure 4.19. f_r is the Resonant Frequency, and BW is the Impedance Bandwidth Determined From 1.5:1 VSWR.

α (°)	f_r (MHz)	BW (%)	Gain (dBi)
0	2044	2.9	8.2
10	2044	5.6	8.1
20	2046	9.0	7.6
30	2044	12.6	7.6
40	2044	16.1	6.1

measured data of the impedance bandwidth and the antenna gain at resonant frequencies are given in Table 4.4. Good agreement between the measurement and the simulation has also been observed (see Fig. 4.20). From the obtained results, the impedance bandwidth for $\alpha = 40°$ reaches 16.1%, which is about 5.5 times that (2.9%) of the case with $\alpha = 0°$. On the other hand, the antenna gain is decreased by about 2.1 dBi, in comparison to the cases of $\alpha = 0°$ and 40°. For the cases with α greater than 40°, it is expected that the antenna gain will be decreased further. However, the experiments were not conducted because the parameters used in this study were not suitable for implementing the design with $\alpha > 40°$ (the two bent sections of the ground plane cannot be bent further because of the presence of the SMA connector).

The measured radiation patterns for $\alpha = 0°$, 20°, and 40° are shown in Figure 4.21. Largely owing to the bending of the ground plane, the XPLs in the H-plane pattern are seen to decrease by about 7 dB (from about 13 dB for $\alpha = 0°$ to about 6 dB for

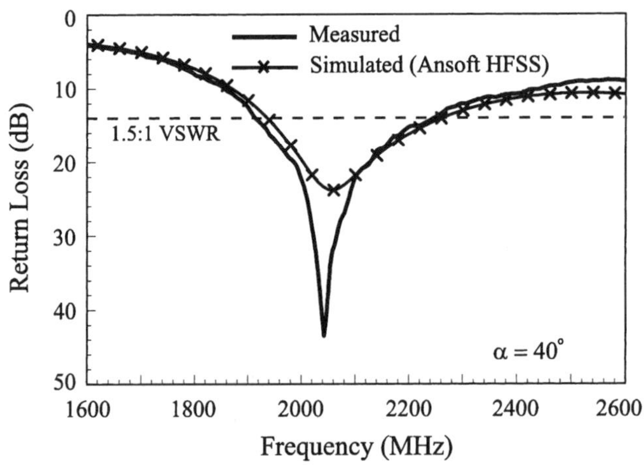

FIGURE 4.20 Measured and simulated return loss of the antenna studied in Figure 4.19 with $\alpha = 40°$.

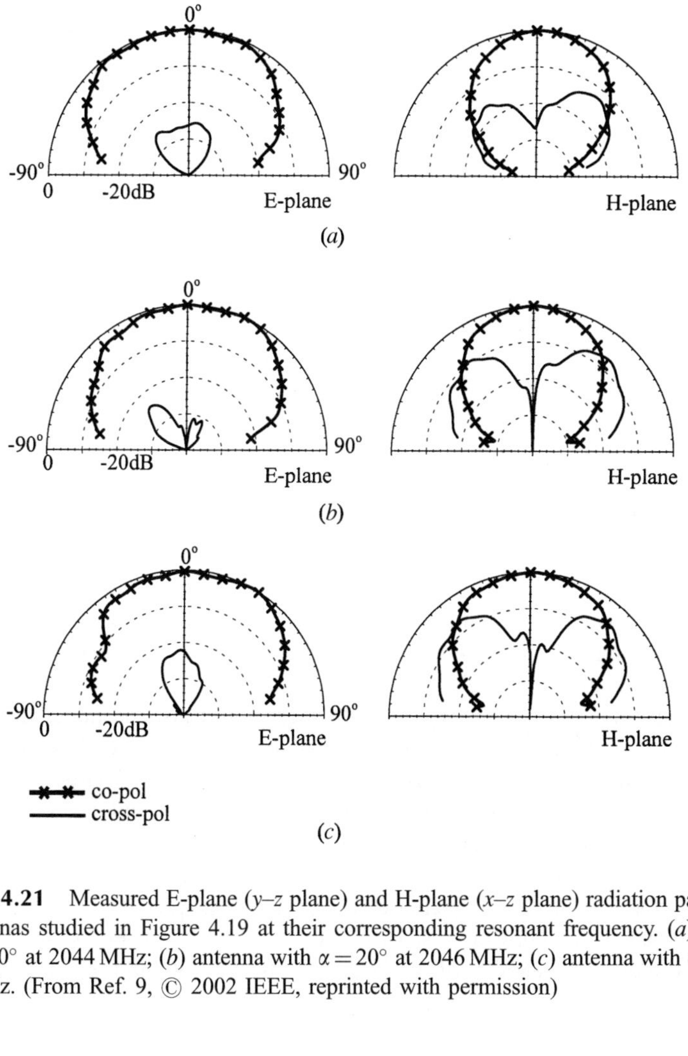

FIGURE 4.21 Measured E-plane (y–z plane) and H-plane (x–z plane) radiation patterns for the antennas studied in Figure 4.19 at their corresponding resonant frequency. (a) Antenna with $\alpha = 0°$ at 2044 MHz; (b) antenna with $\alpha = 20°$ at 2046 MHz; (c) antenna with $\alpha = 40°$ at 2044 MHz. (From Ref. 9, © 2002 IEEE, reprinted with permission)

$\alpha = 40°$). The decrease in the XPL is similar to that observed for the broadband probe-fed antenna designs with a thick air-layer substrate discussed in Sections 4.2.2.1– 4.2.2.3. Also, the beam widths of the H-plane pattern are seen to be widened, which is similar to that observed for a microstrip antenna mounted on a cylindrical surface [18]. As for the E-plane pattern, the XPLs are slightly decreased with an increase of α. However, the XPLs are all well above 20 dB.

It is then found that, when the flange sections are added to the bent ground plane to form a W-shaped ground plane, the antenna gain of the proposed design can be enhanced. Figure 4.22 shows the measured and simulated antenna gain versus the flange height W_f for the case with $\beta = 90°$. The corresponding measured data are given in Table 4.5. It is observed that the antenna gain increases with an increase

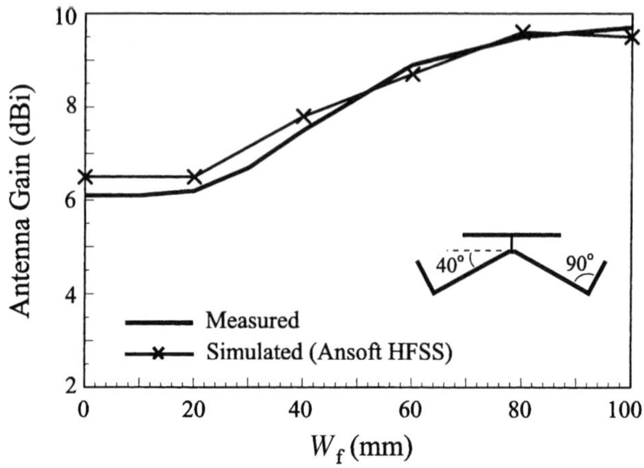

FIGURE 4.22 Measured and simulated antenna gain versus W_f for the antenna shown in Figure 4.18 with $\alpha = 40°$, $\beta = 90°$. Other parameters are given in Figure 4.19. (From Ref. 9, © 2002 IEEE, reprinted with permission)

of the flange height. The measured gain for frequencies across the operating bandwidth is presented in Figure 4.23. When W_f is greater than 60 mm, the antenna gain is larger than that (8.2 dBi) of a conventional probe-fed patch antenna with a thin air substrate (see $\alpha = 0°$ in Table 4.5). It should also be noted that, for the case with $W_f = 80$ mm, the edge of the flange section of the W-shaped ground plane is at about the same level as the radiating patch. For W_f larger than 80 mm, the edge of the flange section becomes higher than the rectangular patch, and the antenna gain is seen to increase slowly.

It is also observed that, although the impedance bandwidth is decreased from 16.1% to about 12% for an increase of the flange height, it is still greater than 4.0 times that of the conventional case with $\alpha = 0°$, $W_f = 0$. Figure 4.24 plots the measured radiation patterns of the cases with $W_f = 60$ and 80 mm studied in

TABLE 4.5 Performance of the Antenna Shown in Figure 4.18 with $\alpha = 40°$, $\beta = 90°$ [9]. Other Parameters are Given in Fig. 4.19. f_r is the Resonant Frequency, and BW is the Impedance Bandwidth Determined From 1.5:1 VSWR.

W_f (mm)	f_r (MHz)	BW (%)	Gain (dBi)
0	2044	16.1	6.1
20	2036	13.1	6.2
40	2012	12.2	7.5
60	2004	12.1	8.9
80	1994	12.5	9.5
100	1990	12.3	9.7

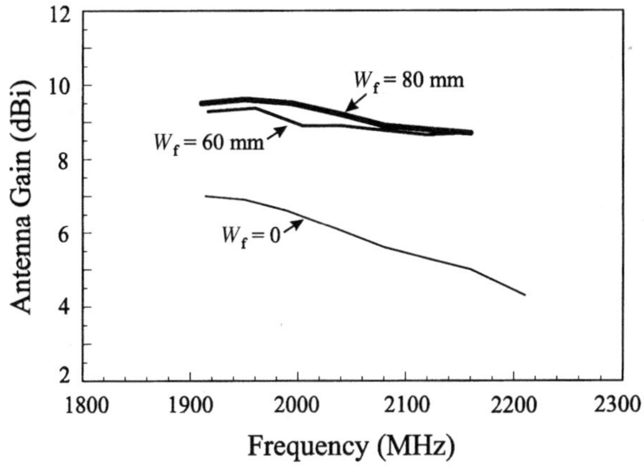

FIGURE 4.23 Measured antenna gain for the antenna shown in Figure 4.18 with $\alpha = 40°$, $\beta = 90°$. Other parameters are given in Figure 4.19. (From Ref. 9, © 2002 IEEE, reprinted with permission)

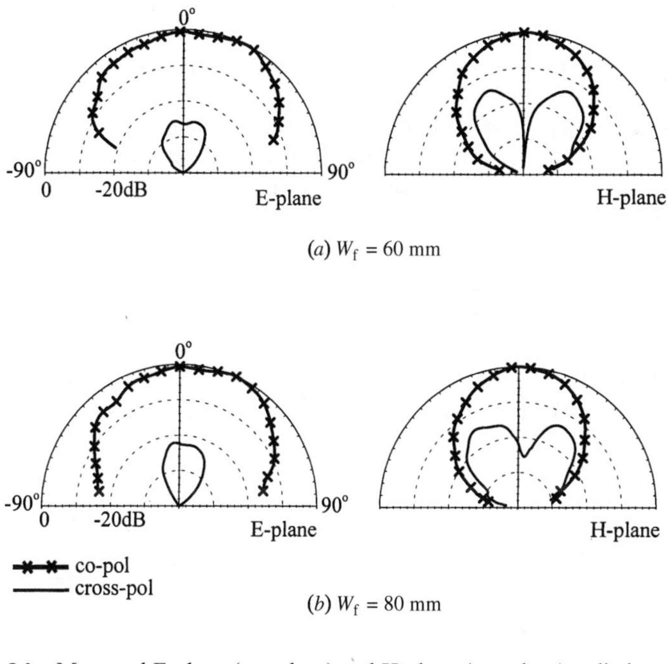

FIGURE 4.24 Measured E-plane (y–z plane) and H-plane (x–z plane) radiation patterns for the antenna shown in Figure 4.18 with $\alpha = 40°$, $\beta = 90°$. (*a*) Antenna with $W_f = 60$ mm at 2004 MHz; (*b*) antenna with $W_f = 80$ mm at 1994 MHz. Other parameters are given in Figure 4.19. (From Ref. 9, © 2002 IEEE, reprinted with permission)

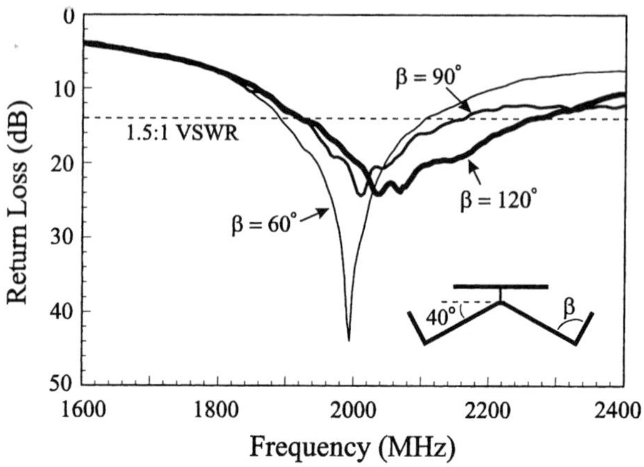

FIGURE 4.25 Measured return loss as a function of β for the antenna shown in Figure 4.18 with $\alpha = 40°$, $W_f = 60$ mm. Other parameters are given in Figure 4.19. (From Ref. 9, © 2002 IEEE, reprinted with permission)

Figure 4.22. It is clearly seen that the XPLs in the H-plane pattern are greatly improved, with comparison to that shown in Figure 4.21*b* and *c* for $\alpha = 20°$ and 40°, and are about 14 dB, which is even better than that shown in Figure 4.21*a* for $\alpha = 0°$ (about 13 dB). The XPLs in the E-plane patterns are also well above 20 dB.

Effects of varying the angle β in the proposed design have also been studied, and the results of measured return loss are shown in Figure 4.25. Table 4.6 lists the corresponding measured data. For obtaining maximum enhanced antenna gain, results indicate that $\beta = 90°$ is the optimal flange angle away from the bent sections of the ground plane. On the other hand, it is seen that decrease in β decreases the impedance bandwidth. This behavior probably occurs because the decrease in β corresponds to the lowering in the effective average substrate thickness in the proposed design, resulting in a smaller impedance bandwidth.

By applying the broadband design described above, a 1×2 W-shaped patch array antenna for the DCS cellular system base station has been implemented. Figure 4.26

TABLE 4.6 Performance of the Antenna Shown in Figure 4.18 with $\alpha = 40°$, $W_f = 60$ mm [9]. Other Parameters are Given in Fig. 4.18. f_r is the Resonant Frequency, and BW is the Impedance Bandwidth Determined from 1.5:1 VSWR.

β (°)	f_r (MHz)	BW (%)	Gain (dBi)
60	1993	10.8	8.5
90	2004	12.1	8.9
120	2038	16.8	7.2

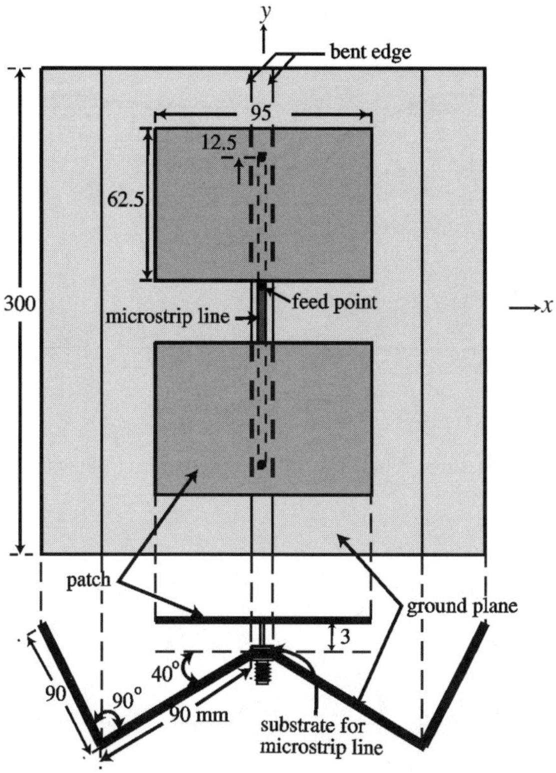

FIGURE 4.26 Geometry of a 1×2 W-shaped patch array antenna for DCS cellular system base station.

shows the antenna geometry. For the center operating frequency to be at about 1800 MHz, the dimensions of the rectangular patch were chosen to be $95 \times 62.5\,\text{mm}^2$. The two rectangular patches were also fed out of phase, and other design dimensions are given in Figure 4.26. Figure 4.27 shows the measured return loss of the constructed prototype. The obtained impedance bandwidth (1.5:1 VSWR) reaches 217 MHz (1681–1898 MHz) or about 12.1% referenced to 1800 MHz. The obtained bandwidth satisfies the operating bandwidth of the DCS system. Figure 4.28 plots the measured radiation pattrens of the array antenna. Good broadside radiation patterns with low cross-polarization radiation has been obtained. The measured antenna gain for the array antenna is shown in Figure 4.29. The antenna gain level is about 12 dBi for frequencies across the DCS band. Also notice that this antenna gain level is about 2.6 dBi larger than that of the case without the flanges ($W_f = 0$ in the geometry shown in Fig. 4.26). With the good antenna performance obtained here, the array antenna shown in Figure 4.26 is very promising for applications in the DCS system.

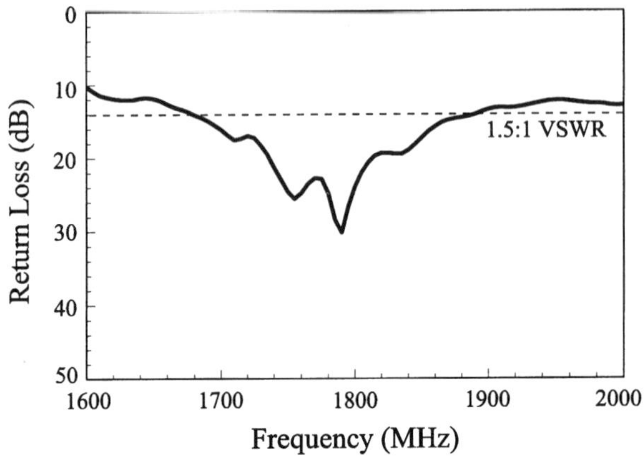

FIGURE 4.27 Measured return loss of the array antenna shown in Figure 4.26.

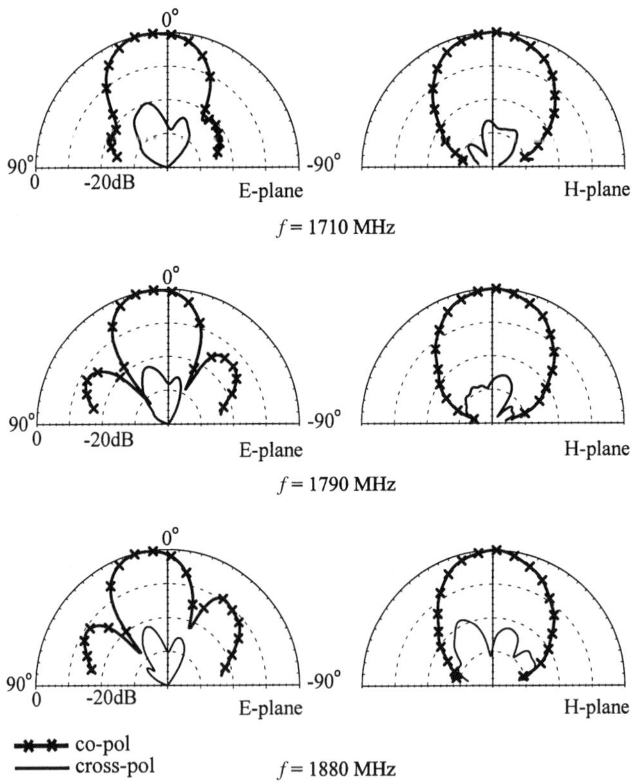

FIGURE 4.28 Measured E-plane (y–z plane) and H-plane (x–z plane) radiation patterns for the array antenna shown in Figure 4.26.

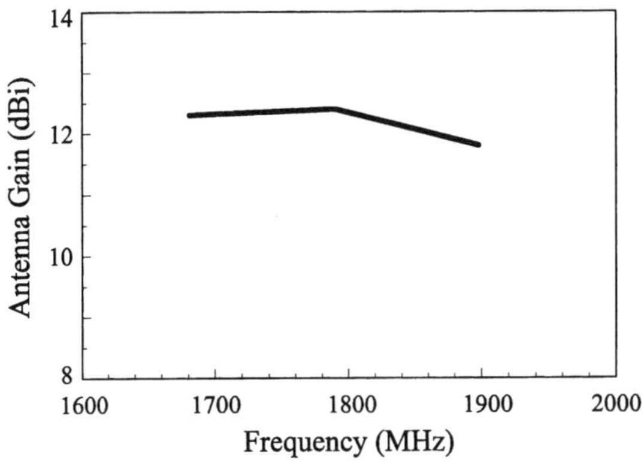

FIGURE 4.29 Measured antenna gain for the array antenna shown in Figure 4.26.

4.2.2.5 With an E-Patch and a U-Shaped Ground Plane Figure 4.30 shows the geometry of the broadband probe-fed patch antenna with an E-shaped patch (E-patch) and a U-shaped ground plane. The E-patch, whose dimensions are chosen to be $60 \times 90\,\text{mm}^2$ in this study, is centered above the U-shaped ground plane. By adjusting proper dimensions (ℓ, w, S) of the two wide slits cut in the E-patch, good impedance matching over a wide impedance bandwidth (about 25–30%) for the proposed antenna can be obtained. The probe feed is within a distance of d_{p} from the patch's radiating edge having the inserted slits. The E-patch is also supported by plastic posts (not shown in the figure) above the U-shaped ground plane with a height of h (i.e., the air-substrate thickness is h).

Along the patch's excitation direction (y-axis), the bottom ground plane has a length of L. The two conducting flanges of the U-shaped ground plane have a height of H and an angle of $(180° - \alpha)$ to the bottom ground plane. Also note that the

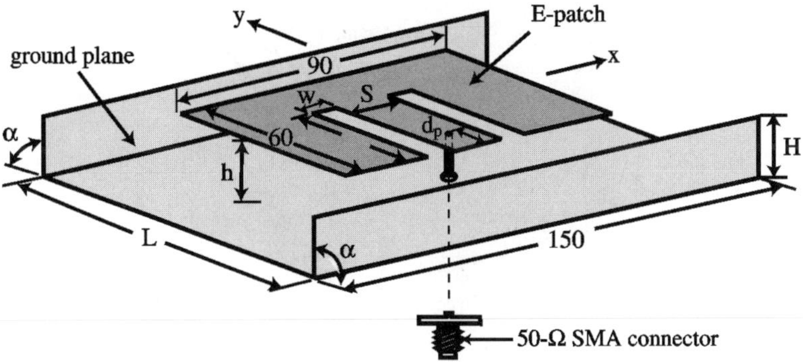

FIGURE 4.30 Geometry of the broadband probe-fed patch antenna with an E-patch and a U-shaped ground plane. (From Ref. 10, © 2002 IEEE, reprinted with permission)

bottom ground plane length in the proposed design should be greater than the E-patch's length but less than the E-patch's length plus two substrate's thicknesses; that is, when $\alpha = 90^\circ$, the distance between the E-patch's radiating edges and the conducting flanges should be less than one substrate thickness. As for the length of the edge of the bottom ground plane in parallel to the patch's radiating edges, its effect on the antenna's cross-polarization radiation is found to be insignificant and it is chosen to be 150 mm in this study (about 0.85 wavelength of the center frequency at 1700 MHz).

Prototypes of the proposed antenna with a center frequency at about 1700 MHz have been constructed and studied. The E-patch had a size of $60 \times 90\ \text{mm}^2$, and the air-substrate thickness was selected to be 14.3 mm (about 8% of the wavelength at the center operating frequency). Figure 4.31 shows the measured return loss of the prototypes with $(H, \alpha) = (30\ \text{mm}, 90^\circ)$, $(30\ \text{mm}, 75^\circ)$, $(15\ \text{mm}, 90^\circ)$, and $(7.5\ \text{mm}, 90^\circ)$. The case with $(H, \alpha) = (0, 0^\circ)$ is also shown for comparison, which represents the antenna with a corresponding planar ground plane of the same size as the bottom ground plane of the prototypes. The bottom ground plane lengths (L) of the prototypes were all chosen to be 80 mm (about 0.45 wavelength of the center operating frequency); that is, when $\alpha = 90^\circ$, the distance between the E-patch's radiating

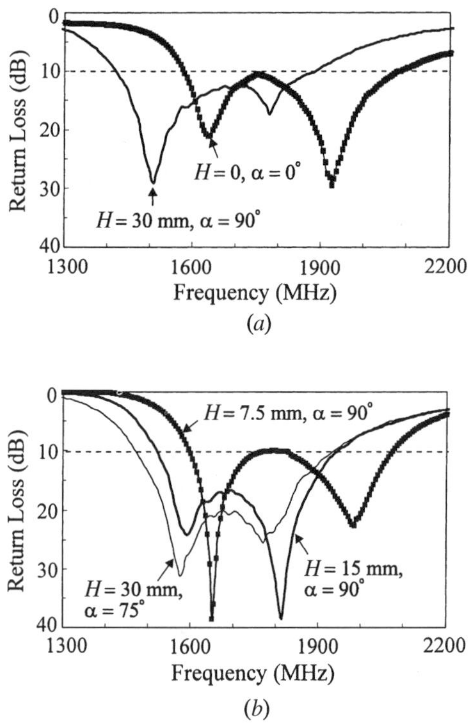

FIGURE 4.31 Measured return loss for the antenna shown in Figure 4.30. (*a*) The cases with $(H, \alpha) = (0, 0^\circ)$ and $(30\ \text{mm}, 90^\circ)$; (*b*) the cases with $(H, \alpha) = (30\ \text{mm}, 75^\circ)$, $(15\ \text{mm}, 90^\circ)$, and $(7.5\ \text{mm}, 90^\circ)$. (From Ref. 10, © 2002 IEEE, reprinted with permission)

TABLE 4.7 Performance of the Antenna Shown in Figure 4.30 [10]; $h = 14.3$ mm, $L = 80$ mm. The Impedance Bandwidth (BW) is Determined from 10 dB Return Loss and Referenced to the Center Frequency at 1700 MHz. f_1 and f_2 are the Antenna's First and Second Resonant Frequencies.

			XPL at f_1		XPL at f_2	
H, α	ℓ, w, S, d_p (mm)	BW (MHz, %)	E-Plane	H-Plane	E-Plane	H-Plane
0, 0°	45, 4, 15, 7	490, 28.8	28.2	9.1	25.8	4.3
30 mm, 90°	53, 4, 15, 5	465, 27.4	32.0	19.6	29.1	19.2
30 mm, 75°	51, 4, 15, 6	445, 26.2	30.7	17.1	29.0	16.2
15 mm, 90°	48, 4, 15, 10	425, 25.0	24.6	16.6	26.6	15.4
7.5 mm, 90°	47, 4, 15, 6	475, 27.9	25.7	11.0	24.0	7.0

edges and the conducting flanges is 10 mm, less than one substrate thickness. When a larger bottom ground plane length was chosen, effects of the cross-polarization reduction due to the U-shaped ground plane become insignificant. For all cases, the dimensions of the inserted wide slits (ℓ, w, S) and the feed position (d_p) were slightly adjusted such that wide impedance bandwidths about 25% or larger were achieved. Also note that, for such broadband probe-fed patch antennas, two resonant modes are usually excited, which forms the wide impedance bandwidth obtained.

The design parameters and experimental data are given in Table 4.7 for comparison. The measured radiation patterns for (H, α) = (0, 0°) and (30 mm, 90°) are shown in Figure 4.32, and the cases for (H, α) = (7.5 mm, 90°) and

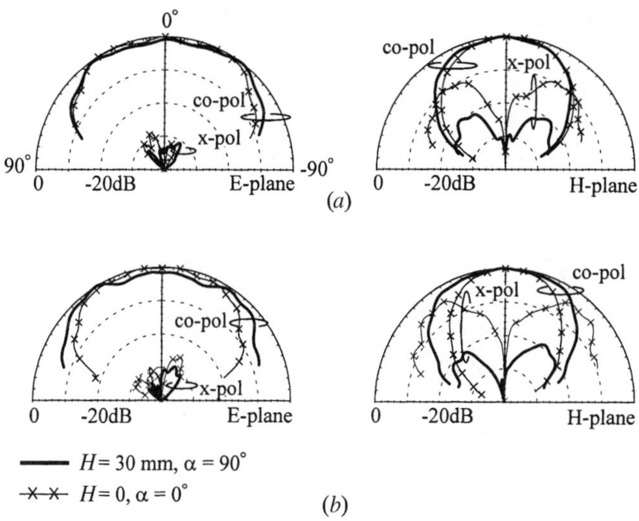

FIGURE 4.32 Measured radiation patterns for (a) the first (f_1) and (b) the second (f_2) resonant frequencies of the antenna shown in Figure 4.30 with (H, α) = (0, 0°) and (30 mm, 90°). (From Ref. 10, © 2002 IEEE, reprinted with permission)

(15 mm, 90°) are presented in Figure 4.33. From the obtained results, it is clearly seen that, when the conducting flanges of the U-shaped ground plane are at $\alpha = 90°$ and have a height of 30 mm (about two times the substrate thickness), the XPLs in the E- and H-plane patterns can be improved, respectively, by about 3.0–4.0 dB (32.0 vs. 28.2 dB and 29.1 vs. 25.8 dB) and 10.0–15.0 dB (19.6 vs. 9.1 dB and 19.2 vs. 4.3 dB), in comparison to the case without conducting flanges [i.e., $(H, \alpha) = (0, 0°)$]. When a larger height of the conducting flanges is chosen ($H > 30$ mm), experimental results show that there is a large dip in the broadside direction of the E-plane pattern of the second resonant frequency (f_2) within the impedance bandwidth, which may become unacceptable for some applications. In addition, a larger height of the conducting flanges increases the antenna height, which means that the proposed antenna no longer has the low-profile advantage like the conventional patch antenna, and thus the related results for the cases with $H > 30$ mm are not studied. As for the effect of varying the angle α on the cross-polarization reduction, experimental results show that the optimal angle is about 90° for a fixed flange height [see the results of $(H, \alpha) = (30\text{ mm}, 90°)$ and (30 mm, 75°) in Table 4.7].

Measured antenna gain for the constructed prototypes is shown in Figure 4.34. It is seen that, for the case with $(H, \alpha) = (30\text{ mm}, 90°)$, the antenna gain has a peak value of 8.0 dBi and small gain variations less than 0.8 dBi, which is better than the case without conducting flanges [$(H, \alpha) = (0, 0°)$]. The relatively lower gain for the case of $(H, \alpha) = (0, 0°)$ is probably due to the smaller ground plane dimension along the antenna's excitation direction, and, by adding conducting flanges to the

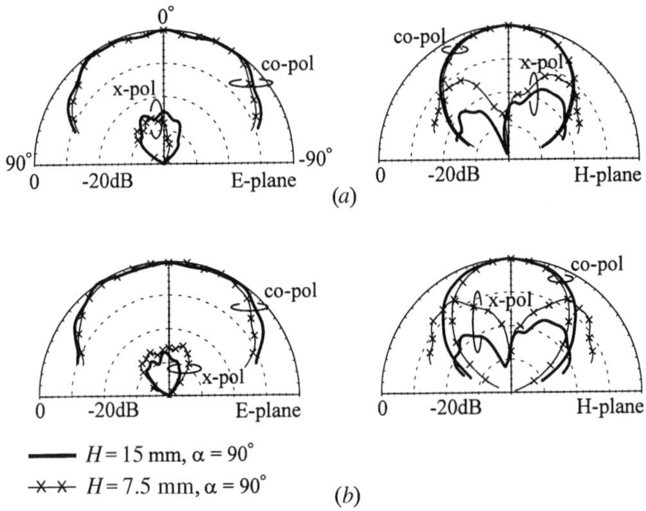

FIGURE 4.33 Measured radiation patterns for (a) the first (f_1) and (b) the second (f_2) resonant frequencies of the antenna shown in Figure 4.30 with $(H, \alpha) = (7.5\text{ mm}, 90°)$ and (15 mm, 90°). (From Ref. 10, © 2002 IEEE, reprinted with permission)

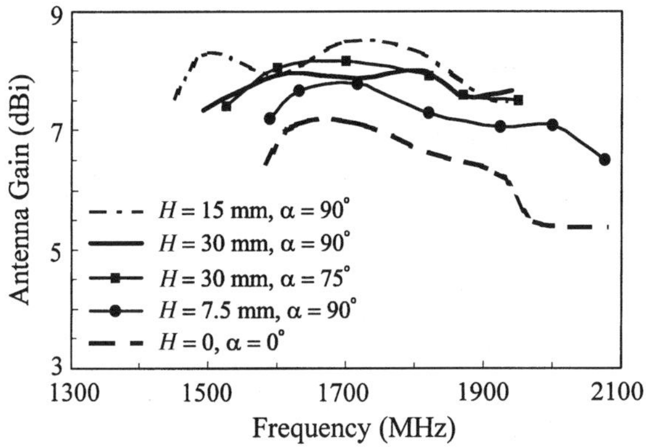

FIGURE 4.34 Measured antenna gain for the antenna shown in Figure 4.30. (From Ref. 10, © 2002 IEEE, reprinted with permission)

smaller ground plane, the effective ground plane size is increased, leading to gain enhancement for the proposed antenna. The present proposed antenna can also be easily applied to an array antenna, as shown in Figure 1.13, for cellular-system base-station applications.

4.2.2.6 *With Two E-Patches Fed Out of Phase*

A 1×2 probe-fed E-patch array antenna fed by using two feeds of a 180° phase shift to achieve a broadband operation with low cross-polarization radiation in the DCS band has been studied recently. Figure 4.35 shows the geometry of the E-patch array antenna studied. The two E-patches have the same dimensions of $60 \times 90\,\text{mm}^2$ and are supported by plastic supports (not shown in the figure) above the ground plane of an inexpensive FR4 substrate (thickness 0.8 mm and relative permittivity 4.4). The distance between the E-patches and the ground plane is 14 mm. The two wide slits in the E-patch have the same length ℓ and the same width w_1 and are inserted at one of the patch's radiating edges. The separation of the two wide slits is w_2, and

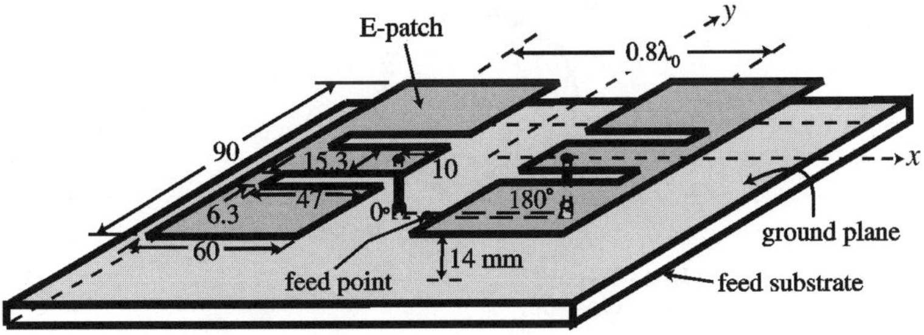

FIGURE 4.35 Geometry of the broadband 1×2 probe-fed E-patch array antenna for DCS operation.

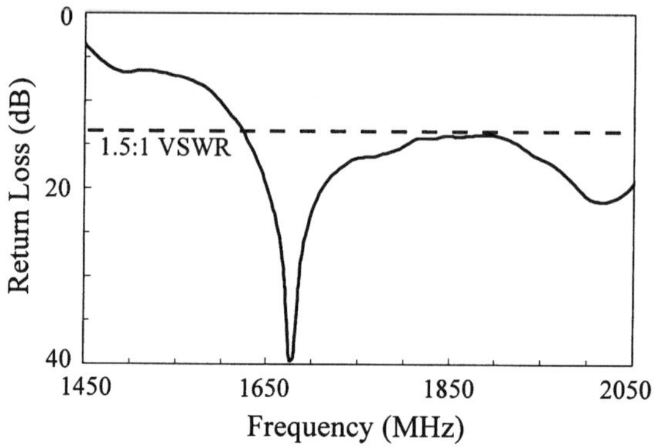

FIGURE 4.36 Measured return loss for the array antenna shown in Figure 4.35.

the two slits are placed symmetrically with respect to the centreline (x-axis) of the E-patch. The probe feed position is selected such that the input impedance seen at the feed position is 100 Ω for the excitation of the array antenna. The spacing is chosen to be about 0.8 wavelength of the center operating frequency so that a better antenna gain can be obtained.

Figure 4.36 shows the measured return loss for the array antenna. The obtained impedance bandwidth (1.5:1 VSWR) can cover the DCS band. Radiation characteristics of the array antenna have also been studied. Figure 4.37 plots the radiation

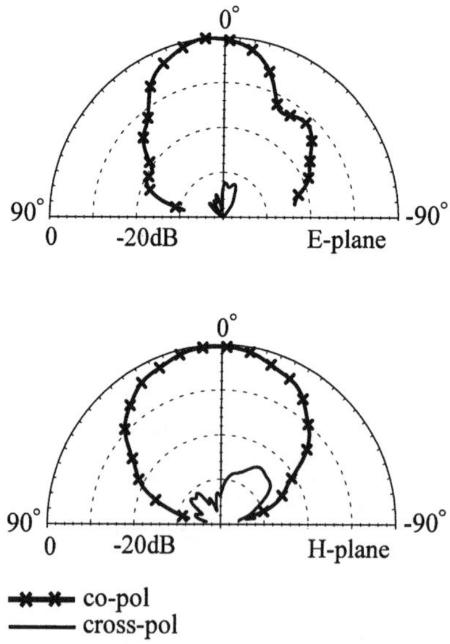

FIGURE 4.37 Measured radiation patterns at 1840 MHz for the array antenna shown in Figure 4.35.

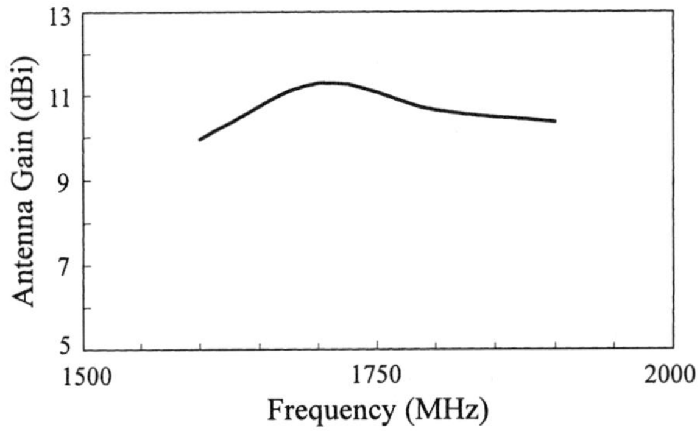

FIGURE 4.38 Measured antenna gain for the array antenna shown in Figure 4.35.

patterns at 1840 MHz, which show good broadside radiation patterns with low cross-polarization radiation (less than −25 dB in both the E- and H-plane patterns). Measured radiation patterns of other frequencies across the impedance bandwidth are similar to those shown here; that is, stable radiation patterns are obtained across the impedance bandwidth. Figure 4.38 presents the measured antenna gain. The peak antenna gain for frequencies across the DCS band reaches about 11.4 dBi, and the gain variations are less than 1.5 dBi.

4.2.3 Coplanar Probe-Fed Design

Figure 4.39 shows the broadband patch antenna edge-fed by a coplanar probe feed. The radiating patch has dimensions of $L \times W$, and the ground plane is L-shaped to facilitate the use of a coplanar probe feed shown in the figure. In this design, the

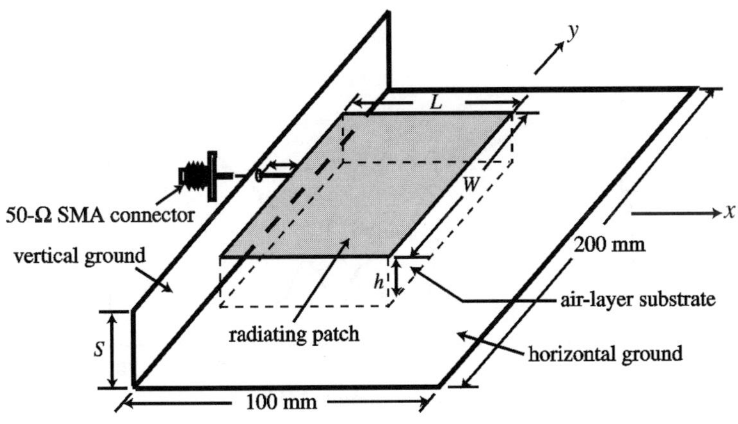

FIGURE 4.39 Geometry of the broadband patch antenna edge-fed by a coplanar probe feed. (From Ref. 11, © 2001 John Wiley & Sons, Inc.)

horizontal portion of the L-shaped ground has dimensions of $100 \times 200\ mm^2$ and the vertical portion of the L-shaped ground has a height of S, which is chosen to be slightly larger than the air-layer substrate thickness h to accommodate the SMA connector of the probe feed. Also note that the probe pin of the probe feed has a length of ℓ, which is chosen to be small (about 2% of wavelength of the operating frequency) so that the introduced probe inductance is small and can be ignored. In addition, it has been found that, simply by placing the probe pin at the center of the patch edge close to the vertical ground, good impedance matching over a wide frequency range can easily be achieved.

Several prototypes of the proposed antenna to operate in the 1800 MHz band have been constructed. Measured return loss for the constructed prototypes with various air-layer substrate thicknesses is presented in Figure 4.40. The radiating patch has a length (L) of 60 mm and a width (W) of 93 mm, and the height (S) of the vertical ground is fixed to be 23 mm, which is slightly larger than the air-layer substrate thickness and is about 14% of the wavelength at 1800 MHz. [Note that, when a larger value of H is chosen, the proposed antenna could have asymmetric E-plane radiation patterns.] The corresponding measured data are also given in Table 4.8 for comparison. It is seen that the obtained impedance bandwidth (1.5:1 VSWR) increases with an increase of the air-layer substrate thickness. For the case of $h = 18$ mm (about 10.8% of the wavelength at 1800 MHz), the impedance bandwidth reaches 186 MHz (1708–1894 MHz) or about 10.3% with respect to the center frequency at 1802 MHz. The obtained impedance bandwidth covers the required bandwidth of the DCS cellular system and can be applied for DCS base station antenna applications.

The radiation characteristics of the constructed prototype have also been studied. Figure 4.41 shows the measured radiation patterns for the constructed prototype studied in Figure 4.40 with $h = 18$ mm. Good broadside radiation patterns have

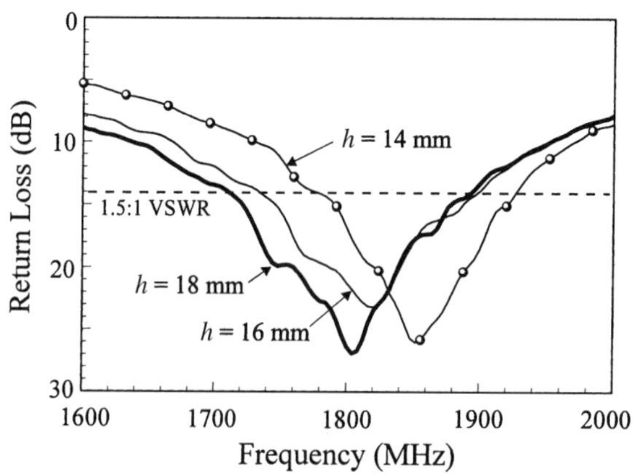

FIGURE 4.40 Measured return loss for the antenna shown in Figure 4.39; $S = 23$ mm, $L = 60$ mm, $W = 93$ mm, $\ell = 3.5$ mm. (From Ref. 11, © 2001 John Wiley & Sons, Inc.)

TABLE 4.8 Performance of the Constructed Prototypes for the Antenna Shown in Figure 4.39 [11]; $S = 23$ mm, $L = 60$ mm, $W = 93$ mm, $\ell = 3.5$ mm. f_c is the Resonant Frequency, and BW is the Impedance Bandwidth Determined From 1.5:1 VSWR.

h (mm)	f_c (MHz)	BW (MHz, %)
14	1850	148, 8.0
16	1818	164, 9.0
18	1802	186, 10.3

been observed, and the cross-polarization radiation is seen to be below −30 dB in the E-plane (x–z plane) patterns for the frequencies within the DCS band. As for the H-plane (y–z plane) patterns, the cross-polarization radiation is below −14 dB for the frequencies within the DCS band. Figure 4.42 shows the measured antenna gain for the constructed prototype studied in Figure 4.40 with $h = 18$ mm. The peak antenna gain reaches about 7.3 dBi, with small gain variations less than about 0.5 dBi.

4.2.4 Microstrip Line-Fed Design

By using a microstrip line feed, a broadband patch antenna suitable for applications in a UMTS cellular-system base station has been implemented. Figure 4.43 shows the antenna geometry. The radiating patch is short-circuited to the ground plane through two identical shorting plates. Because of the short-circuiting, the radiating patch is operated as a quarter-wavelength structure and has a compact size of 20×60 mm^2 for operating in the UMTS band (1920–2170 MHz). In addition, the microstrip feed line in this design has an air-layer substrate, which greatly reduces the fabrication cost of the proposed antenna.

A prototype based on the design dimensions shown in Figure 4.43 has been constructed and tested. Figure 4.44 shows the measured and simulated return loss for the constructed prototype. The measured data agree well with the simulated results. The obtained impedance bandwidth (1.5:1 VSWR) is 390 MHz (1840–2230 MHz), which covers the required bandwidth of the UMTS band. For frequencies across the bandwidth, good broadside radiation patterns have been observed. Figure 4.45 presents the measured antenna gain. The prototype shows an antenna gain level of about 6 dBi, and small gain variations (less than 1 dBi) across the bandwidth are seen.

4.2.5 Capacitively Coupled Feed Design

The broadband patch antenna excited by an L-strip capacitively coupled feed for a DCS base station has been implemented [12]. Figure 4.46 shows the antenna geometry. The radiating patch has a width of W and a projection length of L. Note that, at both of the patch's two radiating edges, the patch is bent downward the antenna's ground plane with a length of ℓ_1, and one of the bent edges is further

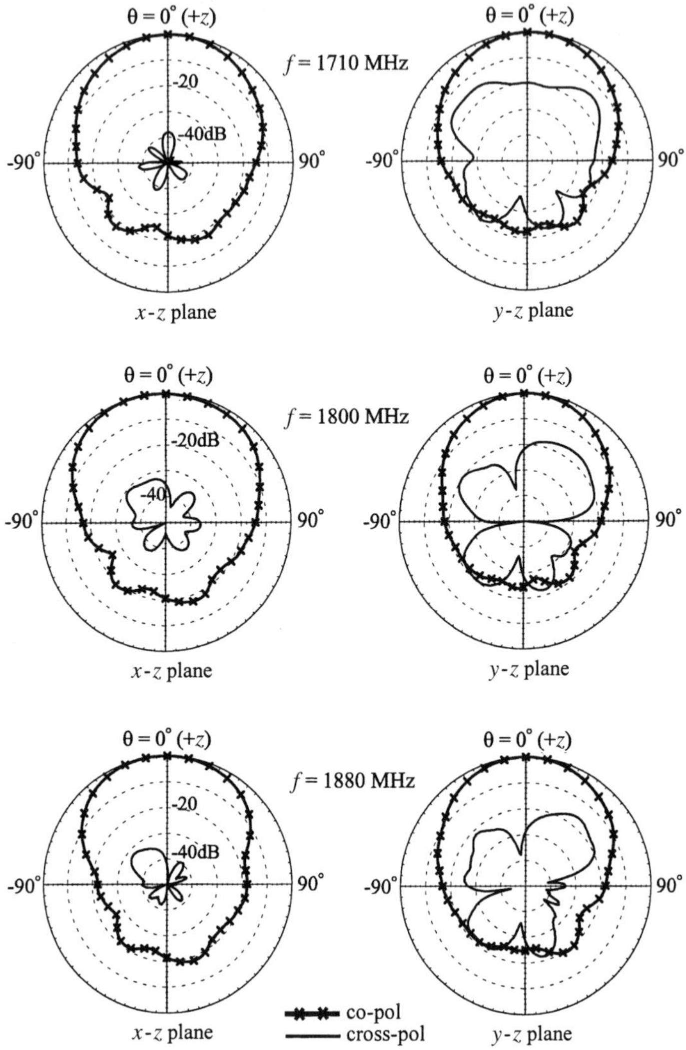

FIGURE 4.41 Measured radiation patterns in the x–z plane (E-plane) and y–z plane (H-plane) for the antenna studied in Figure 4.40 with $h = 18$ mm. (From Ref. 11, © 2001 John Wiley & Sons, Inc.)

bent 90° toward the patch center with a length of ℓ_2. In this case, the total length of the folded radiating patch is $L + 2\ell_1 + \ell_2$. This folded patch is supported by plastic posts (not shown in the figure) above the antenna's ground plane with a height of h. The L-strip capacitively coupled feed is placed along the patch's centerline (x-axis in the figure) and has a distance of d close to the opposite side of the patch's folded bent edge. The L-strip has a width of 2 mm and a length of 20 mm and is below the top

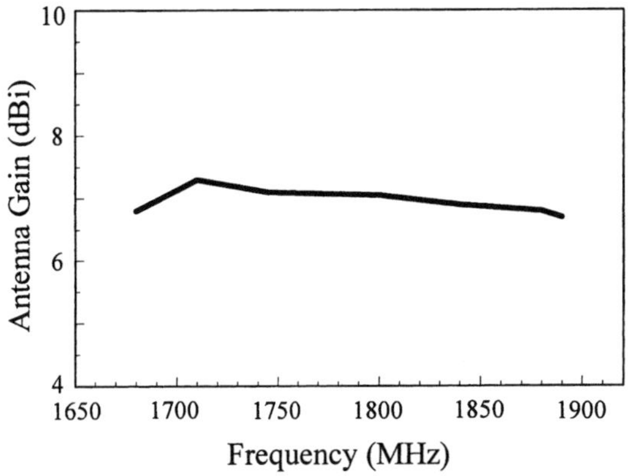

FIGURE 4.42 Measured antenna gain for the antenna studied in Figure 4.40 with $h = 18$ mm. (From Ref. 11, © 2001 John Wiley & Sons, Inc.)

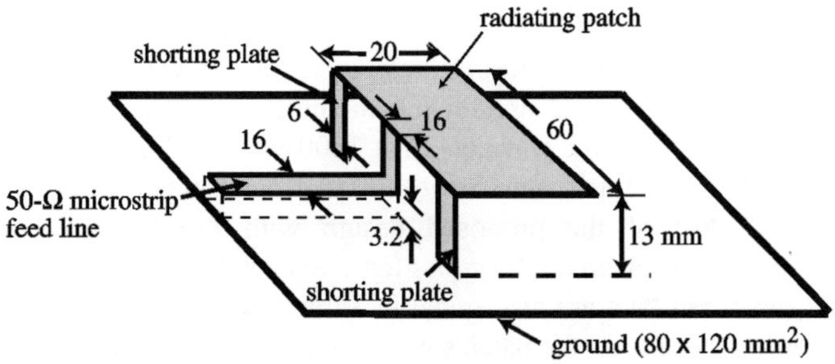

FIGURE 4.43 Geometry of a broadband microstrip-line-fed patch antenna for UMTS operation.

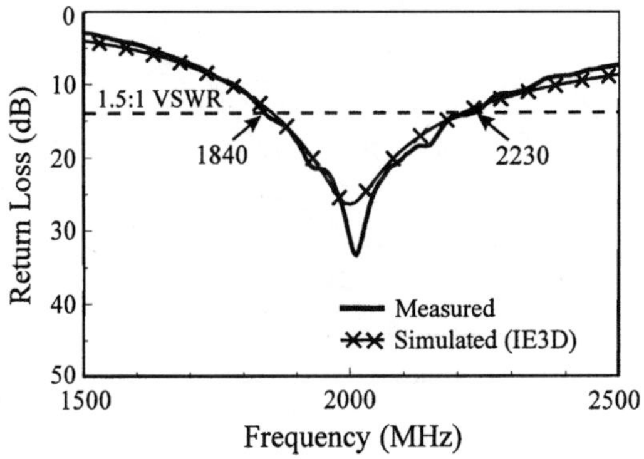

FIGURE 4.44 Measured and simulated return loss for the antenna shown in Figure 4.43.

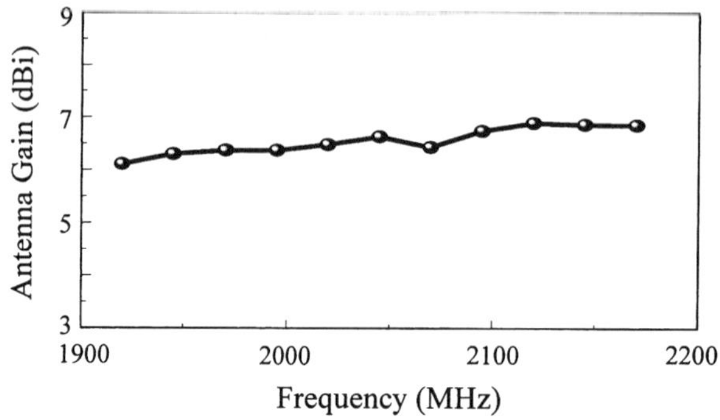

FIGURE 4.45 Measured antenna gain for the antenna shown in Figure 4.43.

portion of the folded patch with a distance of g. In this study, the distance g was chosen to be 4 mm, and by adjusting the distance d, good impedance matching over the frequency range covering the DCS band can be obtained.

In the experiment, the center operating frequency was chosen to be about 1800 MHz, and, to cover the DCS band with an impedance matching better than 1.5:1 VSWR (about 14 dB return loss), a reference antenna with a simple radiating patch of dimensions $60 \times 90\,\text{mm}^2$ and an antenna height (h) of 12 mm, which corresponds to about 7.2% times the wavelength at 1800 MHz, was first constructed and measured. Then, with the total patch length (60 mm) unchanged, several prototypes (antennas A1 to A4) of the proposed design with $\ell_2 = 0$ were constructed. Figure 4.47 shows the measured return loss for antennas A1 to A4 and the reference antenna. The measured data are also given in Table 4.9 for comparison. Note that, with an increase of ℓ_1 (that is, the patch's projection length L is decreased), antennas

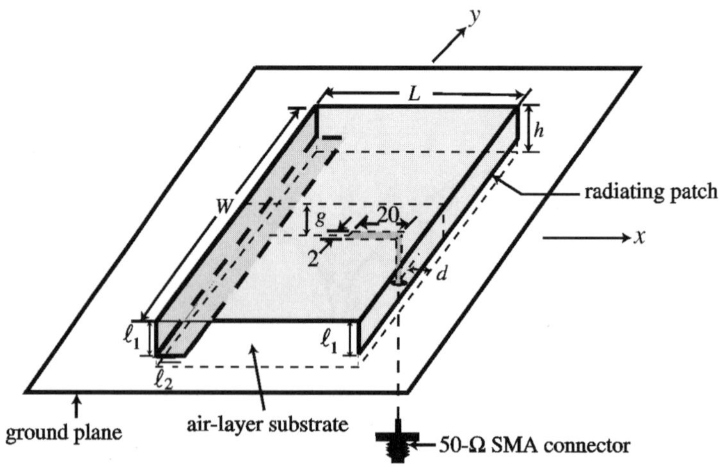

FIGURE 4.46 Geometry of an L-strip capacitively coupled patch antenna for DCS base station. (From Ref. 12, © 2001 John Wiley & Sons, Inc.)

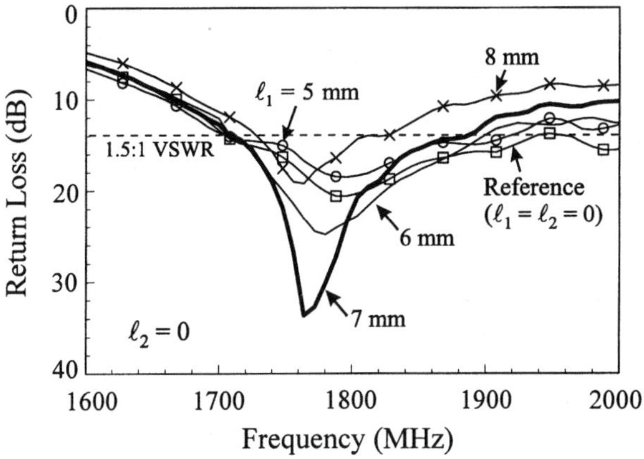

FIGURE 4.47 Measured return loss for the antenna shown in Figure 4.46 with $\ell_2 = 0$ (antennas A1 to A4 shown in Table 4.9) and the reference antenna ($\ell_1 = \ell_2 = 0$); $h = 12$ mm, $W = 90$ mm, $L = 60$ mm–$2\ell_1$, $g = 4$ mm, ground plane size $= 150 \times 220$ mm^2. (From Ref. 12, © 2001 John Wiley & Sons, Inc.)

A1 to A3 can still have a wide impedance bandwidth centered at about 1800 MHz and covering the DCS band. For antenna A4, however, the obtained impedance bandwidth is greatly reduced and cannot meet the bandwidth requirement of the DCS band.

Next, we fixed the length ℓ_1 at 7 mm and increased the length ℓ_2 to be 1 mm (antenna B1), 3 mm (antenna B2), 5 mm (antenna B3), and 7 mm (antenna B4). In this case, the total patch length ($L + 2\ell_1 + \ell_2$) is still fixed to be 60 mm. Measured return loss is presented in Figure 4.48, and the correponding data are also given in Table 4.9. For antennas B1 to B3, the obtained impedance bandwidths can still cover the DCS band. However, for antenna B4, the center operating frequency is

TABLE 4.9 Performance of the Constructed Prototypes Studied in Figures 4.47 and 4.48 [12]; $h = 12$ mm, $W = 90$ mm, $L = 60$ mm $- 2\ell_1 - \ell_2$, $g = 4$ mm, Ground-Plane Size $= 150 \times 220$ mm^2. Impedance Bandwidth (BW) is Determined From 1.5:1 VSWR, and f_L and f_H are the Lower and Upper Frequencies With 1.5:1 VSWR, Respectively.

	ℓ_1 (mm)	ℓ_2 (mm)	d (mm)	f_L, f_H (MHz)	BW (MHz, %)
Reference	0	0	5	1706, 1940	234, 12.8
Antenna A1	5	0	2.5	1706, 1918	212, 11.7
Antenna A2	6	0	2.5	1710, 1900	190, 10.5
Antenna A3	7	0	2.5	1710, 1888	178, 9.9
Antenna A4	8	0	4	1732, 1824	92, 5.2
Antenna B1	7	1	3.5	1710, 1884	174, 9.7
Antenna B2	7	3	3	1710, 1892	182, 10.1
Antenna B3	7	5	2	1710, 1886	176, 9.8
Antenna B4	7	7	1.5	1764, 1940	176, 9.5

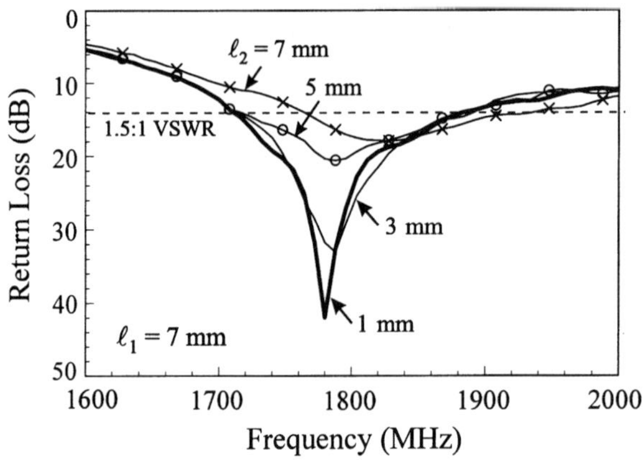

FIGURE 4.48 Measured return loss for the antenna shown in Figure 4.46 with $\ell_1 = 7$ mm (antennas B1 to B4 shown in Table 4.9); $h = 12$ mm, $W = 90$ mm, $L = 46$ mm $- \ell_2$, $g = 4$ mm, ground plane size $= 150 \times 220$ mm^2. (From Ref. 12, © 2001 John Wiley & Sons, Inc.)

increased to about 1852 MHz and it becomes difficult to achieve a wide bandwidth centered at about 1800 MHz suitable for the DCS operation. By comparing antenna B3 ($L = 41$ mm) with the reference antenna ($L = 60$ mm), it can be seen that the patch's projection length for the proposed antenna is reduced by about 31%.

The radiation characteristics of the proposed antenna were also studied. Figures 4.49 and 4.50 plot the radiation patterns for the reference antenna and antenna B3, respectively. Results for the operating frequencies at 1710, 1800, and 1880 MHz are presented. It can be seen that both the proposed antenna and the reference antenna show good broadside radiation patterns. In addition, the proposed antenna shows about the same cross-polarization radiation in both the E-plane (x–z plane) and H-plane (y–z plane) patterns as those of the reference antenna. This behavior is largely because there are no lateral surface current components expected to be generated for lengthening the excited patch surface current path of the proposed antenna, thereby leading to no degradation in the cross-polarization radiation. In Figure 4.51, measured antenna gain for antennas A3, B3, and the reference antenna for frequencies across the DCS band is presented. Small gain variations for the three antennas are observed. However, probably because the top portion of the radiating patch is smaller for the proposed antenna, the gain levels of antenna A3 ($L = 46$ mm) and B3 ($L = 41$ mm) are about 0.6 and 1.0 dBi less, respectively, than that of the reference antenna ($L = 60$ mm).

4.3 ANTENNA FOR DUAL-BAND OR MULTIBAND OPERATION

4.3.1 Design with One Driven Patch and One Parasitic Patch

To obtain dual-band or multiband operation for base station applications, some promising patch antenna designs have also been devised recently. Figure 4.52

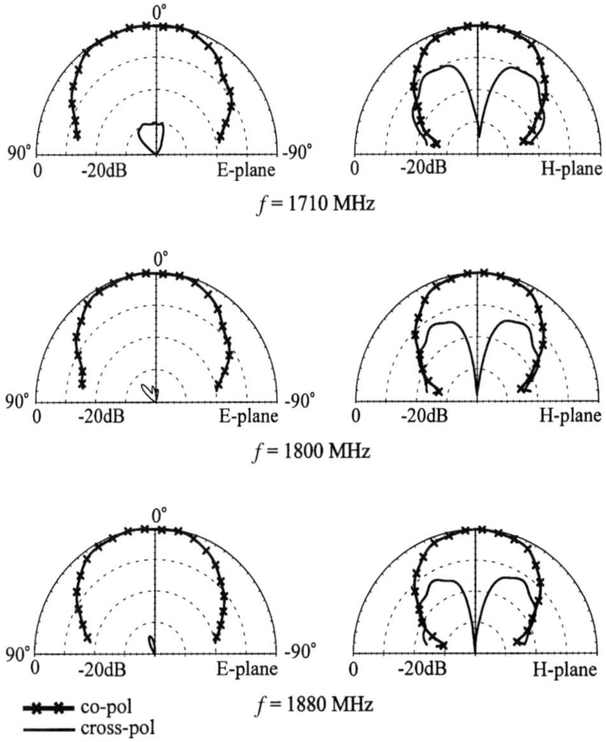

FIGURE 4.49 Measured radiation patterns for the reference antenna ($\ell_1 = \ell_2 = 0$) studied in Figure 4.47. (From Ref. 12, © 2001 John Wiley & Sons, Inc.)

shows a promising patch antenna design suitable for GSM/DCS/PCS triple-band base stations. The proposed antenna comprises a simple rectangular patch for generating the upper band for the DCS and PCS operations and a shorted patch for generating the lower band for the GSM operation. The rectangular patch has dimensions of $55 \times 90\,\text{mm}^2$ and is fed by a coplanar probe feed at the center of the patch edge. To accommodate the coplanar probe feed, a vertical ground of height 31 mm is added and connected to the horizontal ground, which was selected to have dimensions of $170 \times 200\,\text{mm}^2$ in the experiment. The vertical ground was also placed close to the edge of the rectangular patch with a distance of 3.5 mm in the experiment. This small distance makes the required probe pin length of the coplanar probe feed small, although a thick air-layer substrate of thickness 25 mm is used in the proposed antenna to achieve a broadband operation. In this case, the introduced probe inductance is small and its effect on the antenna's impedance matching can be ignored.

The shorted patch is placed at a distance of 27 mm from the rectangular patch and is centered with respect to the coplanar probe feed or the centerline of the rectangular patch. The shorted patch is short-circuited to the horizontal ground through

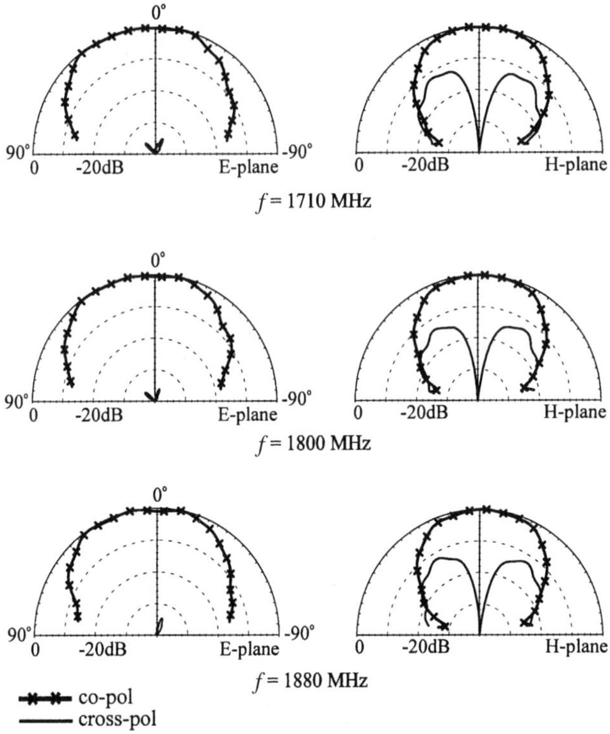

FIGURE 4.50 Measured radiation patterns for antenna B3 ($\ell_1 = 7$ mm, $\ell_2 = 5$ mm) studied in Figure 4.48. (From Ref. 12, © 2001 John Wiley & Sons, Inc.)

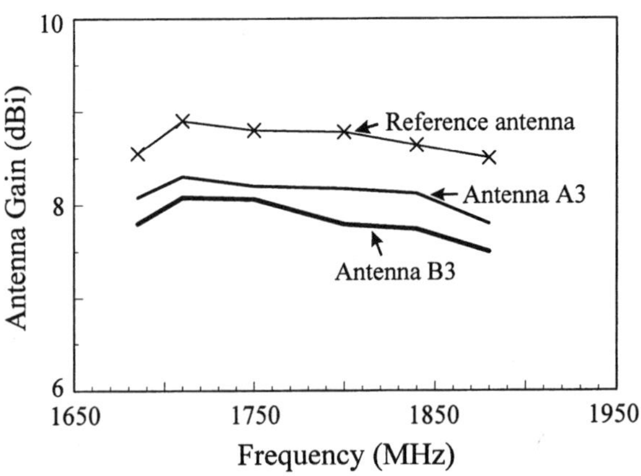

FIGURE 4.51 Measured antenna gain for antennas A3 and B3 and the reference antenna; antenna parameters are given in Figures 4.47 and 4.48. (From Ref. 12, © 2001 John Wiley & Sons, Inc.)

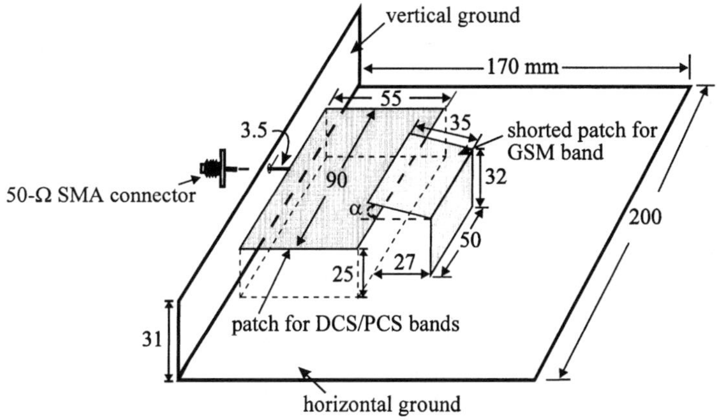

FIGURE 4.52 Geometry of the patch antenna for GSM/DCS/PCS triple-band base station.

a conducting wall of dimensions $32 \times 50\,\text{mm}^2$, and the upper portion of the shorted patch has dimensions of $35 \times 50\,\text{mm}^2$, which is tilted upward with an angle of α. With this arrangement, the shorted patch can be moved closer to the rectangular patch and can be easily excited parasitically. Also note that, for the proposed antenna, the optimal angle for α is 18.5°. When a larger angle is selected, decreasing coupling of the shorted patch to the driven rectangular patch is expected, and the bandwidth of the lower band of the proposed antenna was also found to be decreased. Conversely, when a smaller angle is selected, the shorted patch may be overcoupled, and the impedance matching of the rectangular patch is greatly affected. In this case, the resonant frequency of the rectangular patch was seen to be lowered, and the impedance matching for frequencies across the upper band of the proposed antenna was also greatly affected, resulting in a wide impedance bandwidth difficult to achieve.

It should also be noted that the driven rectangular patch is mainly operated as a half-wavelength structure. However, because of the thick air-layer substrate used, the length (55 mm) of the rectangular patch in the proposed antenna is relatively much less than 50% of the free-space wavelength (only about 34% of the free space wavelength) at the center frequency (1850 MHz) of the upper band. On the other hand, the parasitic shorted patch is operated as a quarter-wavelength structure and, similarly, the total length (67 mm) of the shorted patch is only about 20% of the free space wavelength at the center frequency (900 MHz) of the lower band, which is less than one-quarter wavelength and is also due to the thick air-layer substrate used. Also note that, in the experiment, the optimal dimensions of the proposed design are obtained with the aid of the simulation software Ansoft HFSS (high-frequency structure simulation).

To achieve wide impedance bandwidths for both the lower and upper bands of the proposed antenna, the thicknesses of the air-layer substrates for the driven rectangular patch and the parasitic shorted patch were chosen to be 25 mm (about 15%

of the free space wavelength at 1850 MHz) and 32 mm (about 9.6% of the free space wavelength at 900 MHz), respectively. The measured return loss is presented in Figure 4.53. Two wide impedance bandwidths (1.5:1 VSWR) are obtained. The lower band has a bandwidth of 87 MHz (875–962 MHz), which covers the required bandwidth of the GSM band. On the other hand, a much larger bandwidth of 390 MHz (1610–2000 MHz) for the upper band is obtained, which covers both the DCS and PCS bands.

The radiation characteristics of the proposed antenna have also been studied. In general, broadside radiation patterns are obtained for frequencies across the GSM, DCS, and PCS bands, with slight asymmetry in the E-plane patterns observed, which is largely due to the effect of the added vertical ground plane for accommodating the coplanar probe feed. As for the measured antenna gain, the peak antenna gain for the GSM band is about 5.5 dBi and the gain variations are less than 2 dBi. For the upper band for the DCS and PCS operations, the peak antenna gain reaches about 9.7 dBi and the gain variations are less than 2 dBi.

4.3.2 Design with One Single Patch

Recently, it was demonstrated that GSM/DCS/PCS triple-band operations can be obtained by using a single patch [13]. Figure 4.54 shows the top view and side view of the proposed patch antenna. The radiating patch has a length ℓ and a width w and is fed by a probe feed with a short probe pin of length 3 mm in the same plane. The radiating patch is backed by a U-shaped ground plane, which comprises a horizontal ground (dimensions $L \times W$), a first vertical ground (dimensions $L \times H_1$), and a second vertical ground (dimensions $L \times H_2$). The distance between the radiating patch and the horizontal ground is h, which can be considered as the air-layer substrate thickness, and there are plastic posts (not shown in the figure) supporting the radiating patch above the horizontal ground. In this study, h

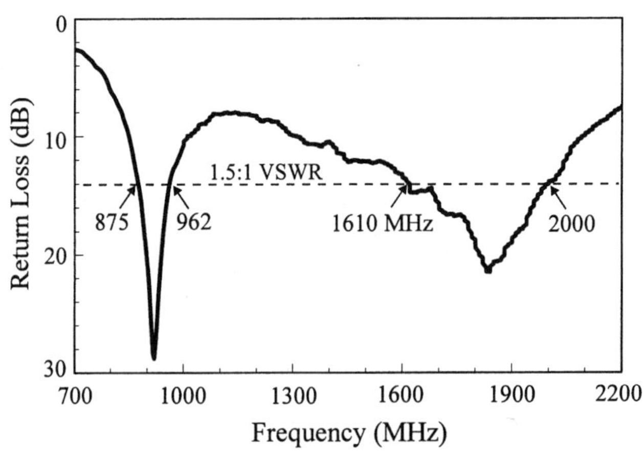

FIGURE 4.53 Measured return loss of the antenna shown in Figure 4.52.

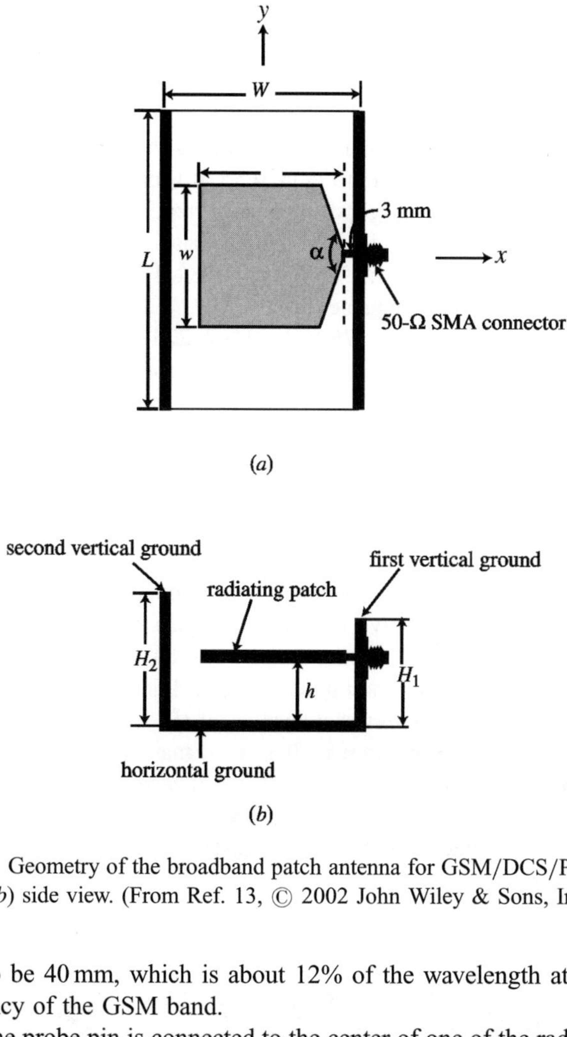

FIGURE 4.54 Geometry of the broadband patch antenna for GSM/DCS/PCS base station. (*a*) Top view; (*b*) side view. (From Ref. 13, © 2002 John Wiley & Sons, Inc.)

was chosen to be 40 mm, which is about 12% of the wavelength at 920 MHz, the center frequency of the GSM band.

Note that the probe pin is connected to the center of one of the radiating edges of the patch and, at that edge, the patch has a flare angle of α. By tuning this flare angle, impedance matching of the proposed antenna can be improved. In this study, the optimal flare angle was selected to be 170°. For the first vertical ground it is mainly used for accommodating the coplanar probe feed, and its height was chosen to be 50 mm (H_1), slightly larger than h. For the second vertical ground its height (H_2) has significant effects on improving the impedance matching of the two desired resonant modes to cover the GSM, DCS, and PCS bands. An optimal height of 70 mm was chosen from experimental and simulation studies. The simulation software Ansoft HFSS is used for the simulation, which makes the optimal dimensions of the proposed antenna easy to determine.

In the experiment, the dimensions of the radiating patch were selected to be $74 \times 65\ \text{mm}^2$ ($\ell \times w$) and the horizontal ground has dimensions $200 \times 100\ \text{mm}^2$

($L \times W$). Note that, in this case, the distance of the radiating patch to the second vertical ground is 23 mm ($= W - \ell - 3$ mm), which is about 30% of the patch length ℓ. If this distance is too small or too large, good impedance matching of the two desired resonant modes will be difficult to obtain. Measured and simulated return loss for the proposed antenna with an optimal height of the second vertical ground ($H_2 = 70$ mm) is presented in Figure 4.55. It is clearly seen that two separate resonant modes are excited with good impedance matching. Reasonable agreement between the simulated results and the measured data is obtained. The lower mode has an impedance bandwidth (1.5:1 VSWR) of 388 MHz (593–981 MHz) or nearly 50% referenced to its center frequency at 787 MHz. On the other hand, the upper band has a bandwidth as large as 575 MHz (1662–2237 MHz) or about 30% referenced to its center frequency at 1950 MHz. The two resonant modes cover the GSM, DCS, and PCS bands. Also note that, in the proposed antenna, the length (74 mm) of the radiating patch is only about 20% of the wavelength of the center frequency (787 MHz) of the lower mode or about 22% of the wavelength at the center frequency (920 MHz) of the GSM band. This resonant patch length is relatively much less than that (usually larger than 35% of the wavelength of the center operating frequency) of the broadband patch antenna studied in Section 4.2. This indicates that, in addition to the broadband dual-frequency operation obtained, the proposed antenna has the attractive feature of compact size in its radiating patch.

To demonstrate the effects of the height H_2 on improving the impedance matching of the two desired resonant modes, Figure 4.56 shows the measured return loss of the proposed antenna with various values of H_2. In Figure 4.56*a*, the cases of $H_2 = 0$, 30, 50, and 70 mm are shown. It is seen that the impedance matching is greatly improved with an increase of H_2. Conversely, results for the cases ($H_2 = 70$, 90, 110, and 130 mm) shown in Figure 4.56*b* indicate that the impedance matching is degraded when H_2 is further increased. These results suggest that there

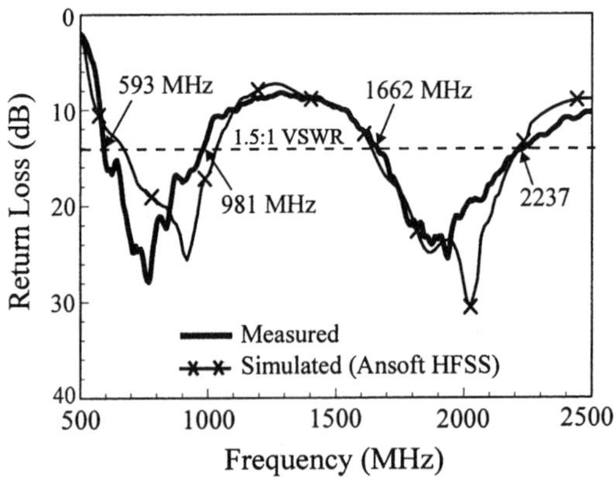

FIGURE 4.55 Measured and simulated return loss for the antenna shown in Figure 4.54; $L = 200$ mm, $W = 100$ mm, $H_1 = 50$ mm, $H_2 = 70$ mm, $h = 40$ mm, $\ell = 74$ mm, $w = 65$ mm, $\alpha = 170°$. (From Ref. 13, © 2002 John Wiley & Sons, Inc.)

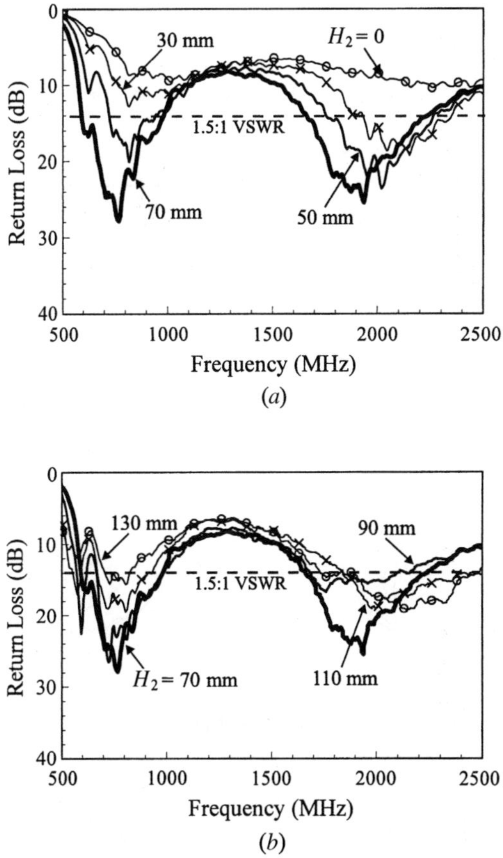

FIGURE 4.56 Measured return loss for the antenna shown in Figure 4.54 with (*a*) $H_2 = 0$, 30, 50, 70 mm and (*b*) $H_2 = 70$, 90, 110, 130 mm. Other parameters are the same as given in Figure 4.55. (From Ref. 13, © 2002 John Wiley & Sons, Inc.)

is an optimal value of H_2 for achieving improved impedance matching of the two desired resonant modes.

Figure 4.57 plots the measured and simulated radiation patterns at center frequencies of the GSM, DCS, and PCS bands for the proposed antenna with $H_2 = 70$ mm. Good broadside radiation patterns are observed. The measurement in general agrees with the simulation. Radiation patterns for other operating frequencies were also measured, and the radiation patterns are stable for frequencies across their respective GSM, DCS, and PCS bands and are similar to those plotted in Figure 4.57. Figure 4.58 presents the measured antenna gain for frequencies across the GSM, DCS, and PCS bands for the proposed antenna with $H_2 = 70$ mm. For the GSM band, the peak antenna gain is about 5 dBi, with very small gain variations

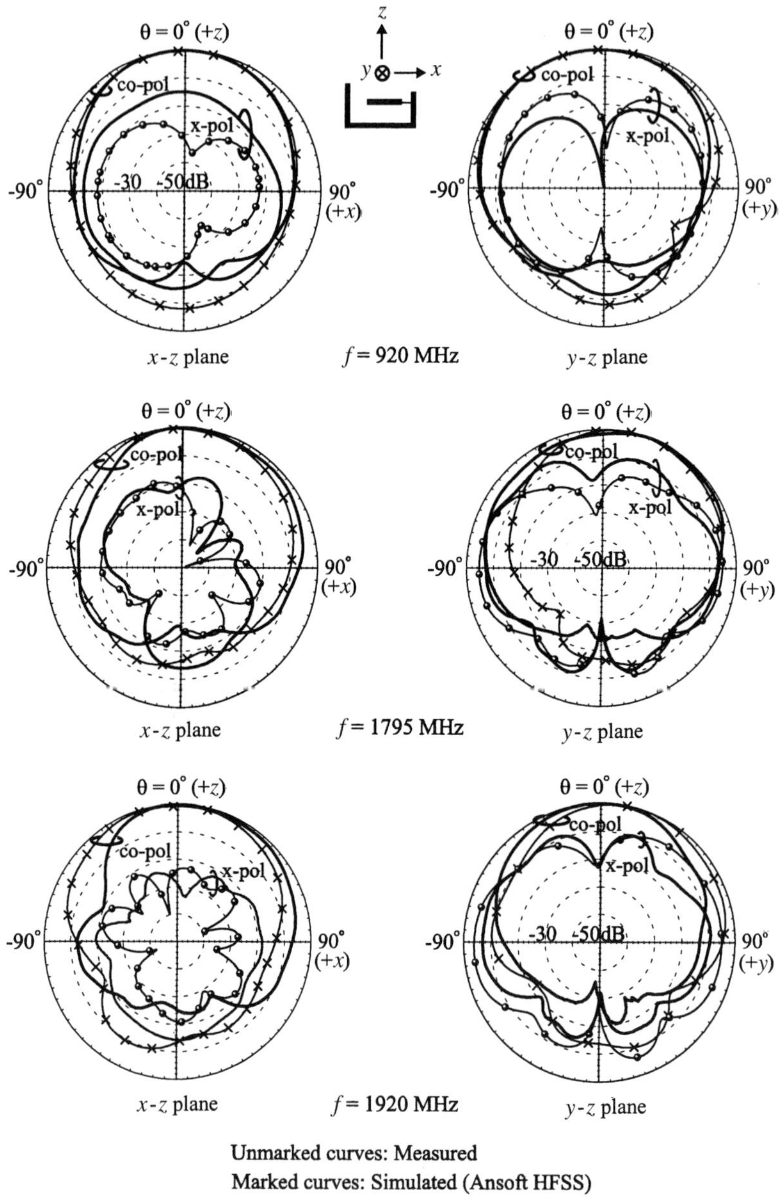

FIGURE 4.57 Measured and simulated radiation patterns for the antenna studied in Figure 4.55. (From Ref. 13, © 2002 John Wiley & Sons, Inc.)

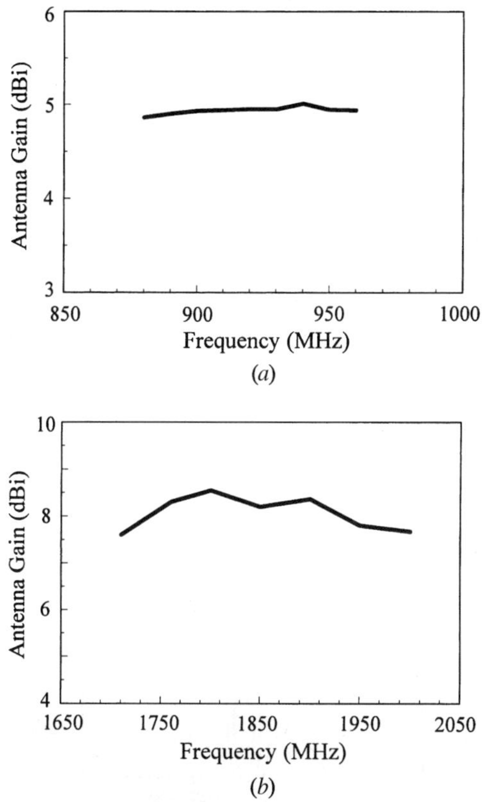

FIGURE 4.58 Measured antenna gain for frequencies across (*a*) the GSM band and (*b*) the DCS and PCS bands for the antenna studied in Figure 4.55. (From Ref. 13, © 2002 John Wiley & Sons, Inc.)

(less than 0.3 dBi) across the band. On the other hand, the DCS and PCS bands have a peak antenna gain of 8.5 dBi and the gain variations are less than 1.0 dBi.

4.4 ANTENNA FOR DUAL-POLARIZED OPERATION USING TWO SAME FEEDS

Good dual-polarized characteristics of slot-coupled patch antennas with H-shaped coupling slots have been obtained [19]. In such dual-polarized patch antennas, highly decoupled feeding ports ($S_{21} < -30$ dB) over a wide operating bandwidth can easily be obtained when the central arm of one H-shaped coupling slot is aligned in the direction of the microstrip feed line of the other H-shaped coupling slot. Several promising arrangements of the two H-shaped coupling slots in the antenna's ground plane have also been described in Reference 19. In this section, another promising design of slot-coupled patch antennas with modified H-shaped coupling slots [14] for achieving dual-polarized radiation with high port isolation over a wide bandwidth is presented.

Figures 4.59 and 4.60 show two designs of the slot-coupled dual-polarized patch antenna with the proposed modified H-shaped coupling slots. The two upper side arms of the modified H-shaped slots are bent inward with an angle α, and the spacing between the tips of the two upper side arms is S_2. (Note that, in the proposed design, if S_2 is less than 3 mm, enhanced isolation between two feeding ports cannot be obtained.) The two lower side arms are perpendicular to the center arm of the modified H-shaped slot and have a length of S_h and a spacing of S_1 (S_1 is also the length of the center arm). The distance between the center arm and the patch center is d. The upper and lower side arms are all chosen to have the same width

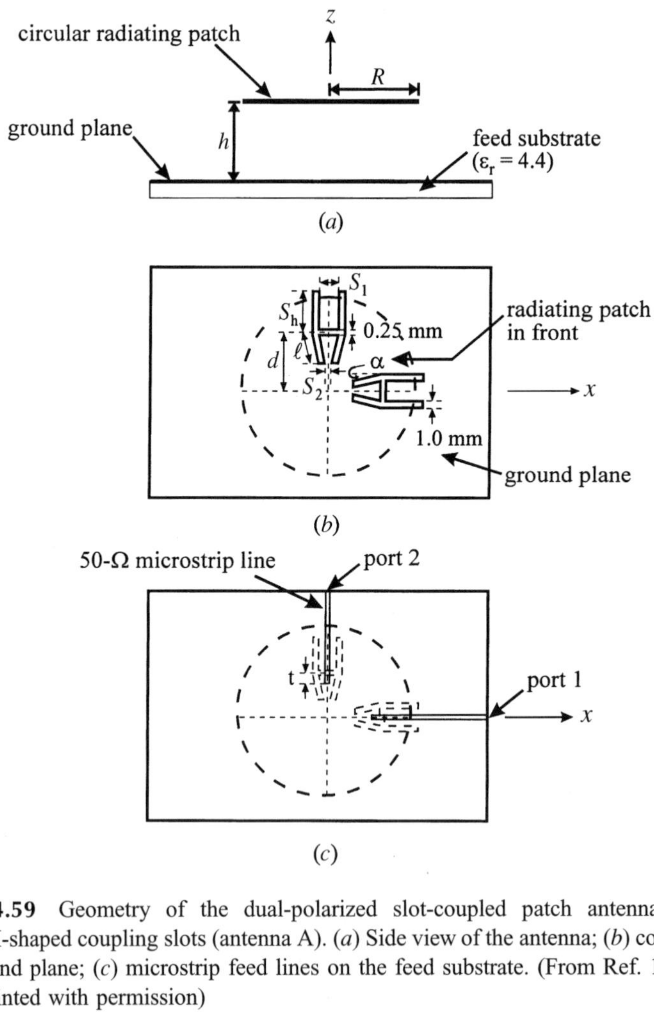

FIGURE 4.59 Geometry of the dual-polarized slot-coupled patch antenna with two modified H-shaped coupling slots (antenna A). (*a*) Side view of the antenna; (*b*) coupling slots in the ground plane; (*c*) microstrip feed lines on the feed substrate. (From Ref. 14, © 2002 IEEE, reprinted with permission)

of 1.0 mm in this study, and a smaller width (0.25 mm) for the center arm is selected. The modified H-shaped slots are fed by 50-Ω microstrip lines, which have a tuning stub length of t and are printed on a feed substrate (an FR4 substrate with relative permittivity 4.4 and thickness 0.8 mm in this study). The radiating patch of a circular shape is used, and the circular patch has a radius of R. The circular patch is also supported by plastic posts (not shown in the figures) of length h and placed above the ground plane of the microstrip feed lines.

In Figure 4.59 (denoted as antenna A in this study), the electromagnetic energy is coupled from the microstrip feed lines to the circular patch via a modified H-shaped slot for each feeding port. The decoupling between the two feeding ports is found to be greatly dependent on the bent angle α of the modified H-shaped slot. There exists an optimal α for achieving optimal decoupling between the two feeding ports. The optimal angle can be obtained with the aid of the simulation software IE3D, and in this study the optimal angle is found to be about 40°. It is also found that, by using a pair of modified H-shaped slots fed by equal amplitudes and phases provided by a Wilkinson power divider for each feeding port (see Fig. 4.60, antenna B), decoupling between the two feeding ports can further be improved and enhanced polarization purity of the two polarizations can also be obtained. This improvement in decoupling and polarization purity is largely because some higher-order modes that contribute to the polarization impurity can be suppressed with the use of two slots fed by equal amplitudes and phases and placed symmetrically with respect to the patch center.

Antenna A was first constructed and measured. The center frequency was designed to be 1800 MHz. For this purpose, the radius of the circular patch was chosen to be 37.25 mm, and the air-substrate's thickness (h) was 9 mm or about 5.4% of the wavelength at the center frequency. The required center arm's length (S_1) was 14 mm, only about 19% of the diameter ($2R$) of the circular patch. Figure 4.61 shows the measured return loss and isolation for antenna A and the reference antenna. Note that, because port 1 and port 2 excitation in this case have almost identical measured results, only the results of port 1 excitation are shown in Figure 4.61. The modified H-shaped coupling slots used for antenna A had an optimal bent angle (α) of 40°, which was found to result in an optimal decoupling between ports 1 and 2. The reference antenna had the same parameters as antenna A, except that the coupling slots used were the conventional H-shaped slots; i.e., $\alpha = 0°$ and $S_2 = S_1$. For both antenna A and the reference antenna, the obtained impedance bandwidths (10 dB return loss) are 7.3 and 8.5%, respectively. The isolation between the two feeding ports for antenna A is less than −31.4 dB over the entire bandwidth, which is about −10 to −35 dB less than that of the reference antenna. The measured radiation patterns in two orthogonal planes at 1800 MHz for port 1 excitation of antenna A are plotted in Figure 4.62. Good broadside radiation patterns are observed, with an antenna gain of 6.8 dBi. However, the cross-polarization levels (XPLs) in both planes are seen to be only about 14 dB, which is worse than that of the reference antenna. As for the backward radiation, both the proposed antenna and the reference antenna are about the same.

Antenna B was also constructed and measured. The parameters of antenna B were selected to be the same as those of antenna A, except that two identical modified H-

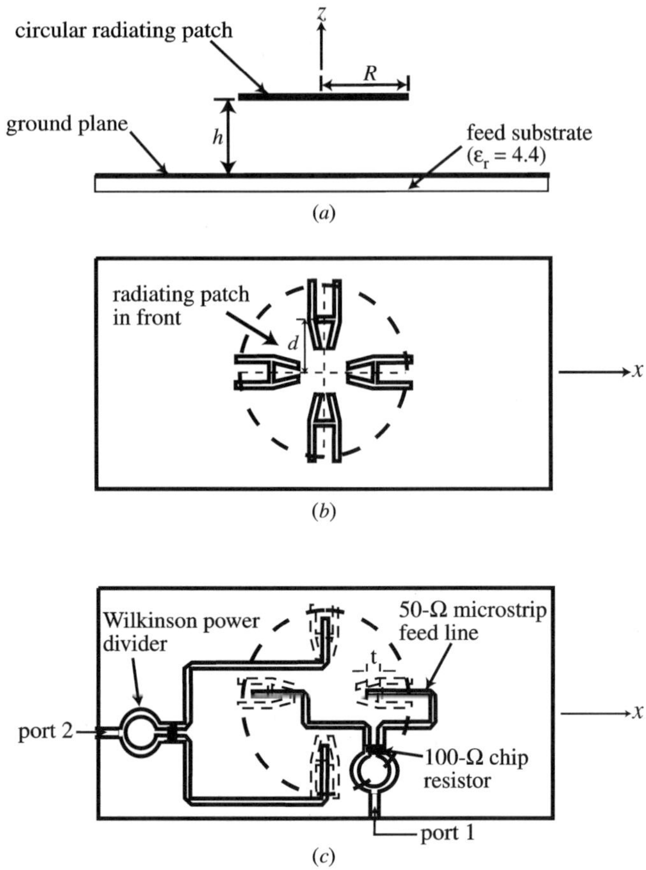

FIGURE 4.60 Geometry of the dual-polarized slot-coupled circular patch antenna with four modified H-shaped coupling slots (antenna B). (*a*) Side view of the antenna; (*b*) coupling slots in the ground plane; (*c*) microstrip feed lines including two Wilkinson power dividers on the feed substrate. (From Ref. 14, © 2002 IEEE, reprinted with permission)

shaped coupling slots, including their associated feed networks, were used for each feeding port of antenna B. Figure 4.63 shows the measured return loss and isolation for antenna B. Both return loss results for port 1 and port 2 excitation are shown in Figure 4.63*a*, and some variations are observed, which are largely because the feed networks for ports 1 and 2 are not symmetrical for antenna B. The obtained impedance bandwidths reach 17.8% (1627–1947 MHz) and 15.2% (1637–1910 MHz) for port 1 and port 2 excitation, respectively, which are about 2.0 times that of antenna A. This increase in the impedance bandwidth is probably because an additional resonance is excited because of the coupling between two modified H-shaped slots and the radiating patch. In addition, the results shown in Figure 4.63*b* indicate that the isolation over the entire bandwidth is further improved, with S_{21} less than −34.4 dB. As for the radiation patterns given in Figure 4.64, good cross-polarization levels (XPL > 20 dB) in two orthogonal planes of the two polar-

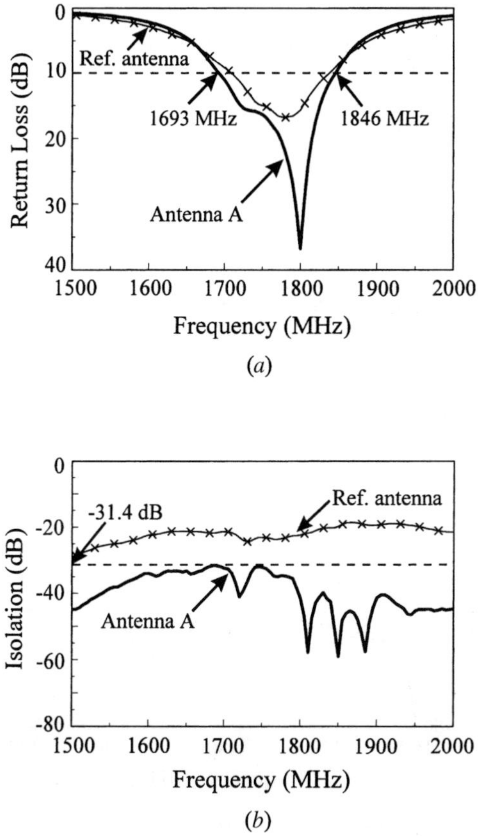

FIGURE 4.61 Measured return loss and isolation for the antenna shown in Figure 4.59 (antenna A) and the reference antenna; $R = 37.25$ mm, $h = 9$ mm, $d = 27$ mm, $S_h = 14.5$ mm, $S_1 = 14$ mm, $S_2 = 6$ mm ($S_2 = 14$ mm for reference antenna), $\alpha = 40°$ ($\alpha = 0°$ for reference antenna), $\ell = 6.4$ mm ($\ell = 14.5$ mm for reference antenna), $t = 3.5$ mm, ground-plane size $= 100 \times 100$ mm^2. (*a*) Return loss for antenna A and reference antenna; (*b*) isolation between ports 1 and 2. (From Ref. 14, © 2002 IEEE, reprinted with permission)

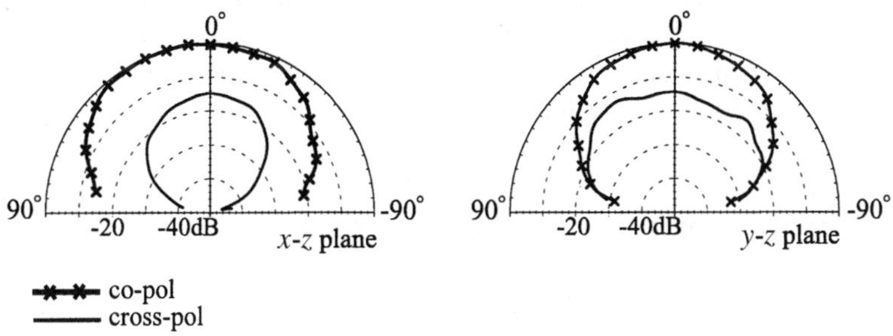

FIGURE 4.62 Measured radiation patterns at 1800 MHz for port 1 excitation of antenna A studied in Figure 4.61. (From Ref. 14, © 2002 IEEE, reprinted with permission)

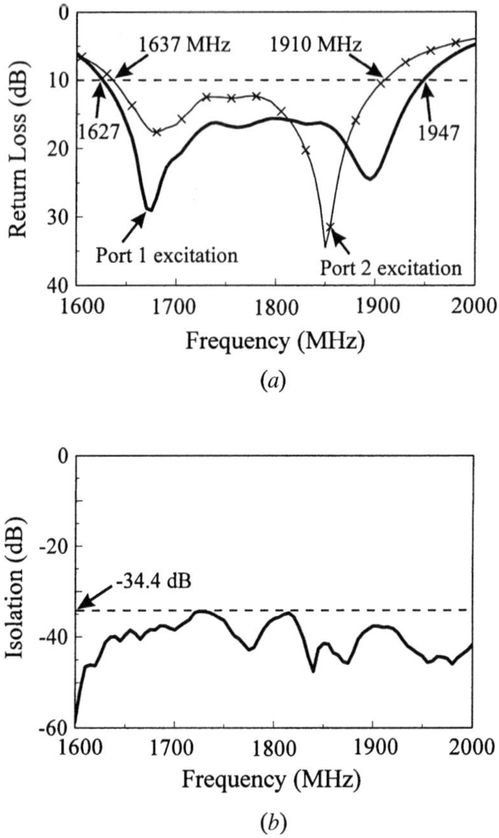

FIGURE 4.63 Measured return loss and isolation for the antenna shown in Figure 4.60 (antenna B); ground-plane size = $110 \times 150\,\text{mm}^2$. Other parameters are the same as given in Figure 4.61. (*a*) Return loss at ports 1 and 2; (*b*) isolation between ports 1 and 2. (From Ref. 14, © 2002 IEEE, reprinted with permission)

izations are obtained. However, the antenna gain of port 1 and port 2 excitation is measured to be 6.6 and 6.5 dBi, respectively, which are 0.3–0.4 dBi less than that of antenna A. This decrease in antenna gain is probably because relatively more complicated feed networks are used for the two feeding ports of antenna B, and increased energy losses in the feed network can be expected.

4.5 ANTENNA FOR DUAL-POLARIZED OPERATION USING HYBRID FEEDS

It has been reported that, when a reflecting plate is added below the ground plane of the dual-polarized slot-coupled patch antennas, the reflecting plate and the ground

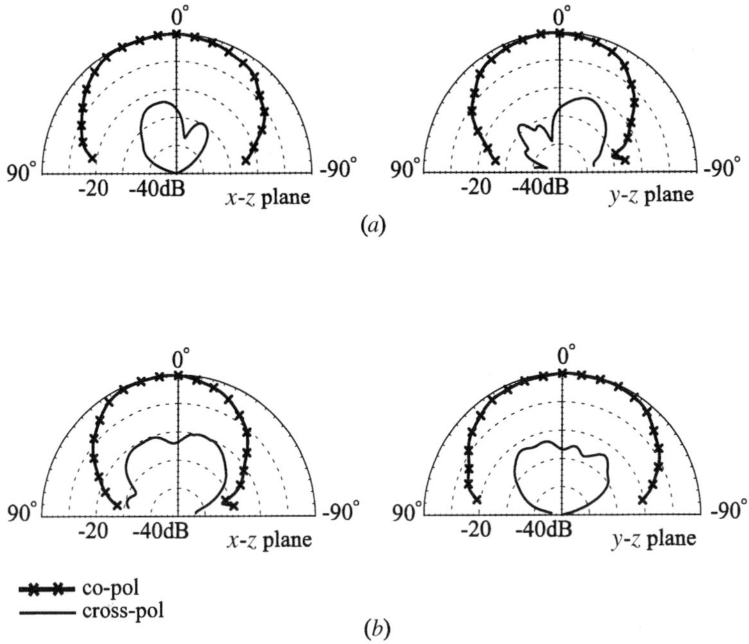

FIGURE 4.64 Measured radiation patterns at 1800 MHz for antenna B studied in Figure 4.63. (*a*) Port 1 excitation; (*b*) port 2 excitation. (From Ref. 14, © 2002 IEEE, reprinted with permission)

plane may form a waveguide and some propagation modes may exist. This behavior may lead to some coupling between the two slot-coupled feeds, and the port decoupling and impedance matching of the two feeding ports will then be affected. By replacing one of the two slot-coupled feeds with a capacitively-coupled feed [15] or a coplanar probe feed or a gap-coupled feed [16] (that is, using two hybrid feeds), this problem can be solved, and good dual-polarized radiation with highly decoupled feeding ports can be obtained. This is because, in this kind of design, possible coupling below the ground plane between the two hybrid feeds can be eliminated or reduced to a minimum, and thus very small or no port decoupling due to the adding of a reflecting plate in practical applications is expected. In the following sections, three promising dual-polarized designs using hybrid feeds are presented.

4.5.1 Design with One Capacitively Coupled Feed and One Slot-Coupled Feed

Figure 4.65 shows the broadband dual-polarized circular patch antenna fed by a capacitively coupled feed and a slot-coupled feed (denoted as antenna A in this study). The circular patch has a radius of R and is supported by nonconducting

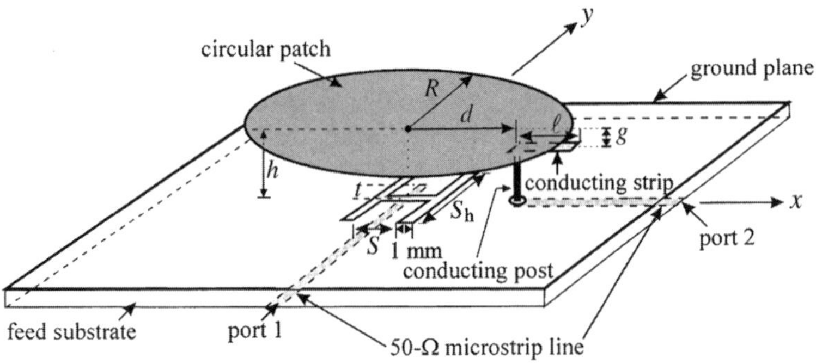

FIGURE 4.65 Geometry of the dual-polarized patch antenna fed by a capacitively coupled feed and a slot-coupled feed (antenna A). (From Ref. 15, © 2002 IEEE, reprinted with permission)

posts (not shown in the figure) with a distance of h above the ground plane of a grounded substrate (feed substrate; its thickness and relative permittivity were 0.8 mm and 4.4, respectively, in this study). Port 1 is the slot-coupled feed with an H-shaped coupling slot cut in the ground plane and centered below the circular patch. The center arm and two side arms of the H-shaped coupling slot have dimensions of $S \times 1$ mm and $S_h \times 1$ mm, respectively. With the use of the H-shaped coupling slot instead of the conventional narrow slot, the required length (S) of the slot's center arm can be reduced, which is found in this study to be effective in improving the isolation between the two feeding ports. Port 1 is then excited through a 50-Ω microstrip line, which has a tuning stub length of t and is printed on the feed substrate.

Port 2 has a capacitively coupled feed. A conducting strip of width 1 mm and length ℓ ($\ell \gg 1$ mm) is placed in parallel to and below the circular patch with a distance of g and couples the electromagnetic energy to the cicrular patch from a 50-Ω microstrip line through a via-hole in the ground plane. The conducting strip is supported by a conducting post, which is away from the center axis of the patch with a distance of d and is connected to the 50-Ω microstrip line at port 2.

A prototype of antenna A has been constructed and measured. The center frequency was designed at 1800 MHz. The radius of the circular patch was chosen to be 35 mm, and the antenna height or the air-substrate's thickness was 13.6 mm or about 8% of the wavelength of the center frequency. Figure 4.66 shows the measured return loss and isolation for antenna A, and the measured E- and H-plane radiation patterns for the two polarizations at 1800 MHz are plotted in Figure 4.67. The obtained impedance bandwidth (10 dB return loss) for port 1 excitation is 304 MHz or about 16.9%, whereas that for port 2 excitation is 17.8%. The isolation between ports 1 and 2 is seen to be less than −30 dB for the entire impedance bandwidths of the two polarizations. From the results shown in Figure 4.67, good cross-polarization levels are in general obtained, except that the XPL in the H-plane (y–z plane) pattern of port 2 excitation is only about 14.4 dB. Measured

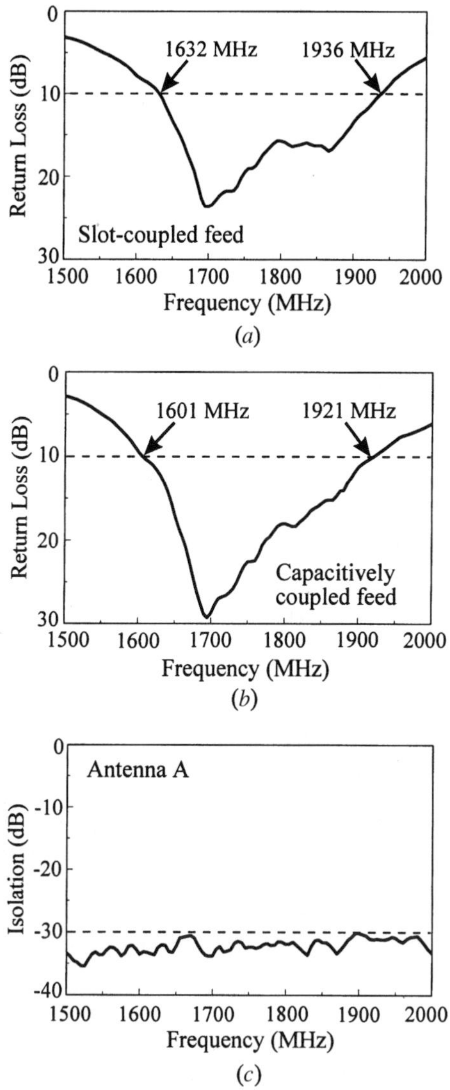

FIGURE 4.66 Measured return loss and isolation for the antenna shown in Figure 4.65 (antenna A); $R = 35$ mm, $h = 13.6$ mm, $d = 22$ mm, $\ell = 20$ mm, $g = 3.2$ mm, $S_h = 19$ mm, $S = 18$ mm, $t = 6.0$ mm, ground plane size $= 100 \times 100$ mm^2. (*a*) Return loss for slot-coupled feed (port 1); (*b*) return loss for capacitively coupled feed (port 2); (*c*) isolation between ports 1 and 2. (From Ref. 15, © 2002 IEEE, reprinted with permission)

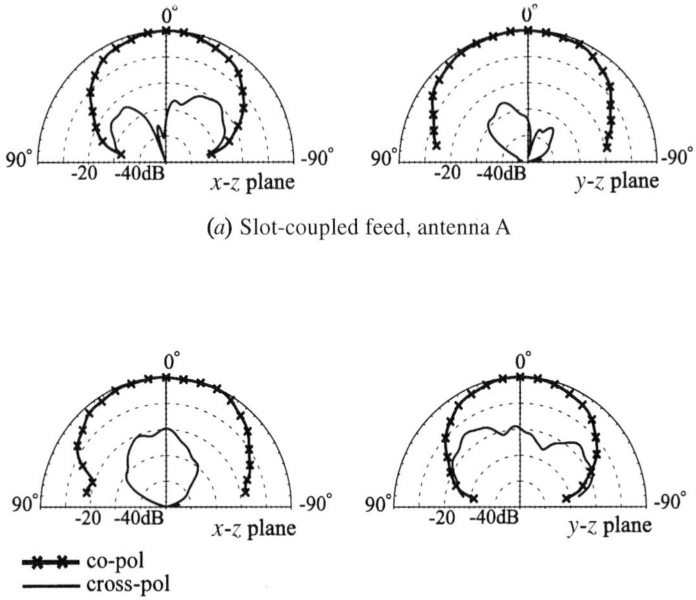

(a) Slot-coupled feed, antenna A

(b) Capacitively-coupled feed, antenna A

FIGURE 4.67 Measured radiation patterns at 1800 MHz for antenna A studied in Figure 4.66. (*a*) Slot-coupled feed (port 1); (*b*) capacitively coupled feed (port 2). (From Ref. 15, © 2002 IEEE, reprinted with permission)

antenna gain in the broadside direction for ports 1 and 2 at 1800 MHz is also observed to be the same (7.4 dBi). These obtained dual-polarization characteristics are listed in Table 4.10 for comparison. Also note that the proposed antenna can be applied to the patch antenna with a square patch. A prototype with a square patch was also fabricated and tested. The obtained results are about the same as

TABLE 4.10 Dual Polarization Performance of Antenna A Shown in Figures 4.66 and 4.67, Antenna B Shown in Figures 4.69 and 4.70, and Antenna C Shown in Figures 4.72 and 4.73 [15]. Impedance Bandwidth (BW) is Determined From 10 dB Return Loss, and E-Plane XPL (XPL_E), H-Plane XPL (XPL_H), and Antenna Gain are Measured Data at 1800 MHz.

	Port 1 excitation				Port 2 excitation				Isolation
	BW (%)	XPL_E (dB)	XPL_H (dB)	Gain (dBi)	BW (%)	XPL_E (dB)	XPL_H (dB)	Gain (dBi)	
Antenna A	16.9	25.6	19.8	7.4	17.8	19.6	14.4	7.4	<−30 dB
Antenna B	13.1	24.5	20.5	7.3	24.7	24.5	20.1	7.0	<−32 dB
Antenna C	16.5	28.8	25.6	10.1	14.8	20.4	21.3	9.2	<−35 dB

those shown in Figures 4.66 and 4.67 for the case with a circular patch. It was also observed that, for antenna A, the measured isolation and return loss of ports 1 and 2 were almost unchanged when a reflecting conducting plate was added below the ground plane to make the proposed antennas usable elements for practical applications.

To improve the XPLs for port 2 (capacitively coupled feed) excitation, the feeding structure shown in Figure 4.68 is proposed (antenna B). Port 2 now has two capacitively coupled feeds of equal input powers and a 180° phase difference, which is provided by a Wilkinson power divider with its two output feedlines having a length difference of a half guided wavelength at the designed center frequency 1800 MHz. The two capacitively coupled feeds are also of the same dimensions and are aligned in the same direction (x-axis in this study). It is then expected that, with comparison to the design of antenna A with a single capacitively coupled feed shown in Figure 4.65, the possible higher-order modes that contribute to the polarization impurity or cross-polarization radiation can be suppressed for port 2 excitation of antenna B, and XPLs can be improved.

Figure 4.69 shows the measured return loss and isolation for antenna B. The radiation patterns of the two polarizations are plotted in Figure 4.70. Related measured data are also given in Table 4.10. It is first observed that the isolation between ports 1 and 2 is improved and is less than −32 dB in the entire impedance bandwidths of the two polarizations, which is probably because the XPLs for port 2 excitation are greatly increased by 4.9 dB (24.5 dB vs. 19.6 dB) and 5.7 dB (20.1 dB vs. 14.4 dB) in the E- and H-plane patterns, respectively. In addition, two resonances in the measured return loss of port 2 excitation are seen (Fig. 4.69*b*) and the impedance bandwidth is increased to be 24.7%, which is probably due to the interaction of the reflections from the antenna feed points and the Wilkinson power divider. On the other hand, probably because the feed circuitry of antenna B is relatively more complex than antenna A, the measured antenna gain of port 2

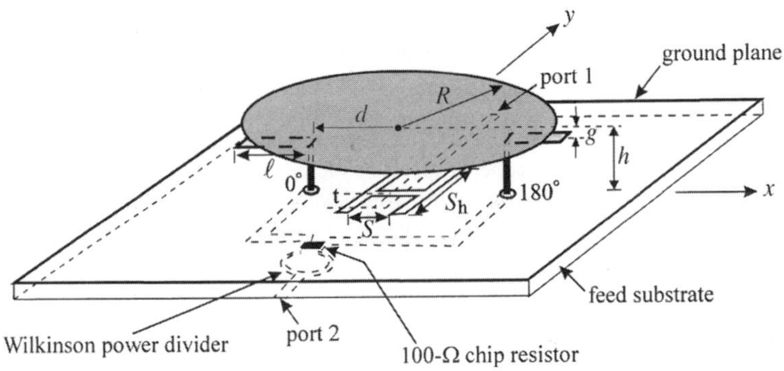

FIGURE 4.68 Geometry of the dual-polarized patch antenna fed by a slot-coupled feed and two capacitively coupled feeds of a 180° phase shift (antenna B). (From Ref. 15, © 2002 IEEE, reprinted with permission)

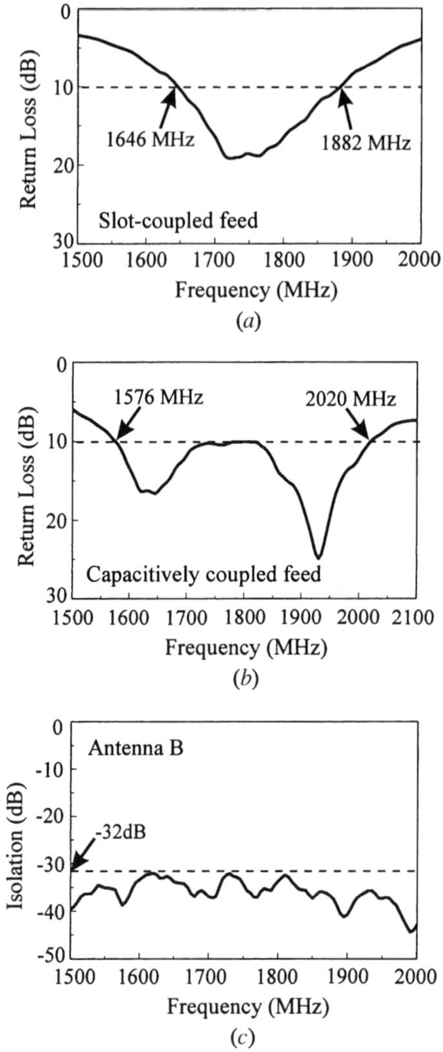

FIGURE 4.69 Measured return loss and isolation for the antenna shown in Figure 4.68 (antenna B); parameters are the same as given in Figure 4.66. (*a*) Return loss for the slot-coupled feed (port 1); (*b*) return loss for the capacitively coupled feed (port 2); (*c*) isolation between ports 1 and 2. (From Ref. 15, © 2002 IEEE, reprinted with permission)

excitation is 0.4 dBi less than that of antenna A. As for port 1 excitation of antenna B, measured characteristics are seen to be about the same as those of antenna A.

The feeding technique studied here has also been applied to a two-element patch antenna for broadband dual-polarized radiation with high isolation and low cross-polarization. The antenna geometry is shown in Figure 4.71 (antenna C). The

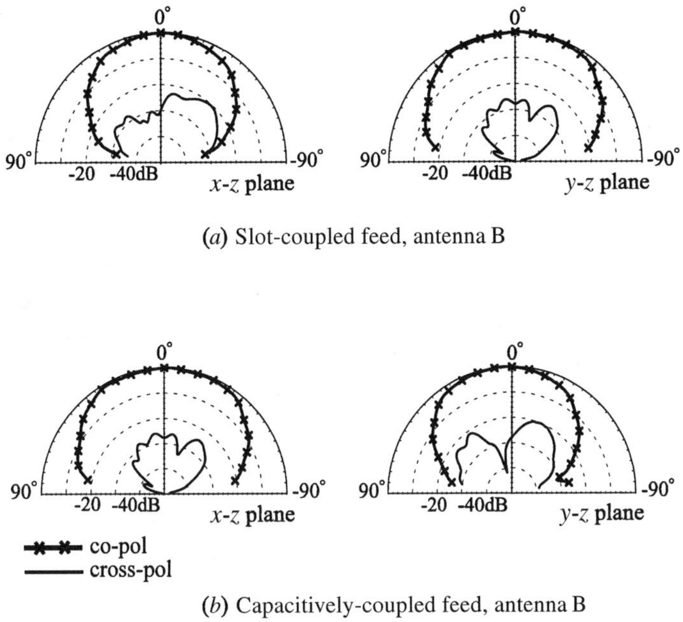

FIGURE 4.70 Measured radiation patterns at 1800 MHz for antenna B studied in Figure 4.69. (*a*) Slot-coupled feed (port 1); (*b*) capacitively coupled feed (port 2). (From Ref. 15, © 2002 IEEE, reprinted with permission)

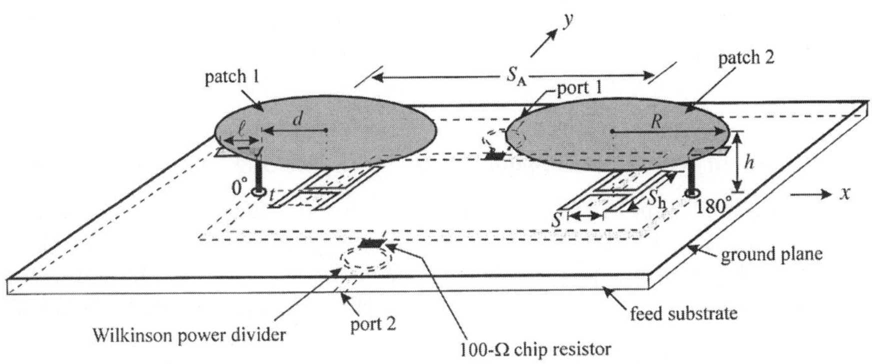

FIGURE 4.71 Geometry of a two-element dual-polarized patch antenna fed by two slot-coupled feeds and two capacitively coupled feeds of a 180° phase shift (antenna C). (From Ref. 15, © 2002 IEEE, reprinted with permission)

spacing between the centers of the two circular patches is S_A, which was selected to be 0.8 times the free space wavelength (λ_0) of the center operating frequency (1800 MHz in this study). Port 2 also has two capacitively coupled feeds of equal input powers and a 180° phase difference, similar to the design of antenna B. However, the two capacitively coupled feeds excite patches 1 and 2, respectively, and are all aligned along the x-axis. In this arrangement, the radiation impurity due to the higher-order modes of the thick air substrate from patch 1 (0° element) can cancel that from patch 2 (180° element), and low cross polarization can be expected, which can also result in the improvement of the isolation between ports 1 and 2. As for port 2, two slot-coupled feeds of equal input powers and same phases are used for exciting patches 1 and 2, respectively, and are centered below the patches. Also note that the parameters of antenna C are the same as those of antennas A and B, and no additional tuning of the dimensions of the capacitively coupled feed and the slot-coupled feed is required for achieving good impedance matching of antenna C.

Figure 4.72 shows the measured return loss and isolation of antenna C, and the measured radiation patterns of the two polarizations at 1800 MHz are presented in Figure 4.73. Measured data are also given in Table 4.10 for comparison. Note that the isolation between the two feeding ports is less than −35 dB for the entire impedance bandwidths of the two polarizations. The XPLs in the two principal planes are also improved to greater than 25 dB for port 1 excitation and 20 dB for port 2 excitation. Cross-polarization radiation in other planes off the main axes (x–z and y–z planes) were also studied. It is found that, in the planes within about ±30° to the x–z or y–z planes, the obtained XPLs are still better than 20 dB for the present design. For other planes, however, no prominent XPL improvement is observed.

The antenna gain is also increased to be 10.1 and 9.2 dBi for the two polarizations. The smaller antenna gain for port 2 excitation is largely because the antenna spacing S_A used in this study is not the optimal spacing for port 2 excitation to achieve an optimal antenna gain. Finally, the impedance bandwidths obtained for ports 1 and 2 are about the same as those of antennas A and B, except that port 2 excitation of antenna B has a much greater bandwidth.

4.5.2 Design with One Coplanar Probe Feed and One Slot-Coupled Feed

The broadband dual-polarized patch antenna using two hybrid feeds of one coplanar probe feed and one slot-coupled feed for operating in the DCS band has been conducted. Figure 4.74 shows the antenna geometry studied. It is first noted that, because different feeds are used for ports 1 and 2 in the proposed design, the resonant lengths are thus different for port 1 and port 2 excitation. A rectangular patch of dimensions $60 \times 70\ \text{mm}^2$, instead of a square patch, is therefore selected to achieve the same desired center operating frequency (at about 1.8 GHz) for the two orthogonal polarizations. The rectangular patch is also centered above a grounded feed substrate of dimensions $100 \times 100\ \text{mm}^2$ (plastic supporters not

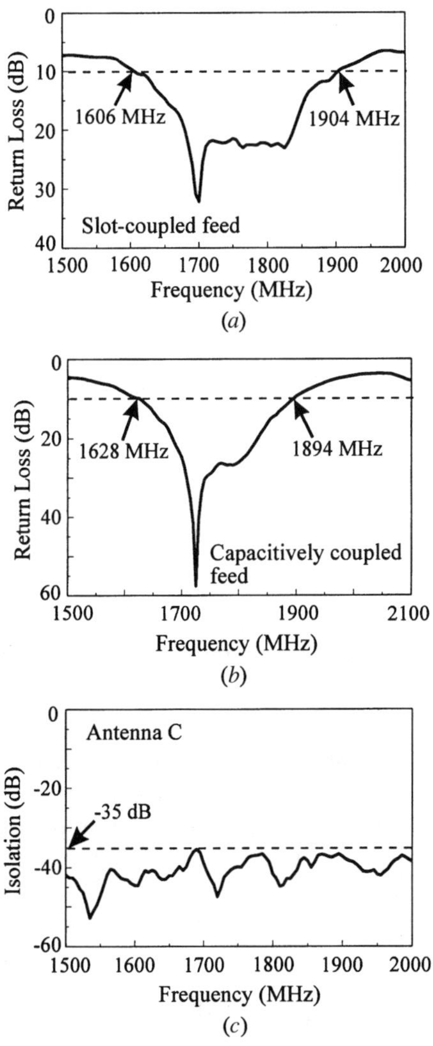

FIGURE 4.72 Measured return loss and isolation for the antenna shown in Figure 4.71 (antenna C); $S_A = 133.3$ mm ($= 0.8\lambda_0$), ground-plane size $= 100 \times 250$ mm^2. Other parameters are the same as given in Figure 4.66. (*a*) Return loss for the slot-coupled feed (port 1); (*b*) return loss for the capacitively coupled feed (port 2); (*c*) isolation between ports 1 and 2. (From Ref. 15, © 2002 IEEE, reprinted with permission)

shown in the figure), and the distance between the rectangular patch and the ground plane is chosen to be 18 mm, which makes possible a wide impedance bandwidth for the proposed antenna. Also notice that, because the excitation with an aperture-coupled feed is usually with a lower resonant frequency than the case with a probe feed, the patch length in the resonant direction of the slot-coupled feed is

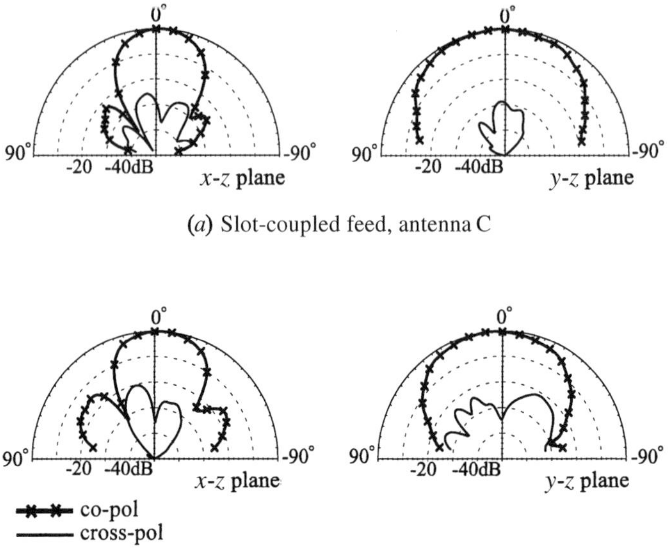

(a) Slot-coupled feed, antenna C

(b) Capacitively-coupled feed, antenna C

FIGURE 4.73 Measured radiation patterns at 1800 MHz for antenna C studied in Figure 4.72. (a) Slot-coupled feed (port 1); (b) capacitively coupled feed (port 2). (From Ref. 15, © 2002 IEEE, reprinted with permission)

chosen to be smaller, that is, 60 mm in the resonant direction for the slot-coupled feed and 70 mm in the resonant direction for the coplanar probe feed.

The slot-coupled feed (port 1 in this design) is with an H-shaped coupling slot cut in the ground plane to couple the electromagnetic energy from the 50-Ω microstrip feed line printed on the feed substrate. The H-shaped coupling slot is centered below the rectangular patch and has a central arm of dimensions $22 \times 1\ \text{mm}^2$ and two

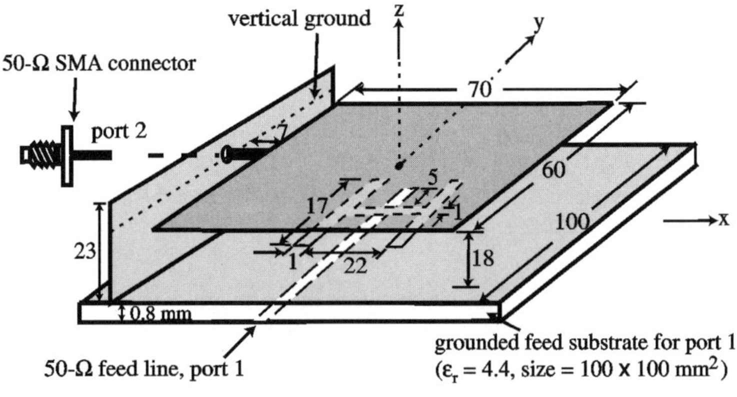

FIGURE 4.74 Geometry of the dual-polarized patch antenna for DCS base station.

identical side arms of dimensions $17 \times 1\,mm^2$. With a tuning stub length of 5 mm for the 50-Ω microstrip feed line, port 1 can be excited with good impedance matching over a wide bandwidth in the 1.8 GHz band.

A coplanar probe feed is used for port 2 excitation. To accommodate the 50-Ω SMA connector for the probe feed, a vertical ground is added (see Fig. 4.74), which has a height of 23 mm and the same width as that (100 mm) of the grounded feed substrate. The SMA connector is also placed in the same plane as the rectangular patch, and a probe pin length of 7 mm is connected, through a via-hole in the vertical ground, to the center of the patch edge. In this case, the impedance matching can be tuned simply by varying the probe pin length of the coplanar probe feed. A wide impedance bandwidth is obtained in this study by selecting the probe pin length to be 7 mm. Also note that the optimal probe pin length in the proposed design is much less than the air-layer thickness. This characteristic is also helpful in decreasing the cross-polarization radiation for port 2 excitation of the proposed antenna.

The proposed design has been constructed and studied. The measured and simulated return loss for port 1 and port 2 excitation is shown in Figure 4.75. The simulated results are obtained from the Ansoft HFSS. Good agreement between the measurement and the simulation is observed. From the measured results, the obtained impedance bandwidths (1.5:1 VSWR) are about 172 and 174 MHz for port 1 and port 2 excitation, respectively. Both the impedance bandwidths cover the required operating bandwidths of the DCS system.

Figure 4.76 shows the measured and simulated isolation (S_{21}) between the two feeding ports. The simulated results in general agree with the measured data. For operating frequency within the DCS band, the isolation is better than −30 dB. Figures 4.77 and 4.78 plot the measured radiation patterns at 1800 MHz for port 1 and port 2 excitation, respectively. Good broadside radiation patterns with cross-polarization radiation less than −20 dB are obtained. For other operating frequencies within the DCS band, measured radiation patterns similar to those shown here are also observed. It is also noted that the backward radiation of port 2 excitation (coplanar probe feed) is smaller than that of port 1 excitation (slot-coupled feed), and the front-to-back ratio is better by about 5 dB for the coplanar probe feed. Figure 4.79 shows the measured antenna gain for port 1 and port 2 excitation. The measured peak antenna gain reaches 8.1 and 7.7 dBi for port 1 and port 2 excitation, respectively, and the gain variations for frequencies within the DCS band are less than 1.3 dBi.

4.5.3 Design with Two Gap-Coupled Probe Feeds and Two Slot-Coupled Feeds

The patch antenna fed by hybrid feeds consisting of two gap-coupled probe feeds and two slot-coupled feeds has also been shown to be very promising for obtaining a broadband dual-polarized operation with highly decoupled ports and low cross-polarization radiation [16]. Figure 4.80 shows the antenna geometry designed for applications in the DCS band. Port 1 has two in-phase slot-coupled feeds and excites

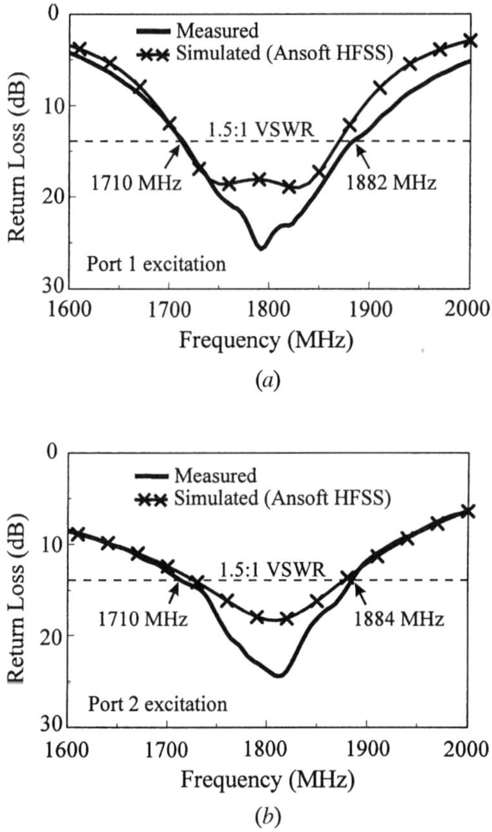

FIGURE 4.75 Measured and simulated return loss for the antenna shown in Figure 4.74. (*a*) Port 1 excitation; (*b*) port 2 excitation.

the linearly polarized radiation along the y–z plane; port 2 has two out-of-phase gap-coupled probe feeds and excites linearly polarized radiation along the x–z plane. The square patch has a side length of L and is supported by plastic posts (not shown in the figure) with a distance of h above the ground plane of a grounded substrate (feed substrate). The feed substrate used is an inexpensive FR4 substrate of thickness 0.8 mm and relative permittivity 4.4, on which the microstrip feed lines, including two Wilkinson power dividers, for ports 1 and 2 are printed. The Wilkinson power divider for port 2 also has one of its two output feed lines with a length difference of a half guided wavelength at the designed center frequency. In this case, a 180° phase shift for the two gap-coupled probe feeds for port 2 can be obtained.

For each gap-coupled probe feed placed at a distance of d away from the patch center, a rectangular ring slot of dimensions $\ell \times 1.5$ mm and width 0.25 mm is embedded around the feed point at the radiating patch. The long side of the rectan-

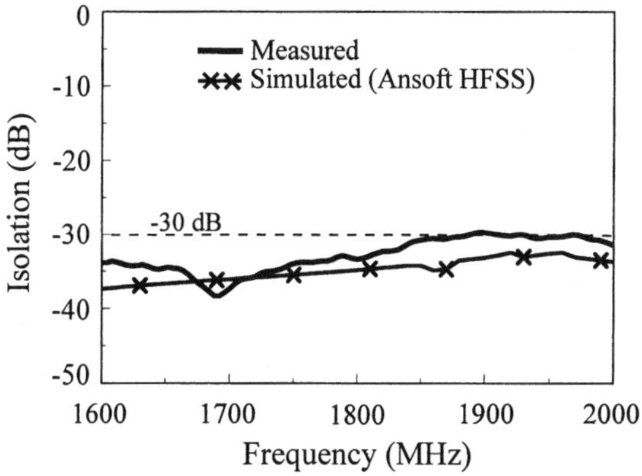

FIGURE 4.76 Measured and simulated isolation (S_{21}) for the antenna shown in Figure 4.74.

gular ring slot is oriented along the excitation direction of port 2, and, by tuning its length (ℓ), good impedance matching over a wide bandwidth for port 2 excitation can be achieved. Also note that, with the use of a narrow rectangular ring slot instead of an annular ring slot, relatively lower cross-polarization radiation for port 2 excitation, especially in the H-plane radiation, is observed in the present study.

The two H-shaped coupling slots for port 1 have the same dimensions, and their center arms and side arms are selected to have dimensions of $S \times 0.25$ mm and $S_h \times 1.0$ mm, respectively. The two H-shaped coupling slots are also placed at the same distance of S_d away from the patch center and are fed by the two output feed lines of a Wilkinson power divider. The output feed lines are 50-Ω microstrip lines and have a tuning stub of length t.

A prototype of the proposed antenna has been successfully implemented. The designed center frequency was 1800 MHz, and the side length of the square patch

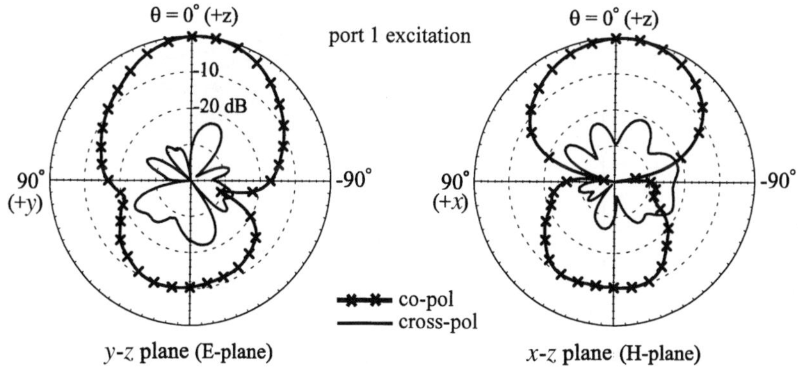

FIGURE 4.77 Measured radiation patterns at 1800 MHz for port 1 excitation of the antenna shown in Figure 4.74.

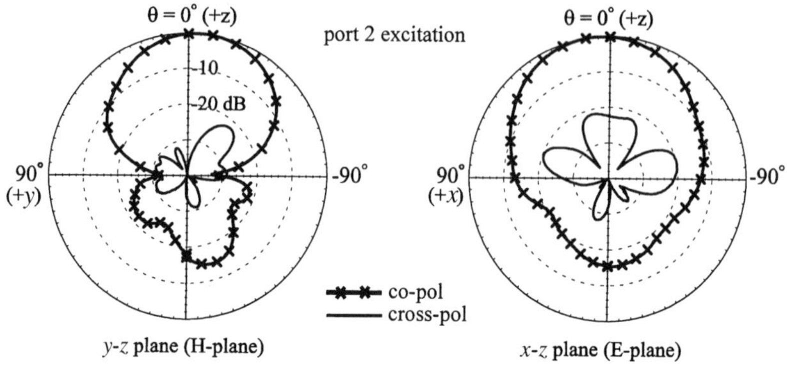

FIGURE 4.78 Measured radiation patterns at 1800 MHz for port 2 excitation of the antenna shown in Figure 4.74.

and the air-substrate's thickness were thus chosen to be 57.5 and 14.2 mm, respectively. The air-substrate's thickness corresponds to about 8.5% of the wavelength of the center operating frequency. The measured return loss for both port 1 and port 2 excitation is presented in Figure 4.81. The IE3D simulation results are also given in the figure for comparison. The measured impedance bandwidths (10 dB return loss) are 261 MHz (or about 14.5%) and 359 MHz (or about 20%) for port 1 and port 2 excitation, respectively. The measured isolation between ports 1 and 2 is shown in Figure 4.82. It can be seen that the isolation for frequencies across the entire bandwidths of ports 1 and 2 is less than −40 dB.

Figure 4.83 plots the measured radiation patterns at 1800 MHz for port 1 and port 2 excitation. Good broadside radiation patterns with low cross-polarization are observed. For port 1 excitation, the cross-polarization radiation for E-plane (*y*–*z*)

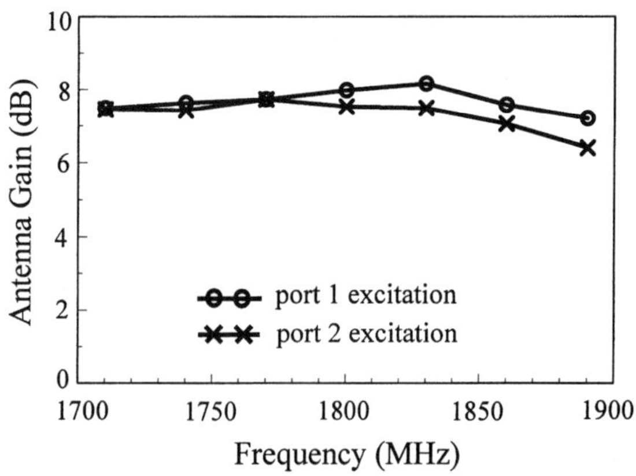

FIGURE 4.79 Measured antenna gain for port 1 and port 2 excitation of the antenna shown in Figure 4.74.

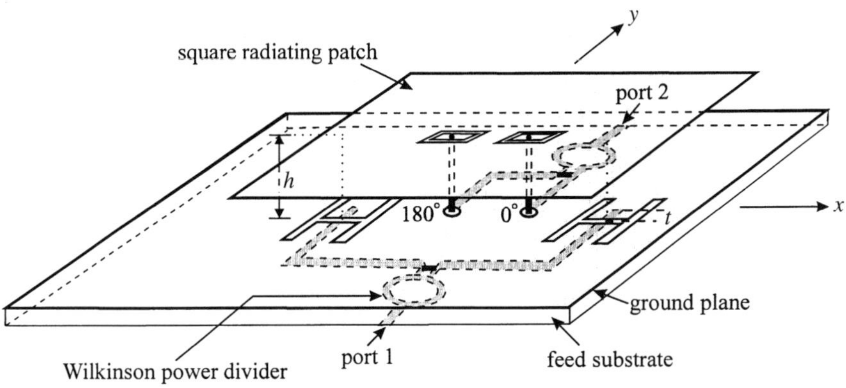

(a)

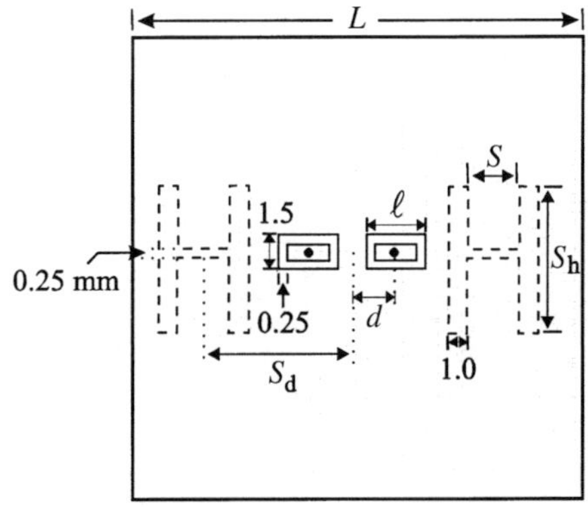

(b)

FIGURE 4.80 (*a*) Geometry of the dual-polarized patch antenna fed by two in-phase slot-coupled feeds (port 1) and two out-of-phase gap-coupled probe feeds (port 2); (*b*) top view of the square patch. (From Ref. 16, © 2002 IEEE, reprinted with permission)

and H-plane (x–z) patterns is less than −25 and −23 dB, respectively. For port 2 excitation, the cross-polarization radiation is less than −25 and −20 dB, respectively, for E-plane (x–z plane) and H-plane (y–z plane) patterns. In addition, the antenna gain in the broadside direction for port 1 and port 2 excitation at 1800 MHz is measured to be 7.3 and 7.0 dBi, respectively, with the gain variations within the bandwidths of ports 1 and 2 being less than 1.5 dBi.

4.6 ANTENNA FOR DUAL-BAND DUAL-POLARIZED OPERATION

A compact and promising dual-polarized antenna suitable for applications in a GSM/DCS dual-band base station has been recently demonstrated [17].

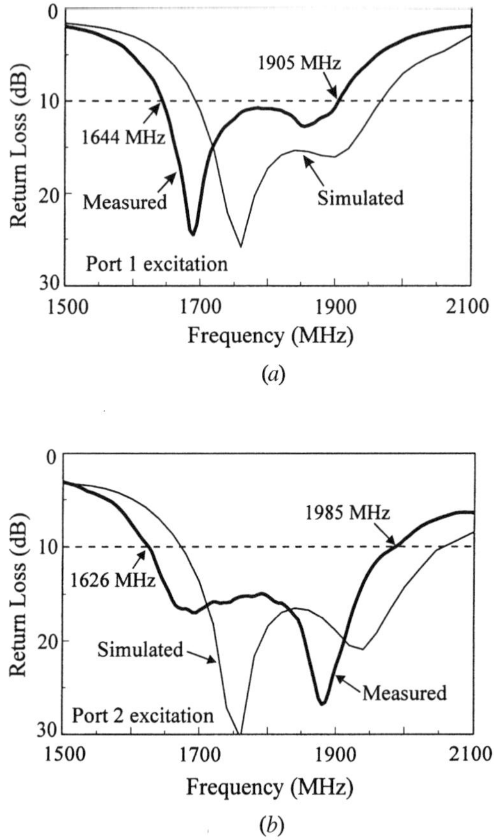

FIGURE 4.81 Measured return loss for the antenna shown in Figure 4.80; $L = 57.5$ mm, $h = 14.2$ mm, $d = 10$ mm, $\ell = 6.5$ mm, $S_h = 30$ mm, $S = 8$ mm, $S_d = 23$ mm, $t = 3.5$ mm, ground plane size $= 100 \times 100$ mm^2. (*a*) Port 1 excitation (slot-coupled feed); (*b*) port 2 excitation (gap-coupled probe feed). (From Ref. 16, © 2002 IEEE, reprinted with permission)

Figure 4.84 shows the antenna geometry. In this antenna design, there are four major parts to be considered. The first two parts are the determination of the proper dimensions of the rectangular ring patch and the notched rectangular patch for the 900 and 1800 MHz operations, respectively. As shown in Figure 4.84*b*, the two patches are printed on the same dielectric substrate (patch substrate) and the notched rectangular patch is placed within the rectangular ring patch to obtain a compact structure. The third part is the arrangement of the coupling slots in the antenna's ground plane (see Fig. 4.84*c*), and the fourth part is the feed network design (Fig. 4.84*d*). The coupling slots and the feed network are printed on two sides of a dielectric substrate (feed substrate). In this study, both the patch and feed substrates used were inexpensive FR4 substrates of thickness 0.8 mm and relative permittivity (ε_r) 4.4. The two substrates are also separated by an air layer of thickness h (see Fig. 4.84*a*; the

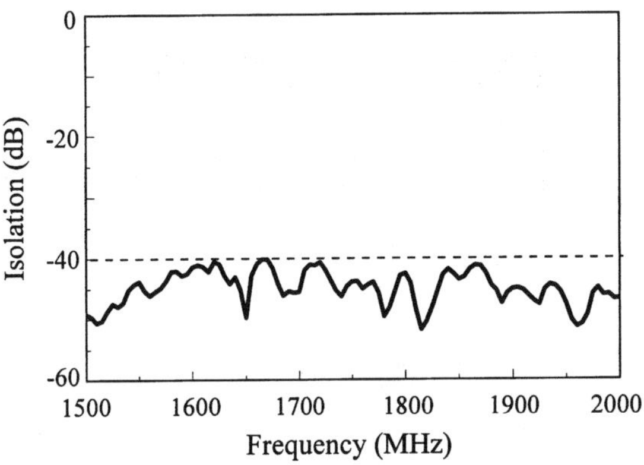

FIGURE 4.82 Measured isolation (S_{21}) for the antenna studied in Figure 4.81. (From Ref. 16, © 2002 IEEE, reprinted with permission)

supporting posts not shown in the figure). The IE3D simulation software was helpful in obtaining proper dimensions of the proposed antenna. The design considerations for the four major parts are described below.

The rectangular ring patch is designed to achieve 900 MHz band operation and is slot-coupled by using two H-shaped coupling slots (slot 1X and slot 2Y in Fig. 4.84*c*) to obtain dual linear polarizations. The two H-shaped slots have a center arm of width 0.5 mm and two side arms of width 1 mm. For the lengths of the slot's center arm and side arm, owing to the width limitation of the rectangular ring patch, slot 1X is designed to be narrow (i.e., a relatively shorter center arm and a relatively longer side arm) and slot 2Y is of a wide H shape, having a relatively longer center arm and a relatively shorter side arm. In addition, because the coupling-slot size can affect the resonant frequency of the antenna, the dimensions of the rectangular ring patch in parallel to the x- and y-axes are chosen to be different to achieve the same operating frequencies around 900 MHz. The designed patch dimensions are given in Figure 4.84*b*.

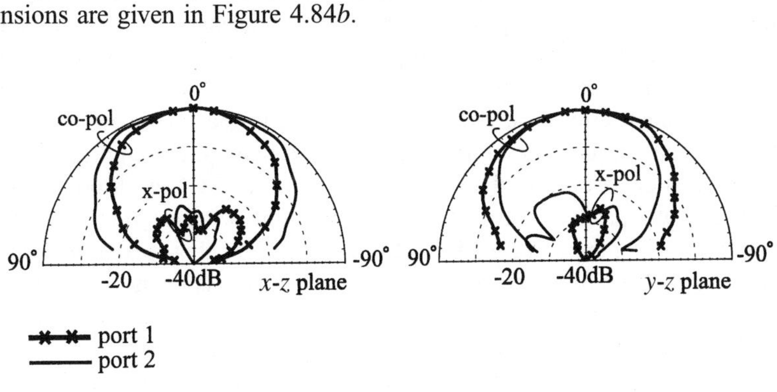

FIGURE 4.83 Measured radiation patterns at 1800 MHz for port 1 and port 2 excitation of the antenna studied in Figure 4.81. (From Ref. 16, © 2002 IEEE, reprinted with permission)

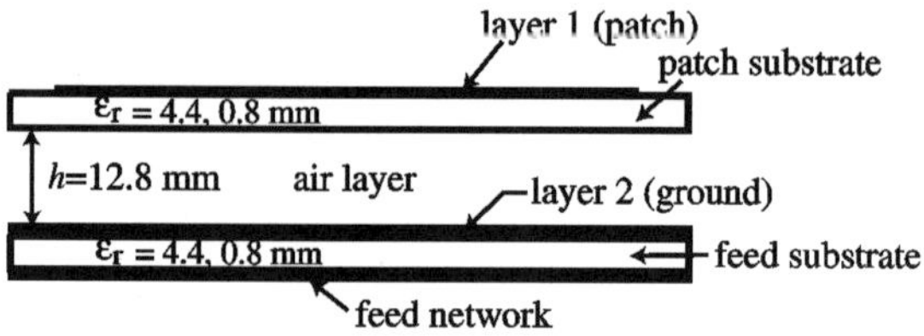

(a)

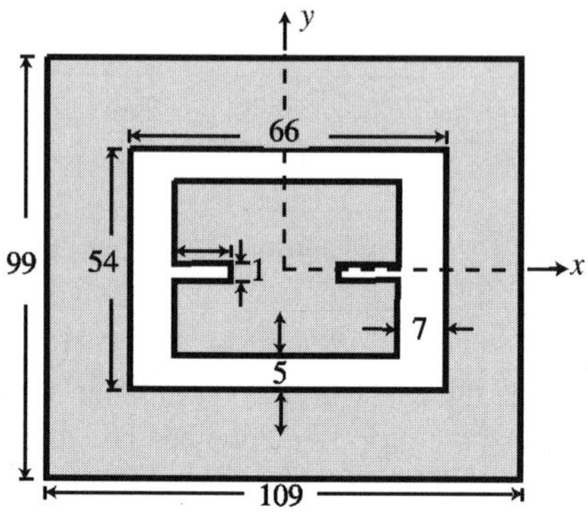

(b)

FIGURE 4.84 Geometry of the GSM/DCS dual-band dual-polarized patch antenna. (a) Side view of the antenna; (b) patches in layer 1; (c) coupling slots in layer 2; (d) microstrip line feed network. (From Ref. 17, © 2001 IEEE, reprinted with permission)

A notched rectangular patch is designed for operating in the 1800 MHz band, and the designed patch dimensions are given in Figure 4.84*b*. By inserting a pair of slits (length ℓ and width 1 mm) along the x-direction (see Fig. 4.84*b*), the side length of the notched rectangular patch along the y-direction can be reduced, which is helpful in obtaining a larger gap between the notched rectangular patch and the rectangular ring patch to reduce the possible coupling between the two radiating patches. In this study, it is found that a gap of 5 mm is required between the two radiating patches. In this case, very small effects on the measured return loss and port isolation of the proposed antenna are observed. Also note that, because the coupling slots (slot 2X and slot 1Y) are arranged along the y-direction, no slits are introduced in the notched rectangular patch in the y-direction to avoid the decrease in the electromagnetic energy coupling from the feed line to the patch. For this reason, the required side length of the notched rectangular patch in the x-direction is larger than that in the y-direction. This also leads to a larger side length of the rectangular ring patch in the x-direction than in the y-direction.

All the H-shaped coupling slots (slots 1X, 1Y, 2X, and 2Y) have the same slot widths (0.5 and 1 mm for the slot's center arm and side arm, respectively), and the lengths of the slots' center arms and side arms are given in Figure 4.84*c*. To

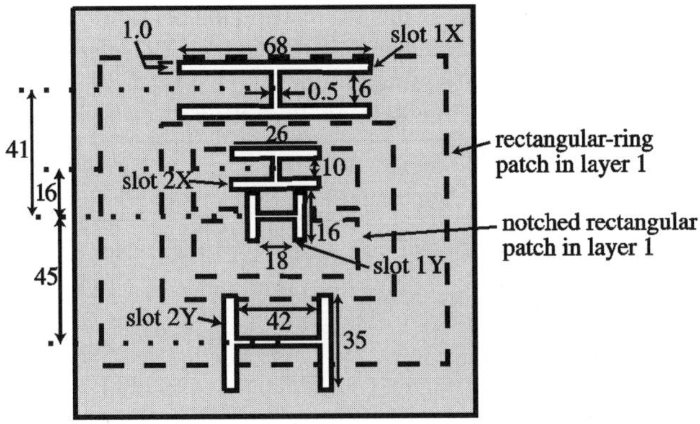

(*c*)

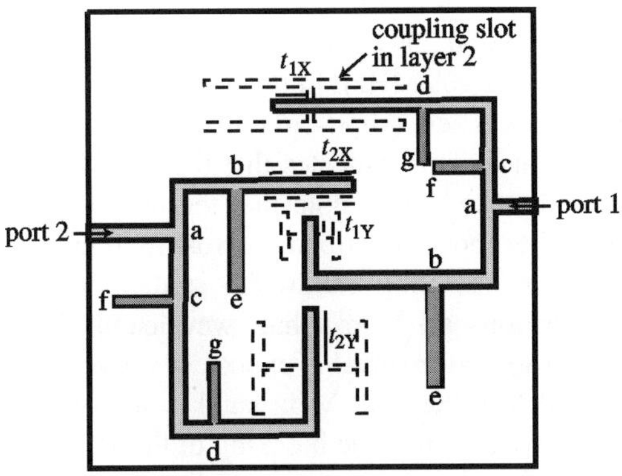

(*d*)

FIGURE 4.84 (*continued*)

obtain high port decoupling, slots 1X and 2Y for the 900 MHz band operation are arranged such that the center arm of slot 1X is in the direction of the microstrip feed line of slot 2Y. For slots 1Y and 2X for the 1800 MHz band operation, the same arrangement is applied. In this case, high isolation between ports 1 and 2 is obtained for the proposed antenna.

In addition to the coupling-slot arrangement described above, the feed network is also designed such that the 1800 MHz (900 MHz) signal into the rectangular ring patch (notched rectangular patch) is blocked. In this condition, the possible excitation of unwanted modes is suppressed, and easy impedance matching of the proposed antenna in the 900 and 1800 MHz bands is also obtained. To achieve this goal, the microstrip line feed network shown in Figure 4.84*d* was designed. The arrangement of the feed networks for ports 1 and 2 is the same. The characteristic impedance of all the microstrip-line sections in the feed network, except $\overline{dg}$ and $\overline{cf}$ sections, is chosen to be 50 Ω (= Z_0). The parameters of the microstrip-line

TABLE 4.11 Lengths and Widths of the Microstrip Line Sections in the Feed Network Shown in Figure 4.84(d). λ_g (=182.4 mm) is the Guided Wavelength in the Feed Substrate at 900 MHz and is Determined From $\lambda_0/\sqrt{\varepsilon_{eff}}$ (λ_0 is the Free-Space Wavelength, and ε_{eff} is the Effective Relative Permittivity).

	Length	Width	Characteristic Impedance
$\overline{ab}$	45.6 mm ($0.25\lambda_g$)	1.53 mm	50 Ω
$\overline{ac}$	45.6 mm ($0.25\lambda_g$)	1.53 mm	50 Ω
$\overline{cd}$	45.6 mm ($0.25\lambda_g$)	1.53 mm	50 Ω
$\overline{be}$	45.6 mm ($0.25\lambda_g$)	1.53 mm	50 Ω
$\overline{cf}$	22.8 mm ($0.125\lambda_g$)	0.17 mm	125 Ω
$\overline{dg}$	22.8 mm ($0.125\lambda_g$)	0.35 mm	100 Ω

sections in the feed network are given in Table 4.11. The $\overline{\text{be}}$ section has a length of $0.25\lambda_g$ and is placed at a distance of $0.25\lambda_g$ away from point a (λ_g denotes the guided wavelength in the feed substrate at 900 MHz). In this case, the impedance at point b seen into the antenna at 900 MHz is zero (short-circuit condition), and thus is infinite (open-circuit condition) at point a. This indicates that the 900 MHz signal into the notched rectangular patch is blocked. Also note that the bandwidth of the single-stub band-stop filter using the above-described microstrip line section is about 25%, which is larger than the operating bandwidth studied here. On the other hand, because $\overline{\text{ab}}$ and $\overline{\text{be}}$ sections all have a half wavelength at 1800 MHz, no impedance mismatch for the 1800 MHz signal is expected; and thus good excitation of the notched rectangular patch in the 1800 MHz band is still obtained.

For the case of the 900 MHz excitation, the blocking of the 1800 MHz signal into the rectangular-ring patch is achieved by the adding of the microstrip line section $\overline{\text{cf}}$ with a length of $0.125\lambda_g$ (λ_g is again the guided wavelength in the feed substrate at 900 MHz). However, the $\overline{\text{cf}}$ section will also affect the 900 MHz signal into the rectangular-ring patch. To solve this problem, the microstrip line section $\overline{\text{dg}}$ is added at a distance $0.25\lambda_g$ away from point c. In this case, the 1800 MHz signal into the rectangular ring patch is still blocked, and the impedance at point c seen into the antenna at 900 MHz can be derived to be

$$\frac{1}{Z_c} = \frac{Z_0}{Z_0^2 + (Z_0^2/Z_{dg})^2} + j\left[\frac{-Z_0^2/Z_{dg}}{Z_0^2 + (Z_0^2/Z_{dg})^2} + \frac{1}{Z_{cf}}\right], \tag{4.1}$$

where Z_c is the impedance at point c seen into the antenna, and Z_{dg} and Z_{cf} are the characteristic impedance of $\overline{\text{dg}}$ and $\overline{\text{cf}}$ sections, respectively; $Z_0 = 50\,\Omega$. From Equation 4.1, it can be found that, by choosing Z_{dg} and Z_{cf} to be 100 and 125 Ω, respectively, the impedance Z_c has a real value of 62.5 Ω. To achieve an impedance very close to 50 Ω for Z_c, larger values of Z_{dg} and Z_{cf} must be selected, which quickly decreases the widths of the microstrip line sections. Because smaller microstrip line widths require more care in the fabrication of the feed network, Z_{dg} and Z_{cf} in this study are chosen to be 100 and 125 Ω, respectively. In this case, the impedance mismatch for the 900 MHz signal is slight, and good measured return loss for the proposed antenna operated in the 900 MHz band is obtained. Also, as shown in Figure 4.84*d*, t_{1X}, t_{2X}, t_{1Y}, and t_{2Y} are tuning stub lengths for slots 1X, 2X, 1Y, and 2Y, respectively. In this design, t_{1X}, t_{2X}, t_{1Y}, and t_{2Y} are selected to be 7.0, 3.5, 5.5, and 6.5 mm, respectively. It should also be noted that the feed network design described here is not a unique solution for the proposed antenna. However, the present feed network provides good results for the proposed antenna.

A prototype of the proposed antenna has been constructed and studied. Figure 4.85 shows the measured return loss for port 1 and port 2 excitation. The total height of the radiating patch to the antenna's ground plane is 13.6 mm, corresponding to about 4% of the wavelength at 900 MHz. The simulated results obtained from the IE3D simulation software are also shown for comparison. Good agreement

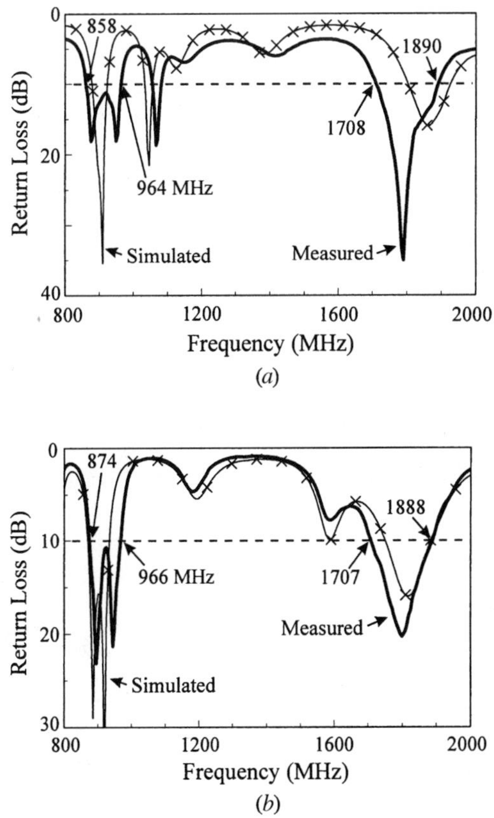

FIGURE 4.85 Measured return loss for the antenna shown in Figure 4.84; $\ell = 13$ mm ground plane size $= 150 \times 150$ mm^2. (*a*) Port 1 excitation; (*b*) port 2 excitation. (From Ref. 17, © 2001 IEEE, reprinted with permission)

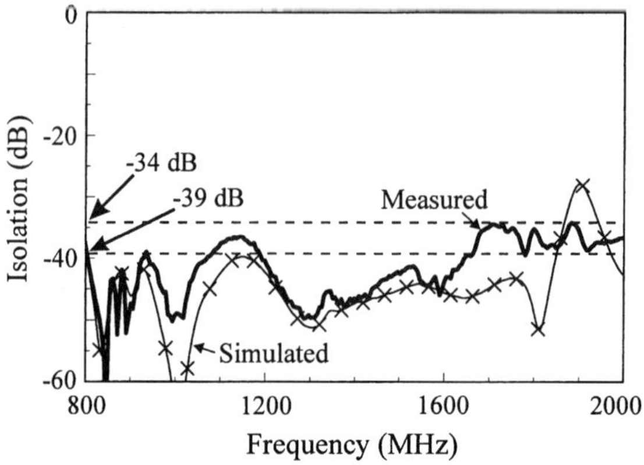

FIGURE 4.86 Measured isolation for the antenna shown in Figure 4.84. (From Ref. 17, © 2001 IEEE, reprinted with permission)

is observed. The obtained impedance bandwidths (10 dB return loss) for the 900 and 1800 MHz bands are about 10% and cover the GSM and DCS bands, respectively. The measured and simulated isolation between ports 1 and 2 is presented in Figure 4.86. From the measured results, the isolation (S_{21}) across the 900 and 1800 MHz bands is less than −39 and −34 dB, respectively. Very good port decoupling is obtained for the proposed antenna.

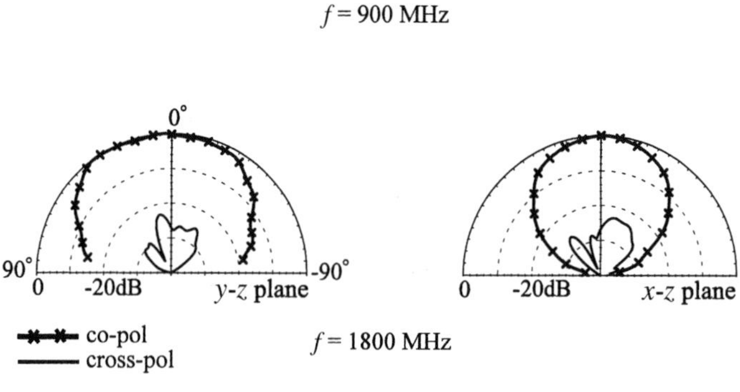

FIGURE 4.87 Measured radiation patterns at 900 and 1800 MHz for port 1 excitation of the antenna shown in Figure 4.84. (From Ref. 17, © 2001 IEEE, reprinted with permission)

Radiation characteristics of the proposed antenna have also been studied. Measured radiation patterns in two principal planes at 900 and 1800 MHz for port 1 excitation are plotted in Figure 4.87. Good broadside radiation patterns are obtained, and the obtained antenna gain is about 6.5–7.3 dBi. The estimated antenna efficiency is about 60%. Similar radiation characteristics at 900 and 1800 MHz for port 2 excitation as seen in Figure 4.87 have also been observed. For both the 900 and 1800 MHz band operations, the dual linearly polarized waves excited by ports 1 and 2 have also been seen to have orthogonal polarizations. Finally, it should be noted that it is possible for the proposed antenna to have 2 + 2 ports instead of 2 ports studied here, if preferred. In addition, for practical applications in a GSM/DCS cellular system base station, the impedance matching of the operating frequencies of the proposed antenna should be further enhanced to be better than 1.5:1 VSWR.

REFERENCES

1. K. L. Wong, Compact and Broadband Microstrip Antennas, Chap. 7, John Wiley & Sons, New York, NY, 2002.
2. F. Yang, X. X. Zhang, X. Ye, and Y. Rahmat-Samii, "Wide-band E-shaped patch antennas for wireless communications," IEEE Trans. Antennas Propagat., vol. 49, pp. 1094–1100, July 2001.
3. K. L. Wong and W. H. Hsu, "A broadband rectangular patch antenna with a pair of wide slits," IEEE Trans. Antennas Propagat., vol. 49, pp. 1345–1347, Sept. 2001.
4. B. L. Ooi and Q. Shen, "A novel E-shaped broadband microstrip antenna," Microwave Opt. Technol. Lett., vol. 27, pp. 348–352, Dec. 5, 2000.
5. H. C. Tung and K. L. Wong, "A compact dual-polarized patch antenna for 1800 MHz band operation," Microwave Opt. Technol. Lett., vol. 29, pp. 1–2, April 5, 2001.
6. F. S. Chang and K. L. Wong, "A broadband probe-fed planar patch antenna with a short probe pin and a conducting cylinder transition," Microwave Opt. Technol. Lett., vol. 31, pp. 282–284, Nov. 20, 2001.
7. F. S. Chang and K. L. Wong, "A broadband probe-fed patch antenna for a DCS base station," Microwave Opt. Technol. Lett., vol. 30, pp. 341–343, Sept. 5, 2001.
8. P. L. Teng, C. L. Tang, and K. L. Wong, "A broadband planar patch antenna fed by a short probe feed," in Proc. 2001 Asia-Pacific Microwave Conf., pp. 1243–1246.
9. K. L. Wong, C. L. Tang, and J. Y. Chiou, "Broadband probe-fed patch antenna with a W-shaped ground plane," IEEE Trans. Antennas Propagat., vol. 50, June 2002.
10. W. H. Hsu and K. L. Wong, "Broadband probe-fed patch antenna with a U-shaped ground plane for cross-polarization reduction," IEEE Trans. Antennas Propagat., vol. 50, pp. 352–355, March 2002.
11. F. S. Chang and K. L. Wong, "Broadband patch antenna edge-fed by a coplanar probe feed," Microwave Opt. Technol. Lett., vol. 31, pp. 287–289, Nov. 20, 2001.
12. F. S. Chang and K. L. Wong, "Compact L-strip capacitively coupled patch antenna with a folded radiating patch for a DCS base station," Microwave Opt. Technol. Lett., vol. 31, pp. 60–62, Oct. 5, 2001.

13. F. S. Chang and K. L. Wong, "Broadband dual-frequency coplanar probe fed patch antenna for GSM/DCS/PCS base stations," Microwave Opt. Technol. Lett., vol. 33, pp. 330–332, June 5, 2002.
14. K. L. Wong, H.C. Tung, and T.W. Chiou, "Broadband dual-polarized aperture-coupled patch antennas with modified H-shaped coupling slots," IEEE Trans. Antennas Propagat., vol. 50, pp. 188–191, Feb. 2002.
15. K. L. Wong and T. W. Chiou, "Broadband dual-polarized patch antennas fed by capacitively coupled feed and slot-coupled feed," IEEE Trans. Antennas Propagat., vol. 50, pp. 346–351, March 2002.
16. T. W. Chiou and K. L. Wong, "Broadband dual-polarized single microstrip patch antenna with high isolation and low cross polarization," IEEE Trans. Antennas Propagat., vol. 50, pp. 399–401, March 2002.
17. T. W. Chiou and K. L. Wong, "A compact dual-polarized aperture-coupled patch antenna for GSM 900/1800-MHz systems," in Proc. 2001 Asia-Pacific Microwave Conf., pp. 95–98.
18. K. L. Wong, Design of Nonplanar Microstrip Antennas and Transmission Lines, John Wiley & Sons, New York, NY, 1999.
19. K. L. Wong, Compact and Broadband Microstrip Antennas, Chap. 8, John Wiley & Sons, New York, NY, 2002.

CHAPTER FIVE

Antennas for WLAN Applications

5.1 INTRODUCTION

In this chapter, a variety of advanced planar antenna designs suitable for WLAN operations are presented. Some promising patch antenna designs to achieve CP radiation [1] and dual-polarized (DP) operation in the 2.4 GHz band are first shown in Section 5.2. These CP and DP antennas are suitable for WLAN access point applications and can overcome the multipath fading problem, and enhanced system performance can be obtained. A dual-band patch antenna design for obtaining WLAN operations in the 2.4 and 5.2 GHz bands [2] is also demonstrated. The design of a promising printed dipole array for obtaining omnidirectional radiation with an enhanced gain is also addressed.

In Section 5.3, surface-mountable antennas such as the ceramic chip antenna (using a ceramic chip as the base for printing the radiating metal lines) [3–7], FR4 chip antenna (using an FR4 chip as the base for printing the radiating metal lines), and folded strip antenna (no ceramic or FR4 chips are required) [8] are introduced. Advanced designs of these surface-mountable antennas, especially for the latter two cases for achieving 2.4/5.2 GHz operations, are presented.

By printing and integrating a monopole antenna on the circuit board of the system, the overall antenna volume can be greatly reduced [9]. Several promising dual-band monopole antennas [10–12] and diversity monopole antennas [13] have recently been reported. Instead of being integrated on the circuit board of the system, the printed monopole antenna designed to be fed by using a coaxial line [14] is also very suitable for applications in a notebook computer or a personal digital assistant (PDA) or the like for WLAN operations. Such printed monopole antennas can be installed in the corners [15] or the hinge [16] of the notebook computer. In this case, a concealed antenna is obtained. These related designs are studied in Sections 5.4 and 5.5.

The printed dipole antenna [17] is also a promising candidate for use in a notebook computer or a PDA for WLAN operations [15, 16]. Several novel designs capable of dual-band operations are introduced in Section 5.6. The PIFA designs discussed in Chapter 2 for 900/1800 MHz operations can also be applied to obtain 2.4/5.2 GHz operations. Some novel designs [18, 19] other than those discussed in Chapter 2 are presented in Section 5.7. Finally, in Section 5.8, a low-cost, compact dual-polarized microstrip antenna with a slotted ground plane [20, 21] suitable for WLAN operation in the 2.4 GHz band is shown. This antenna utilizes an inexpensive FR4 substrate, and good antenna performance in the 2.4 GHz band has been obtained.

5.2 WLAN ACCESS POINT ANTENNA

5.2.1 Broadband CP Design

Circularly polarized antennas are attractive for wireless communication applications, because no strict orientation between the access point and the mobile unit are required. In this section, we demonstrate a novel single-feed broadband circularly polarized patch antenna [1]. This CP antenna has a simple structure and can be constructed with low cost. The antenna comprises a corner-truncated square patch, an L-shaped ground plane, and a probe feed (see Fig. 5.1). Through a via-

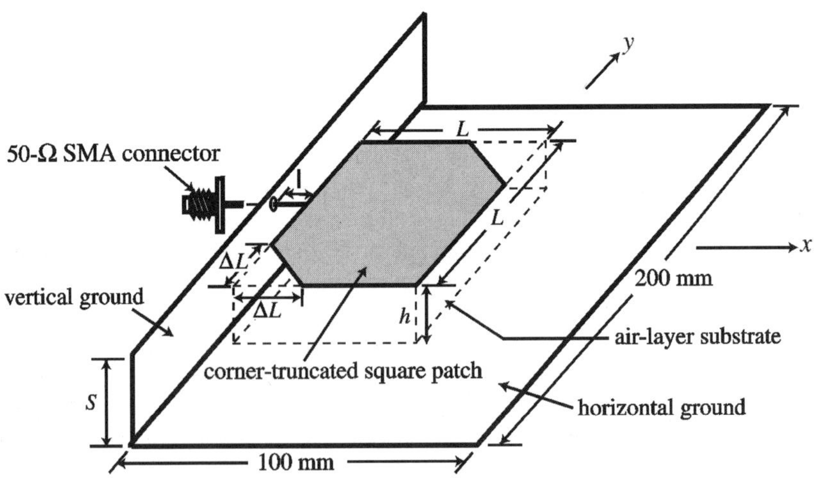

FIGURE 5.1 Geometry of the broadband circularly polarized patch antenna fed by a coplanar probe feed. (From Ref. 1, © 2002 IEEE, reprinted with permission)

hole in the vertical ground of the L-shaped ground plane, the radiating patch can be easily excited by a coplanar probe feed (see Section 4.2.3), and CP radiation over a wide frequency range (>10%) can be achieved. The obtained CP performance is among the best reported for single-feed single-element patch antennas.

As shown in Figure 5.1, the square patch has a side length of L, and two opposite corners having a side length of ΔL are cut to provide as a perturbation for obtaining two orthogonal near-degenerate resonant modes for CP radiation. This corner-truncated square patch is placed above an L-shaped ground plane, which consists of a horizontal ground ($100 \times 200\,\text{mm}^2$ in this study) and a vertical ground ($S \times 200\,\text{mm}^2$). The additional vertical ground used in the design is for the purpose of feeding the square patch with a coplanar probe feed. In this case, although a thick air-layer substrate of thickness h is used between the square patch and the horizontal ground, the probe pin length (ℓ) can be kept to be small to eliminate the problem of large probe inductance effects on the antenna's impedance matching.

In addition, it should be noted that the optimal feed position for this CP design is at the center of the patch edge facing the vertical ground. As for the height (h) of the vertical ground, it is chosen to be slightly larger than the air-layer substrate thickness to accommodate the SMA connector and to minimize its affect on the radiation characteristics of the antenna. If a large vertical ground is used, it could result in prominent asymmetric radiation patterns for the proposed antenna, especially in the x–y plane (the plane in perpendicular to the vertical ground) radiation.

A prototype suitable for operating in the 2.4 GHz band was constructed and studied. The side length (L) of the square patch and the air-layer substrate thickness (h) were selected to be 43 mm (about 36% of the wavelength at resonance) and 18 mm (about 15% of the wavelength at resonance), respectively, and the optimal side length of the truncated corners was determined to be 3.1 mm (about $0.072L$). [Note that, because of the thick air-layer substrate thickness used, the patch's resonant length is much less than one-half wavelength and will approach one-quarter wavelength if a very large value of h is used or the horizontal ground is removed. In this case, the proposed antenna will behave as a monopole antenna, and the CP radiation, however, cannot be obtained by truncating the corners of the square patch.] In addition, the probe pin length (ℓ) was selected to be 3.5 mm, which is much less than the air-layer substrate thickness. Thus the probe inductance is minimized, and its effect on the antenna's impedance matching can be ignored. As for the height (S) of the vertical ground, it was chosen to be 23 mm, slightly larger than the air-layer substrate thickness.

Measured input impedance of the constructed prototype is presented in Figure 5.2. It is seen that a very small loop occurs close to the center of the Smith chart. The small loop indicates that two resonant modes are excited at close frequencies, which makes CP radiation possible. The measured and simulated return loss is also shown in Figure 5.3. Reasonable agreement between the measured data and the simulated results is obtained. The obtained impedance bandwidth, determined by 1.5:1 VSWR, is as large as 740 MHz (2270–3010 MHz), or about 30% referenced to 2.45 GHz.

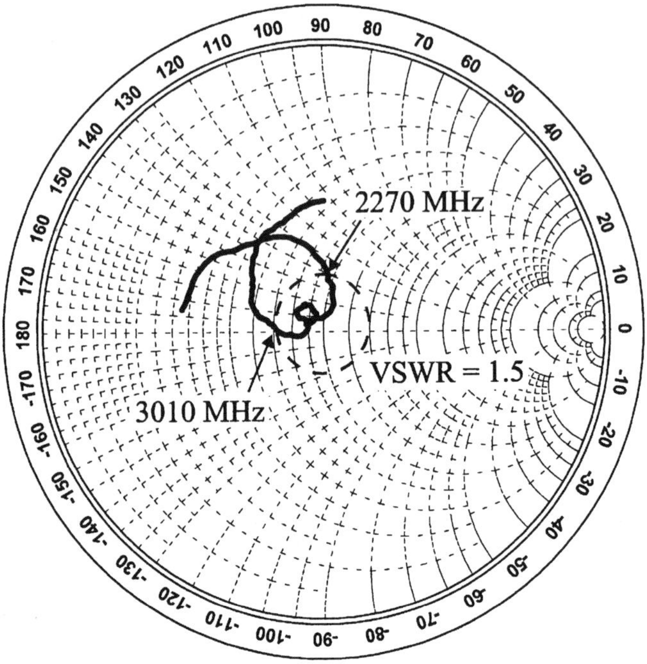

FIGURE 5.2 Measured input impedance on a Smith chart for the antenna shown in Figure 5.1 with $S = 23$ mm, $L = 43$ mm, $\Delta L = 3.1$ mm, $h = 18$ mm, $\ell = 3.5$ mm. (From Ref. 1, © 2002 IEEE, reprinted with permission)

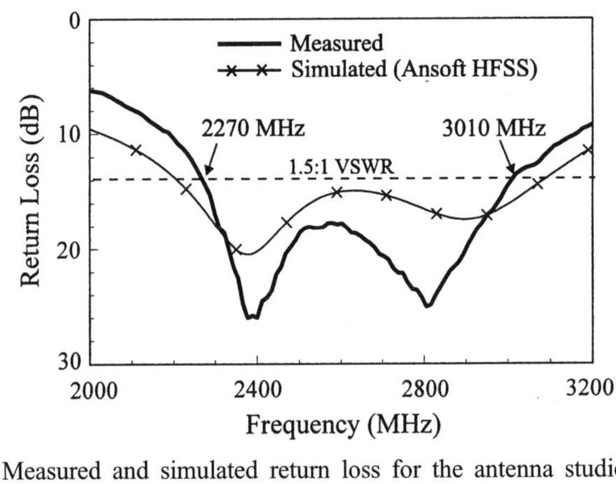

FIGURE 5.3 Measured and simulated return loss for the antenna studied in Figure 5.2. (From Ref. 1, © 2002 IEEE, reprinted with permission)

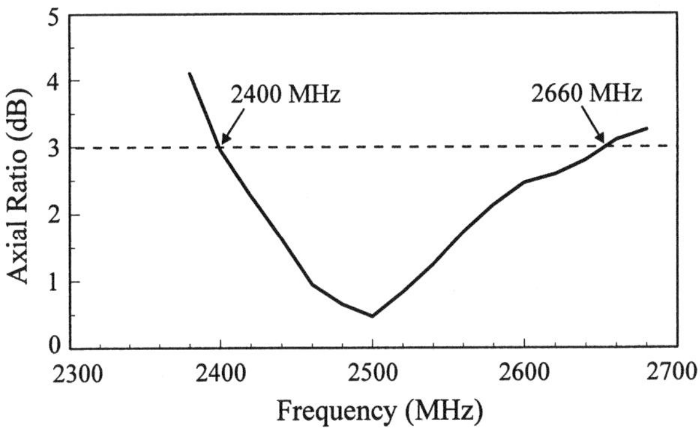

FIGURE 5.4 Measured axial ratio for the antenna studied in Figure 5.2. (From Ref. 1, © 2002 IEEE, reprinted with permission)

Figure 5.4 shows the measured axial ratio in the broadside direction of the constructed prototype. The obtained 3-dB axial-ratio bandwidth reaches 260 MHz (2400–2660 MHz), or about 10.6% with respect to 2.45 GHz. From the measured return loss shown in Figure 5.3, the frequencies within the obtained CP bandwidth are also seen to have a return loss better than about 17 dB. Measured spinning linear radiation patterns at 2500 MHz are plotted in Figure 5.5. Slight asymmetry of the radiation patterns is seen, which is probably due to the presence of the vertical ground in this CP design. As for the backward radiation, it is seen to be less than −20 dB, relative to the main lobe in the forward direction. Figure 5.6 presents the measured antenna gain for frequencies within the obtained CP bandwidth. The peak antenna gain reaches about 9.2 dBi, and the average gain level is about 8.5 dBi or larger.

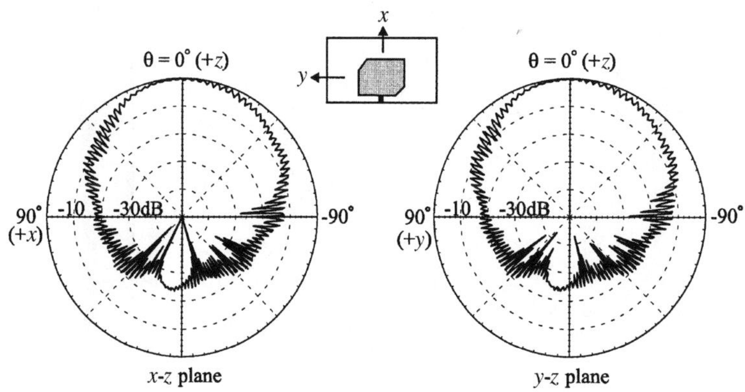

FIGURE 5.5 Measured spinning linear radiation patterns at 2500 MHz for the antenna studied in Figure 5.2. (From Ref. 1, © 2002 IEEE, reprinted with permission)

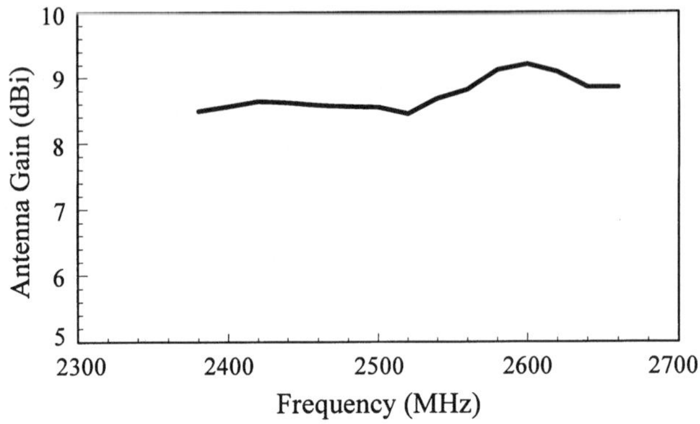

FIGURE 5.6 Measured antenna gain for the antenna studied in Figure 5.2. (From Ref. 1, © 2002 IEEE, reprinted with permission)

5.2.2 Dual-Polarized Design

To obtain dual-polarized operation in the 2.4 GHz band for WLAN applications, the broadband dual-polarized designs shown in Sections 4.4 and 4.5 are also promising candidates. In this section, a dual-polarized probe-fed patch antenna fed by using two probe feeds with high port decoupling for operating in the 2.4 GHz band is presented. Figure 5.7 shows the antenna geometry. The radiating patch is square in shape and has a side length of L. Two identical probe feeds having a short probe pin of length 3 mm and a triangular transition patch with a flare angle α are placed at the centers of two adjacent patch edges. The two probe feeds generate $\pm 45°$ slanted linear polarizations with respect to the y-axis shown in Figure 5.7

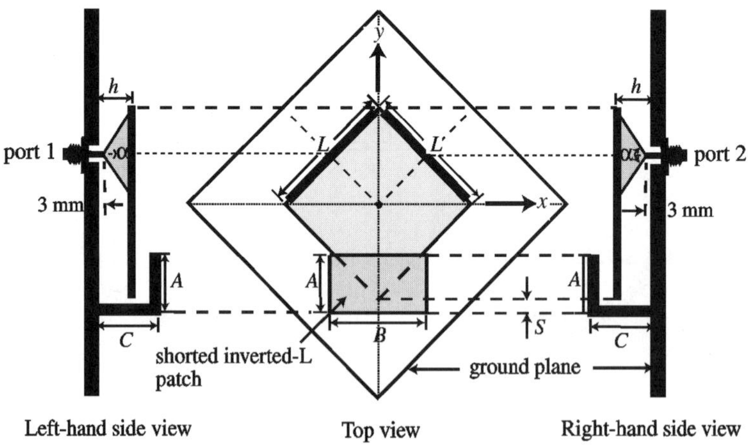

FIGURE 5.7 Geometry of the dual-polarized probe-fed patch antenna for WLAN application.

and also support the square patch above the ground plane with a height of h (h much larger than 3 mm in the design).

It should first be noted that, with the two probe feeds, the port decoupling is usually no better than −20 dB. To improve the port decoupling, a shorted inverted-L patch comprising a vertical portion of dimensions $B \times C$ ($C > h$) and a horizontal portion of dimensions $A \times B$ is placed with a distance of S to the patch corner facing the two probe feeds. The shorted inverted-L patch is parasitically coupled to the excited square patch. In addition, through capacitive coupling between the square patch and the horizontal portion of the shorted inverted-L patch, it is expected that some excited patch surface currents flowing from the excited port toward the other one can be reduced such that port decoupling can be enhanced. It is found that, for WLAN applications in the 2.45 GHz band, the optimal value of B is about one patch's side length (L), and A is about half the patch's side length ($0.5L$); the height C is about 1.5 times the air-layer thickness, and S is about $0.2L$.

Designs without (the reference antenna) and with (the proposed antenna) the shorted inverted-L patch have been constructed and studied. The measured and simulated results for the reflection coefficient (S_{11}) and isolation (S_{21}) are shown in Figures 5.8 and 5.9. The measured data are in good agreement with the simulated results, especially for the proposed antenna shown in Figure 5.9. The obtained impedance bandwidths (1.5:1 VSWR) for both designs cover the 2.4 GHz WLAN band (2400–2484 MHz). From the results, it is seen that the presence of the shorted inverted-L patch has significant effects on S_{21}. For the proposed antenna studied in Figure 5.9, the isolation for frequencies within the WLAN band is less than −30 dB, which is about −15 dB less than that of the reference antenna studied in Figure 5.8. From the simulated excited patch surface currents shown in Figure 5.10, the effect of

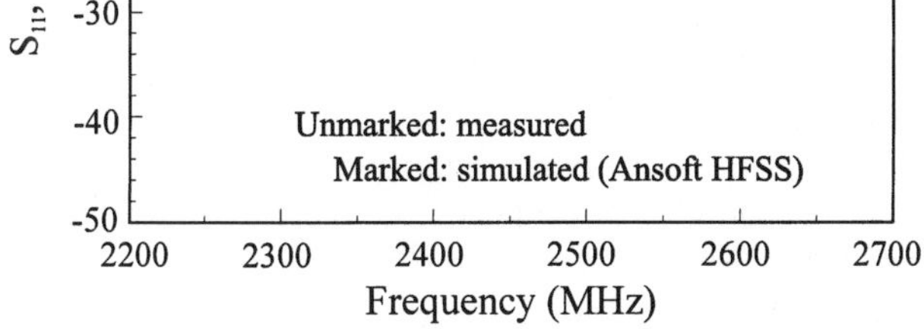

FIGURE 5.8 Measured and simulated reflection coefficient (S_{11}) and isolation (S_{21}) for the reference antenna (the geometry shown in Figure 5.7 without the shorted inverted-L patch); $h = 14$ mm, $L = 53$ mm, $\alpha = 135°$, ground plane size $= 150 \times 150$ mm^2.

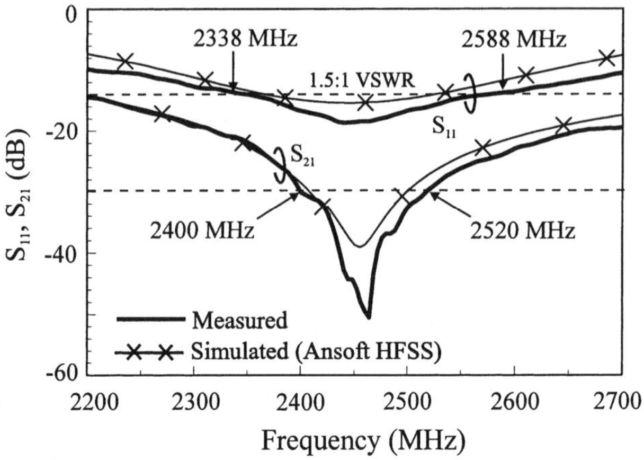

FIGURE 5.9 Measured and simulated reflection coefficient (S_{11}) and isolation (S_{21}) for the antenna shown in Figure 5.7 with $h = 14$ mm, $L = 53$ mm, $\alpha = 135°$, $A = 25$ mm, $B = 48$ mm, $C = 22$ mm, $S = 10$ mm, ground plane size $= 150 \times 150$ mm^2.

the shorted inverted-L patch on improving the port decoupling can be seen much more clearly. The results for the reference antenna (Fig. 5.10*a*) show that, for port 2 excitation, a large portion of the excited patch surface currents are in parallel to the resonance direction for port 1, which degrades the isolation between ports 1 and 2. On the other hand, for the proposed antenna (Fig. 5.10*b*), owing to the presence of the shorted inverted-L patch close to the patch corner facing both ports 1 and 2, the excited patch surface currents are in general in the resonance direction of port 2. This behavior is probably the reason for the better isolation obtained for the proposed antenna.

The radiation characteristics of the proposed antenna for port 1 and port 2 excitation have been measured and presented in Figures 5.11 and 5.12, respectively. Simulated results are also shown, which in general agree with the measured data. Although there is a shorted inverted-L patch presence, the proposed antenna still has good broadside radiation patterns. This is probably because the coupling between the shorted inverted-L patch and the driven square patch is not so strong (the distance between the shorted patch and the driven patch is as large as 8 mm in the proposed antenna), and thus no strong effects on the radiation patterns are observed. Relatively large cross-polarization radiation is also seen. However, this is not a disadvantage for WLAN applications, where the wave propagation usually experiences multiple reflections between the transmitter and the receiver, especially in indoor operations. The measured antenna gain for port 1 and port 2 excitation of the proposed antenna is shown in Figure 5.13. Because of the symmetric structures for ports 1 and 2, their antenna gain is about the same (the small differences are largely due to the manufacturing errors), and the antenna gain level is about 7.5 dBi.

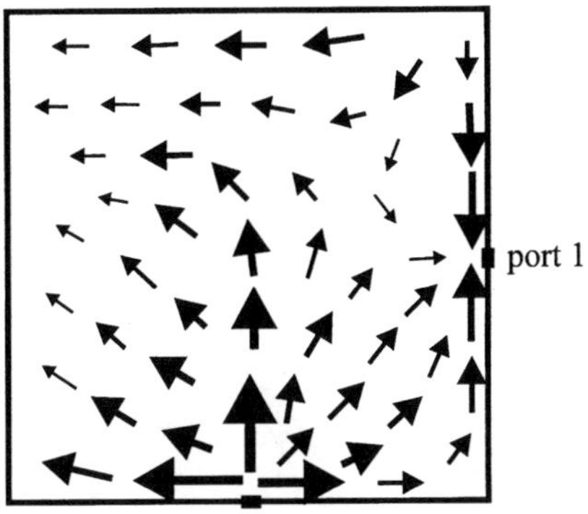

(*a*)

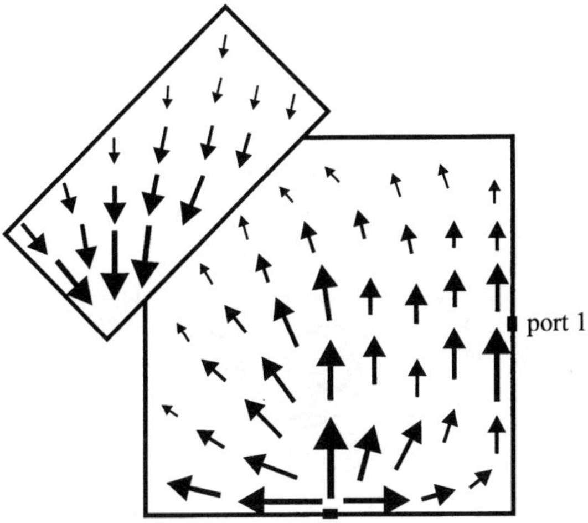

(*b*)

FIGURE 5.10 Simulated patch surface current distributions for port 2 excitation of (*a*) the reference antenna studied in Figure 5.8 and (*b*) the proposed antenna studied in Figure 5.9.

5.2.3 Dual-Band Design

In this section, a dual-band planar inverted-L patch antenna suitable for WLAN applications in the 2.4 and 5.2 GHz bands is demonstrated [2]. The antenna design is shown in Figure 5.14. The inverted-L patch is placed above a grounded FR4 substrate of thickness 0.8 mm and relative permittivity 4.4. To achieve a low profile for the proposed antenna, the vertical portion of the inverted-L patch is chosen to have a small height of 3.5 mm, which makes the total antenna height 4.3 mm, including the thickness of the FR4 substrate. The horizontal portion of the inverted-L patch, whose dimensions are given in the figure, is chosen to be trapezoid

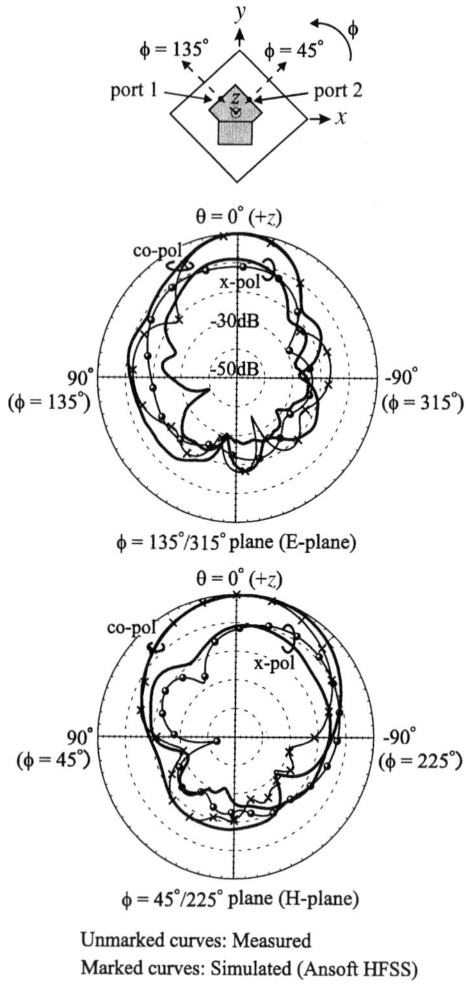

FIGURE 5.11 Measured and simulated radiation pattern at 2450 MHz for port 1 excitation of the antenna studied in Figure 5.9.

or with a widened patch width to achieve a larger frequency ratio between the two operating frequencies. Also note that the inverted-L patch antenna is operated as a quarter-wavelength structure, and thus the total length of the inverted-L patch can be roughly determined by the quarter-wavelength (about 31 mm) of the lower operating frequency (2.4 GHz).

In this design, it was also found that, by placing a pair of narrow slits close to and in parallel to the patch's nonradiating edges, the excited patch surface currents of the upper resonant frequency can be effectively meandered, whereas that of the lower resonant frequency is slightly affected. This characteristic makes it very convenient

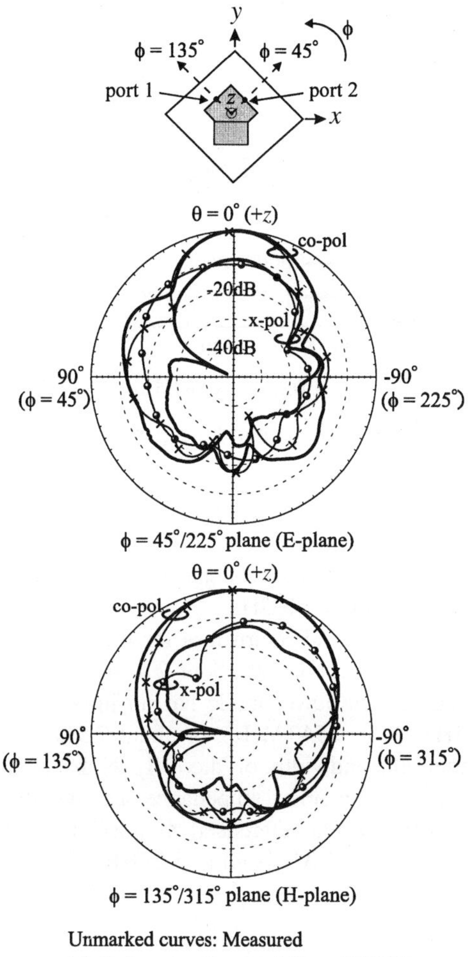

FIGURE 5.12 Measured and simulated radiation pattern at 2450 MHz for port 2 excitation of the antenna studied in Figure 5.9.

in fine-tuning the lower and upper operating frequencies in this dual-band design to about 2.4 and 5.2 GHz, respectively. To achieve good impedance matching for the two operating frequencies, a probe feed with a microstrip tuning stub printed on the FR4 substrate was used to feed the antenna. Through a via-hole in the grounded FR4 substrate, the probe feed is placed at the center of the vertical portion of the inverted-L patch. The width of the microstrip tuning stub was chosen to be 1.53 mm in the study, which gives it a 50-Ω characteristic impedance, and the optimal length of the microstrip tuning stub in the design was found to be 12 mm,

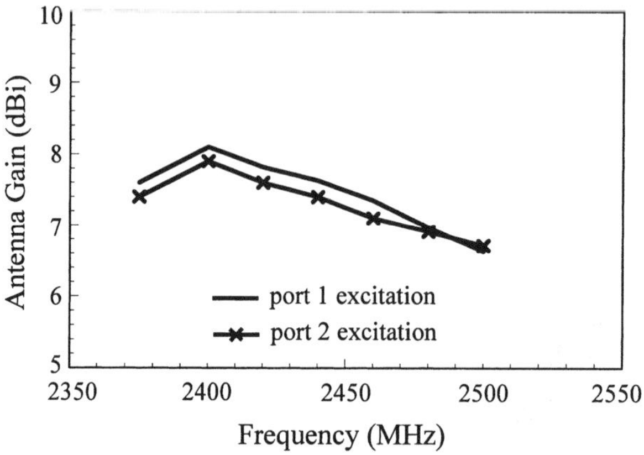

FIGURE 5.13 Measured antenna gain for port 1 and port 2 excitation of the antenna studied in Figure 5.9.

which is about 10% of the wavelength of the operating frequency at 2.4 GHz or about 0.2 times the wavelength at 5.2 GHz.

Figure 5.15 shows the measured return loss for the constructed prototype. It is clearly seen that two resonant modes at about 2.4 and 5.2 GHz are excited with good impedance matching. The lower mode has an impedance bandwidth (10 dB return loss) of 180 MHz (2360–2540 MHz), or about 7.5% referenced to 2.4 GHz, which covers the required bandwidth of the WLAN band at 2.4 GHz. For the upper resonant mode, the obtained impedance bandwidth reaches 420 MHz (5060–5480 MHz), or about 8.1% referenced to 5.2 GHz, which also meets the bandwidth requirement of the WLAN band at 5.2 GHz.

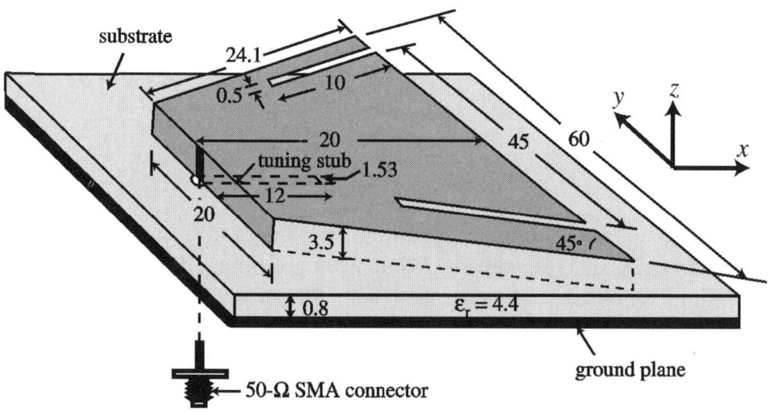

FIGURE 5.14 Geometry of the dual-band planar inverted-L patch antenna for 2.4/5.2 GHz operations. (From Ref. 2, © 2001 John Wiley & Sons, Inc.)

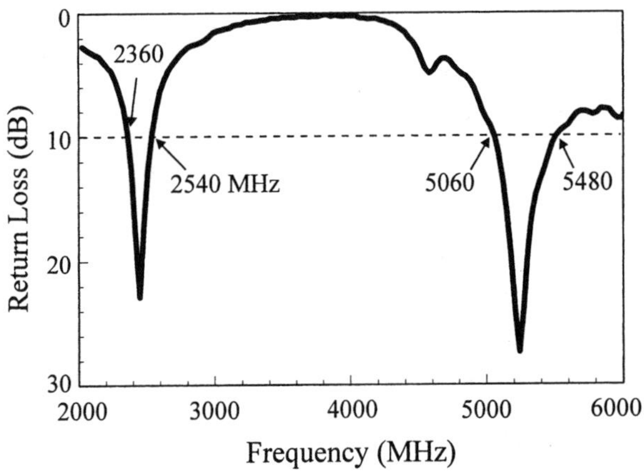

FIGURE 5.15 Measured return loss for the antenna shown in Figure 5.14; ground plane size = $120 \times 120\,\text{mm}^2$. (From Ref. 2, © 2001 John Wiley & Sons, Inc.)

Simulated patch surface current distributions, obtained from IE3D simulation, for the lower and upper bands of the constructed prototype are shown in Figure 5.16. From the simulated results, because of the inserted slits in the inverted-L patch, large meandering of the patch surface current paths for the operating frequency in the upper band is observed. On the other hand, small effects of the inserted slits on the patch surface current path for the operating frequency in the lower band is seen. This characteristic facilitates the fine-tuning of the two operating frequencies of 2.4 and 5.2 GHz in this dual-band design.

Figure 5.17*a* and *b* plots the measured radiation patterns in the x–z plane (E-plane) and y–z plane (H-plane) for the operating frequencies at 2.45 and 5.25 GHz. In general, broadside radiation patterns are seen. Large cross-polarization radiation in the H-plane pattern for both the two operating frequencies is also observed. The measured antenna gain for the frequencies within the lower and upper bands is presented in Figure 5.18*a* and *b*, respectively. The peak antenna gain for the lower and upper bands reaches 5.7 and 6.4 dBi, respectively, and small gain variations less than 0.5 dBi are observed for both bands.

5.2.4 Printed Dipole Array Design for Omnidirectional Radiation

Printed dipole linear arrays, which show omnidirectional radiation, have been applied in the WLAN access points. Figure 5.19 shows the geometry of a promising dipole array design, and Figure 5.20 shows its practical application in an access point for WLAN operations. In this design, the feed network (its layout is not shown in the figure) is designed to provide equal powers at the four outputs (ports 1 to 4) and a 180° phase difference between the lower two ports (ports 1 and 2) and the upper two ports (ports 3 and 4). In this case, the lower and upper two dipoles can have the same polarization to achieve a constructive array effect.

(*a*)

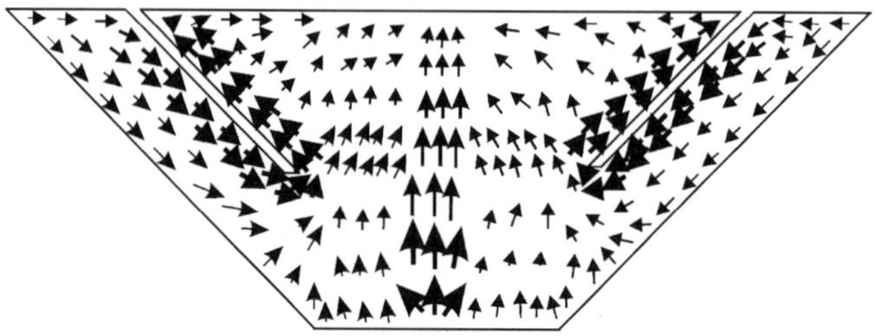

(*b*)

FIGURE 5.16 Simulated patch surface current distributions, obtained from IE3D simulation, in the horizontal portion of the inverted-L patch for the antenna shown in Figure 5.14. (*a*) $f = 2.45$ GHz; (*b*) $f = 5.25$ GHz. (From Ref. 2, © 2001 John Wiley & Sons, Inc.)

For WLAN applications in the 5.2 GHz band (5150–5350 MHz) with 1.5:1 VSWR impedance matching, this dipole array can have a narrow width of about 12 mm and an antenna height of less than 90 mm. Very good omnidirectional radiation with small variations (can be less than 2 dBi) can be obtained for this dipole array, and the peak antenna gain is about 6 dBi.

5.3 SURFACE-MOUNTABLE ANTENNA

5.3.1 Ceramic Chip Antenna

Ceramic chip antennas usually comprise a ceramic base and metal line patterns printed on or embedded within the base [3–7]. Such chip antennas are compact in size and can be directly mounted on the system circuit board of a communication device, leading to reduced packaging cost. There are two major types of the ceramic chip antennas: one is with a ground plane printed on the bottom of the ceramic base, and the other is without a ground plane. For the former type, the chip antenna can be directly mounted on the ground plane of the system circuit board, leading to smaller limitations in practical applications. A typical design is described in Figure 1.16 in Section 1.5. However, because of the large permittivity of the ceramic base (usually

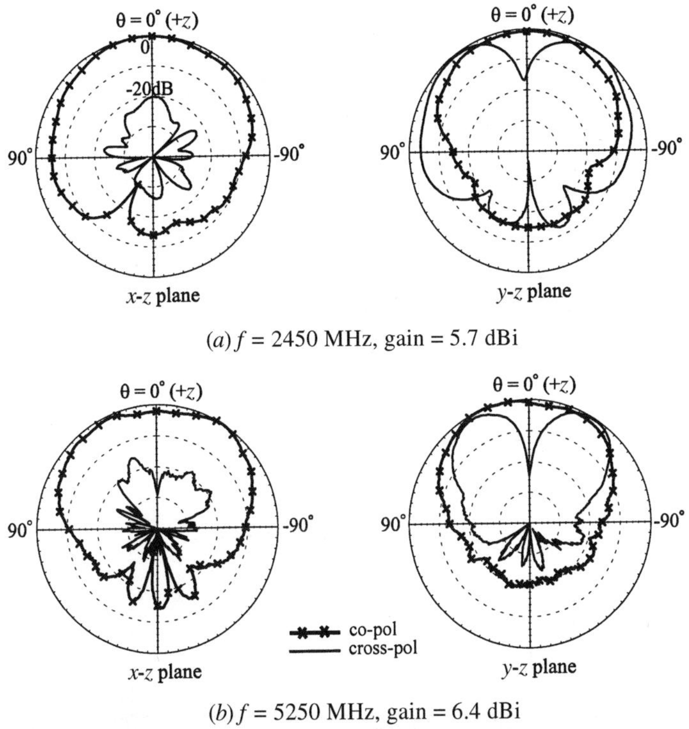

FIGURE 5.17 Measured radiation patterns in the x–z plane (E-plane) and y–z plane (H-plane) for the antenna shown in Figure 5.14. (*a*) $f = 2.45$ GHz; (*b*) $f = 5.25$ GHz. (From Ref. 2, © 2001 John Wiley & Sons, Inc.)

larger than 7), the impedance bandwidth and radiation efficiency of the chip antenna are usually greatly decreased.

On the other hand, for the latter type of ceramic chip antenna without a ground plane, the ceramic base mainly serves as a supporting base for the printed metal lines. In this case, the radiating metal lines are printed on the surfaces of the ceramic base or embedded within the ceramic base as shown in Figure 5.21. In this condition, the ceramic chip antenna is operated as a monopole antenna and should be placed at a certain distance from the ground plane of the system circuit board, which causes some limitations in practical applications. However, there is an advantage of such ceramic chip monopole antennas; that is, relatively much smaller effects of the ceramic base on the antenna's impedance bandwidth and radiation efficiency are expected, with comparison to the type with a ground plane.

For both kinds of ceramic chip antenna, however, there is a common disadvantage; that is, they tend to break easily because the ceramic materials are fragile in nature. To overcome this problem, one can use an FR4 base instead of the ceramic base or simply use a folded metal strip with an air or a plastic base. The related designs are described in Sections 5.3.2 and 5.3.3.

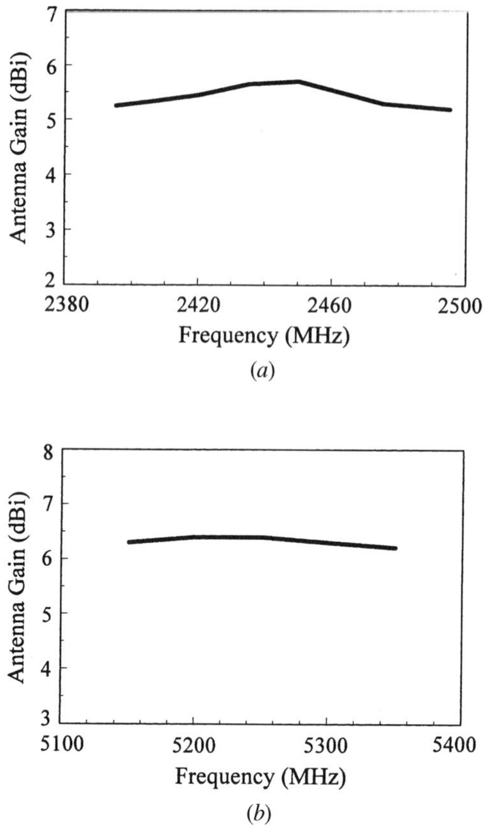

FIGURE 5.18 Measured antenna gain for the antenna shown in Figure 5.14. (*a*) The 2.4 GHz band; (*b*) the 5.2 GHz band. (From Ref. 2, © 2001 John Wiley & Sons, Inc.)

5.3.2 Dual-Band FR4 Chip Antenna

This section introduces a promising dual-band FR4 chip antenna for 2.4/5.2 GHz operations. The proposed antenna mounted on a microwave substrate is shown in Figure 5.22. The grounded microwave substrate has a size of $35 \times 100\,\text{mm}^2$ and can be considered as the circuit board of a practical cellular phone. That is, the proposed dual-band FR4 chip antenna can be considered to be applied in a cellular phone to achieve WLAN or Bluetooth operations. Also note that the ground plane printed on the back of the microwave substrate does not cover the portion (an area of $5 \times 9\,\text{mm}^2$ in this design) underneath the dual-band FR4 chip antenna; that is, the FR4 chip antenna is operated as a monopole antenna. In addition, the proposed antenna can be easily excited by a 50-Ω microstrip line printed on the same substrate and is also surface mountable like the ceramic chip antennas.

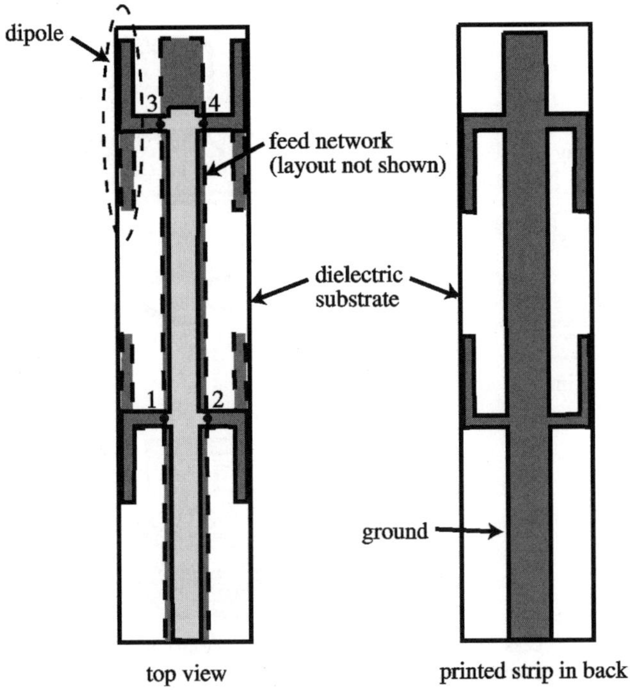

FIGURE 5.19 Geometry of a dipole array with omnidirectional radiation for a WLAN access point.

The FR4 chip antenna comprises an FR4 base and metal lines printed on the surfaces of the FR4 base. Note that the FR4 base has a compact size of $1.6 \times 6 \times 6\,\text{mm}^3$, and the printed metal lines have a uniform width of 1 mm and are meandered on the upper and bottom surfaces of the FR4 base to achieve a reduced antenna size. More importantly, this meandering can fine-tune the antenna's first two resonant frequencies to the desired 2.4 and 5.2 GHz in this design. Also note that the total length of the metal line is roughly one-quarter wavelength of the lower operating frequency (2.4 GHz) in this dual-band design, and other design dimensions can be determined with the aid of the simulation software Ansoft HFSS.

A prototype based on the design shown in Figure 5.22 has been constructed and tested. Figure 5.23 shows the measured and simulated return loss. The measured data are in reasonable agreement with the simulated results. Two resonant modes at about 2.4 and 5.2 GHz are excited. The obtained impedance bandwidth (2.5:1 VSWR) for the lower (2.4 GHz) band is 107 MHz (2395–2502 MHz), and that for the upper (5.2 GHz) band is as large as 820 MHz (5115–5935 MHz). The two obtained bandwidths cover the required bandwidths of the WLAN operations in the 2.4 and 5.2 GHz bands. Good radiation characteristics have also been obtained. The measured antenna gain levels in the 2.4 and 5.2 GHz bands are about 1.6 and 1.2 dBi, respectively.

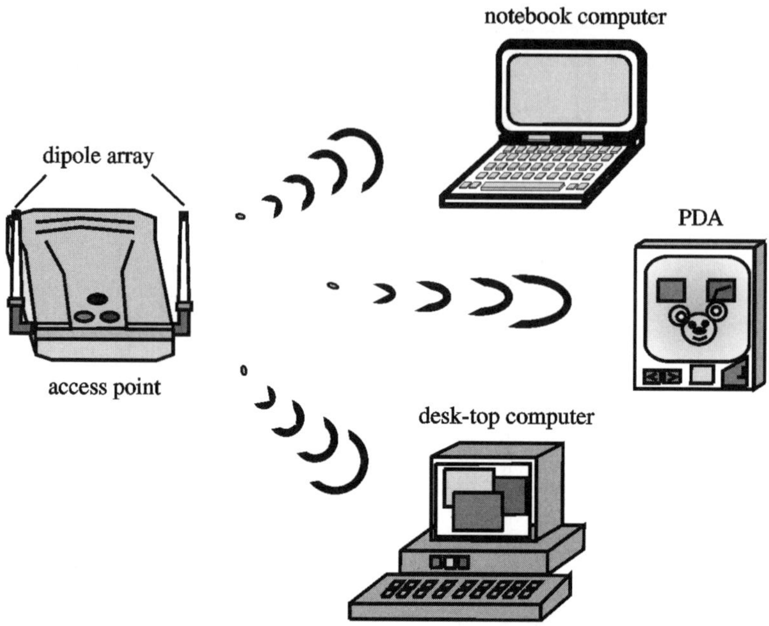

FIGURE 5.20 An access point with two identical dipole arrays shown in Figure 5.19 for WLAN operations with spatial diversity.

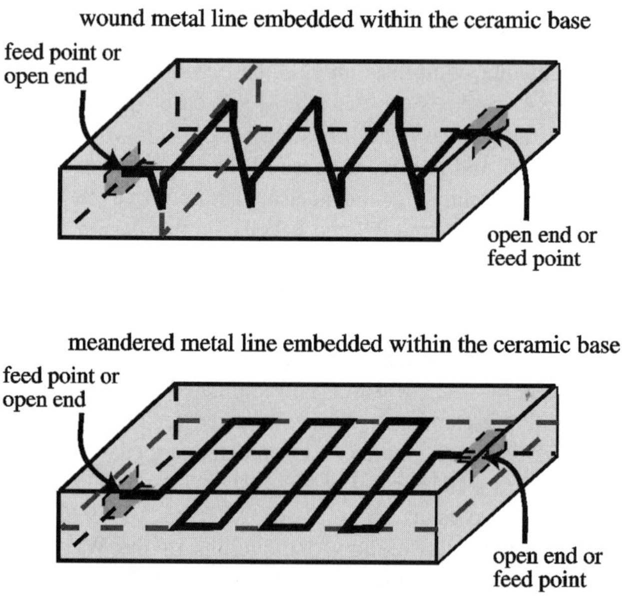

FIGURE 5.21 Geometries of the ceramic chip monopole antenna with its radiating metal lines embedded within the ceramic chip base.

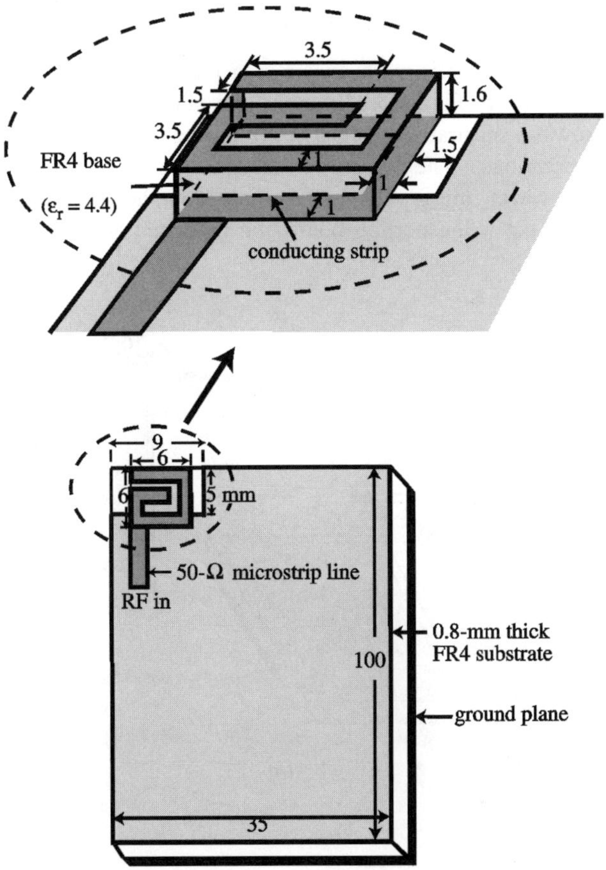

FIGURE 5.22 Geometry of the dual-band FR4 chip antenna mounted on a microwave substrate for 2.4/5.2 GHz operations.

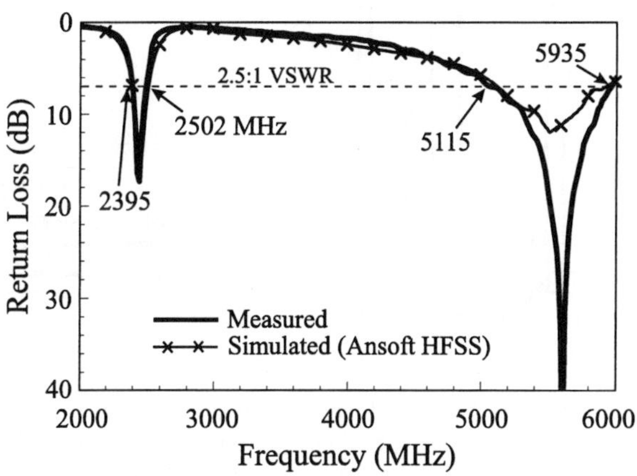

FIGURE 5.23 Measured and simulated return loss for the antenna shown in Figure 5.22.

5.3.3 Dual-Band Folded Strip Monopole Antenna

By using a folded metal strip with an air or a plastic base, a surface-mountable antenna can be obtained. Figure 5.24*a* shows the proposed antenna mounted on a grounded microwave substrate (thickness 0.8 mm and relative permittivity 4.4) for 2.4/5.2 GHz operations. The size of the grounded microwave substrate is $40 \times 100\,\text{mm}^2$. However, the ground plane printed on the back of the microwave substrate does not cover the portion below the proposed antenna; the size of this

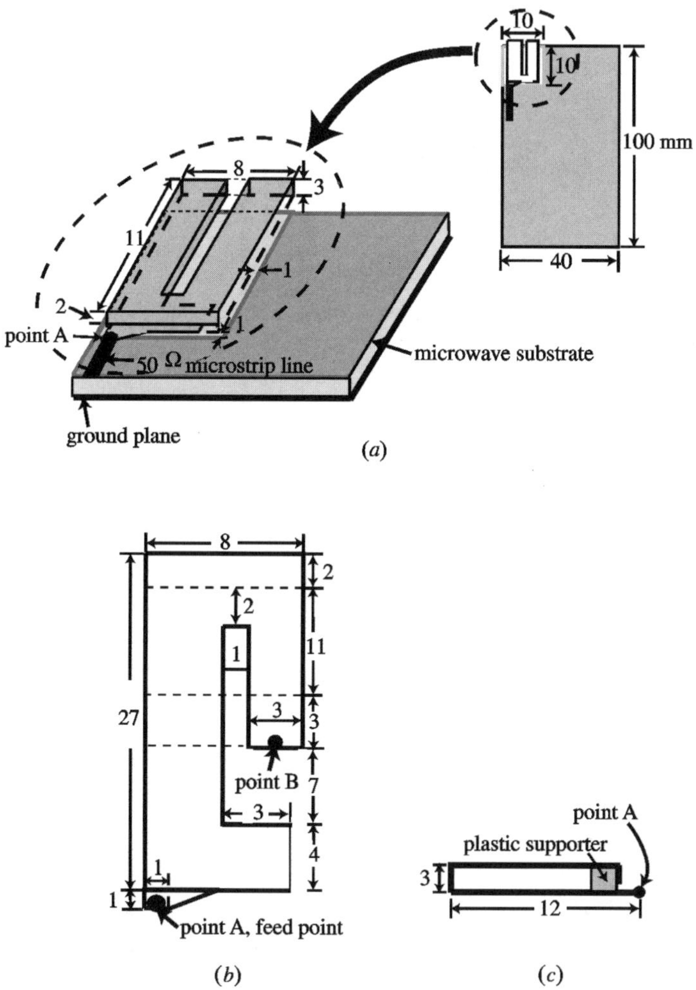

FIGURE 5.24 (*a*) Geometry of the dual-band folded strip monopole antenna mounted on a microwave substrate for 2.4/5.2 GHz operations; (*b*) dimensions of the antenna unfolded into a planar structure; (*c*) side view of the antenna with a plastic supporter placed at the side edge near the feed point. (From Ref. 8, © 2002 John Wiley & Sons, Inc.)

portion is $10 \times 10\,mm^2$ in this study. As discussed in Section 5.3.2, the microwave substrate used can also be considered as the circuit board of a practical cellular phone or a communication device.

The proposed antenna in general looks like a rectangular disk in shape, has a low profile of 3 mm, and occupies an area of $8 \times 12\,mm^2$. The proposed antenna also has a small trapezoid protruded portion (about 1 mm in the protruded length, see the region comprising point A in the figure), which is used for easy connection with the feed, preferably a 50-Ω microstrip line printed on the same microwave substrate. Furthermore, by tuning the size of this protruded portion, improved impedance matching at the two desired operating frequencies for the proposed antenna can be obtained.

The proposed antenna can be easily constructed by folding a planar monopole made of a copper plate, whose dimensions are given in Figure 5.24*b*; the dashed lines shown in the figure are the folding lines. The copper plate used in this study has a thickness of 0.2 mm and, by incorporating placing a plastic supporter in the form of a square rod $3 \times 3\,mm^2$ in cross section and 8 mm in length at the side edge near the feed point (see Fig. 5.24*c*), good rigidity is obtained for the proposed antenna.

To provide dual-frequency operation at two desired operating frequencies, an L-shaped slit is inserted in the proposed monopole (see the proposed monopole in the planar structure shown in Fig. 5.24*b*). The inserted slit in part meanders the antenna's main resonant path (starting from point A, then along the long edge of the copper plate and passing the end of the inserted slit inside the plate, then ending at point B), which reduces the required size of the proposed antenna for a fixed operating frequency and in part is used to fine-tune the antenna's first two resonant frequencies to be two desired operating frequencies for the proposed antenna. In this study, the length of the antenna's main resonant path is about 42 mm, corresponding to about 34% of the wavelength at 2.45 GHz, which is larger than one-quarter wavelength at resonance for a conventional straight monopole antenna. This relatively larger required resonant length is probably due to the coupling effect between adjacent sections of the proposed folded monopole, which reduces the antenna's effective resonant length. Also note that the rectangular stub, having an area of $3 \times 4\,mm^2$ in this study, near the feed point can greatly affect the impedance matching of the two desired operating frequencies. That is, by tuning the size of this rectangular stub, improved impedance matching for the proposed dual-frequency operation can be obtained. This effect is similar to that of the protruded trapezoid portion at the feed point.

Figure 5.25 shows the measured and simulated return loss of the constructed prototype. Reasonable agreement between the measurement and the simulation is observed. It is clearly seen that, at the desired operating frequencies at 2.4 and 5.2 GHz, two separate resonant modes are excited. The lower mode has an impedance bandwidth (10 dB return loss) of 130 MHz (2380–2510 MHz), and the upper mode has an even larger bandwidth of 418 MHz (4940–5358 MHz). The obtained bandwidths cover WLAN operations in both the 2.4 GHz (2400–2484 MHz) and 5.2 GHz (5150–5350 MHz) bands.

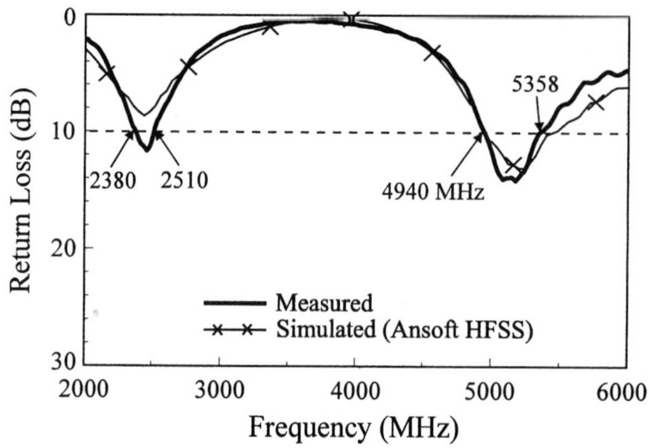

FIGURE 5.25 Measured and simulated return loss for the antenna shown in Figure 5.24. (From Ref. 8, © 2002 John Wiley & Sons, Inc.)

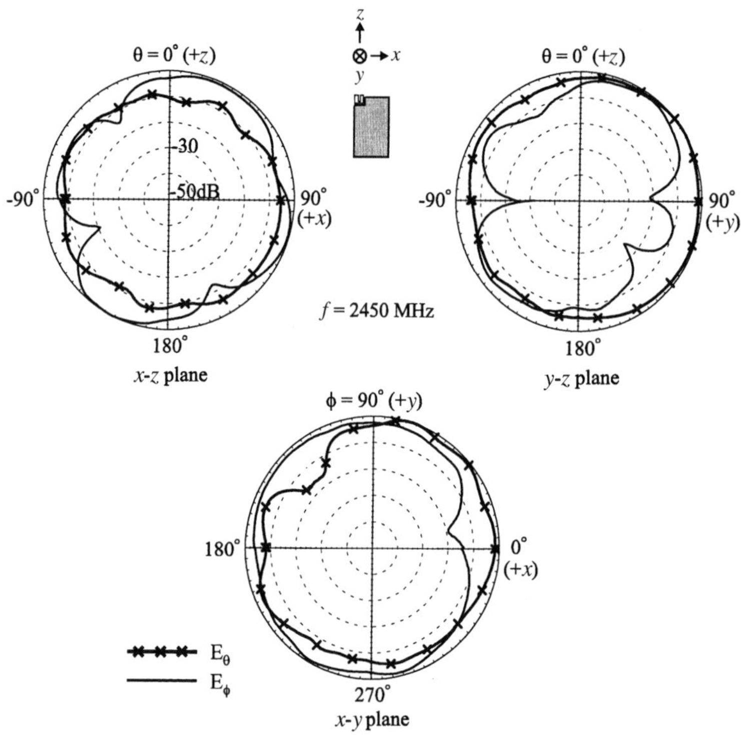

FIGURE 5.26 Measured radiation patterns at 2450 MHz for the antenna shown in Figure 5.24. (From Ref. 8, © 2002 John Wiley & Sons, Inc.)

Figures 5.26 and 5.27 plot the measured radiation patterns at 2.45 and 5.25 GHz, respectively. Measurements at other operating frequencies across the 2.4 and 5.2 GHz bands also show radiation patterns similar to those shown in Figures 5.26 and 5.27, respectively, which indicates that the radiation patterns are stable in the two desired operating bands. Also note that both the E_θ and E_φ components have comparable intensities. This characteristic is an advantage for the proposed antenna for WLAN operations, because their wave propagation environment is usually complex. In this case, comparable E_θ and E_φ radiation can decrease the effect of multiple-path fading problem and enhance the system performance. Figure 5.28 shows the measured antenna gain for frequencies across the lower and upper bands. The 2.4 GHz band has a peak antenna gain of about 2.8 dBi, and the gain variations are less than about 1.3 dBi. On the other hand, the peak antenna gain of the 5.2 GHz band reaches about 3.1 dBi, and small gain variations (less than about 0.5 dBi) are seen.

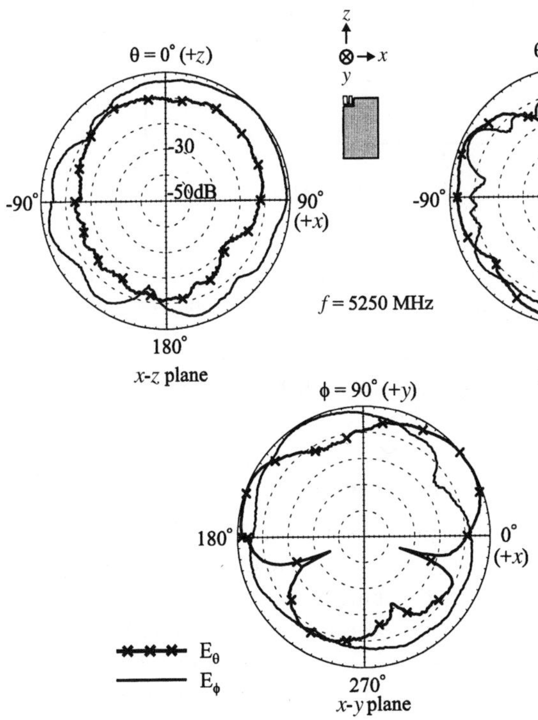

FIGURE 5.27 Measured radiation patterns at 5250 MHz for the antenna shown in Figure 5.24. (From Ref. 8, © 2002 John Wiley & Sons, Inc.)

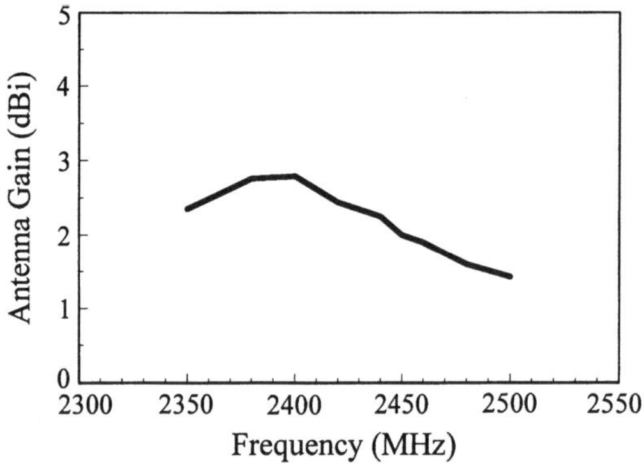

(*a*)

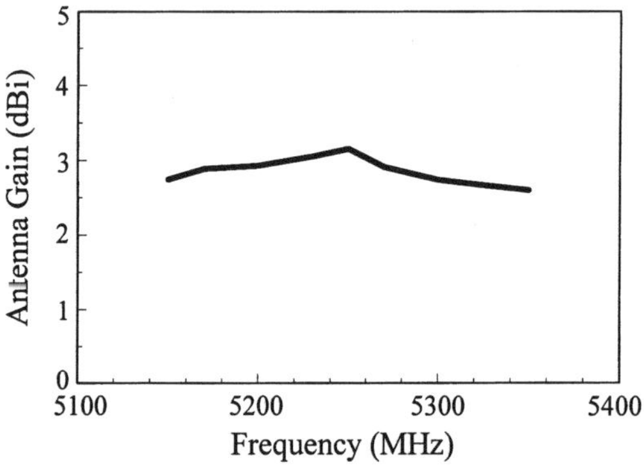

(*b*)

FIGURE 5.28 Measured antenna gain across (*a*) the 2.4 GHz band and (*b*) the 5.2 GHz band for the antenna shown in Figure 5.24. (From Ref. 8, © 2002 John Wiley & Sons, Inc.)

5.4 PRINTED MONOPOLE ANTENNA FOR DUAL-BAND OPERATION

5.4.1 Stacked Inverted-F Monopole Antenna

The inverted-F monopole antenna printed on a dielectric substrate is capable of integration with the associated circuitry on the same substrate, which can reduce the fabrication cost and the required volume of the whole system. In this section, a dual-band design of the printed inverted-F monopole antenna for applications in the 2.4 and 5.2 GHz bands is described. Figure 5.29 shows the antenna design [10]. The antenna consists of two stacked inverted-F metal strips. The two metal strips and the 50-Ω microstrip feed line are all printed on the same dielectric substrate (in this study, the FR4 substrate of thickness 1.6 mm and relative permittivity 4.4 was used.). For design convenience, the width of the two metal strips

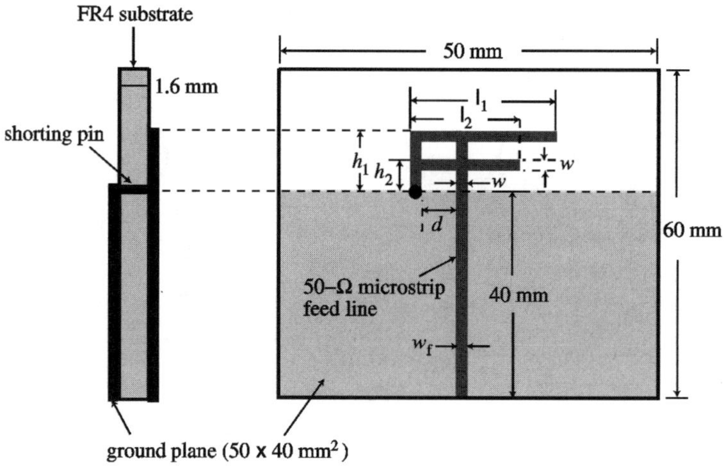

FIGURE 5.29 Geometry of the dual-band printed inverted-F antenna. (From Ref. 10, © 2001 John Wiley & Sons, Inc.)

was chosen to be the same as that of the 50-Ω microstrip feed line. The two metal strips are also short-circuited, with a common shorting pin, through a via-hole in the dielectric substrate to the ground plane printed on the other side of the dielectric substrate.

The larger metal strip has a horizontal length of ℓ_1 and a vertical height of h_1 and is placed on top of the smaller strip having a horizontal length of ℓ_2 and a vertical height of h_2. Also note that the two shorted metal strips are both designed to operate as a quarter-wavelength structure in the proposed design and the total length $(\ell_1 + h_1)$ of the larger inverted-F metal strip can be determined from one-quarter wavelength of the lower operating frequency. However, probably because of the coupling of the two shorted metal strips, the resonant length of the smaller inverted-F strip is increased, and the total length $\ell_2 + h_2$ was found to be about 0.35 wavelength of the upper operating frequency. It has also been found that, by adjusting the distance d between the shorting-pin position and the microstrip feed line, good impedance matching for both operating frequencies can be obtained.

A prototype based on the proposed dual-band design has been constructed and studied. For operating in the 2.4 GHz band, the length $\ell_1 + h_1$ was chosen to be about 31 mm, which corresponds to about 25% of the wavelength at 2.4 GHz. As for operating in the 5.2 GHz band, the length $\ell_2 + h_2$ was found to be 20 mm, which corresponds to about 35% of the wavelength at 5.2 GHz. Also, to achieve a lower height for the proposed antenna, the height or the length of the vertical portion of the inverted-F strips was chosen to be less than 10% of the wavelength of the operating frequencies; thus, in the proposed design, h_1 and h_2 were chosen to be 10 mm (8% of the wavelength at 2.4 GHz) and 5 mm (8.7% of the wavelength at 5.2 GHz), respectively, which makes ℓ_1 and ℓ_2 21 and 15 mm, respectively. Also, the optimal feed position d for achieving good impedance matching of the 2.4

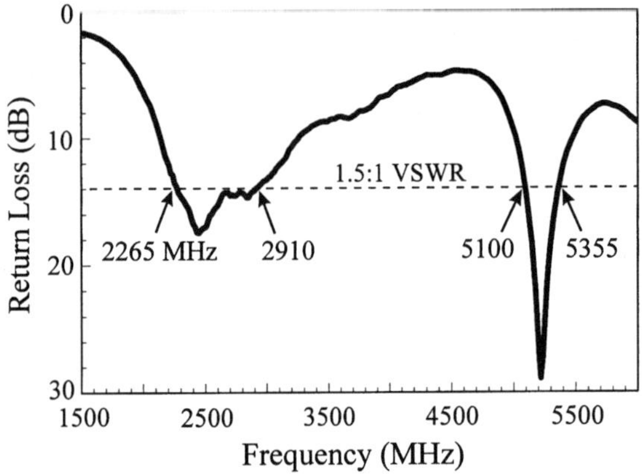

FIGURE 5.30 Measured input impedance for the antenna shown in Figure 5.29 with $\ell_1 = 21$ mm, $h_1 = 10$ mm, $\ell_2 = 15$ mm, $h_2 = 5$ mm, $w = w_f = 3.05$ mm, $d = 3$ mm. (From Ref. 10, © 2001 John Wiley & Sons, Inc.)

and 5.2 GHz bands was found to be 3 mm (about $0.15\ell_1$) for the constructed prototype.

Figure 5.30 shows the measured return loss of the constructed prototype. The obtained impedance bandwidth (1.5:1 VSWR) reaches 645 MHz (2265–2910 MHz) and 255 MHz (5100–5355 MHz) for the lower and upper operating bands, respectively. The two operating bands cover the required bandwidths of the WLAN operations in the 2.4 and 5.2 GHz bands. The radiation characteristics of the frequencies across the lower and upper bands have also been studied. Figures 5.31 and 5.32 plot the measured radiation patterns at 2.45 and 5.25 GHz, respectively. It is seen that, in the azimuthal plane (x–y plane), the radiation patterns for E_θ are close to omnidirectional, especially for the lower operating frequency at 2.45 GHz. The measured antenna gain for the frequencies within the 2.4 and 5.2 GHz bands is presented in Figure 5.33. The peak antenna gain for the 2.4 GHz band is about 0.8 dBi, and that for the 5.2 GHz band is about 0.4 dBi. For both bands, small gain variations have also been observed.

5.4.2 Integrated F-Shaped Monopole

Another promising dual-frequency design of the printed monopole antenna has also been devised [11], and the antenna geometry is shown in Figure 5.34. The printed monopole antenna has an F shape and is integrated on a microwave substrate of dimensions $45 \times 80\ \text{mm}^2$, which can be considered as a circuit board of a communication device. The microwave substrate used here has a thickness of 0.8 mm and a relative permittivity of 4.4. The dimensions given in the figure are for achieving the 2.4 and 5.2 GHz operations. The proposed monopole is printed on a small region (size $15 \times 15\ \text{mm}^2$ in this study) at one of the corners of the microwave substrate;

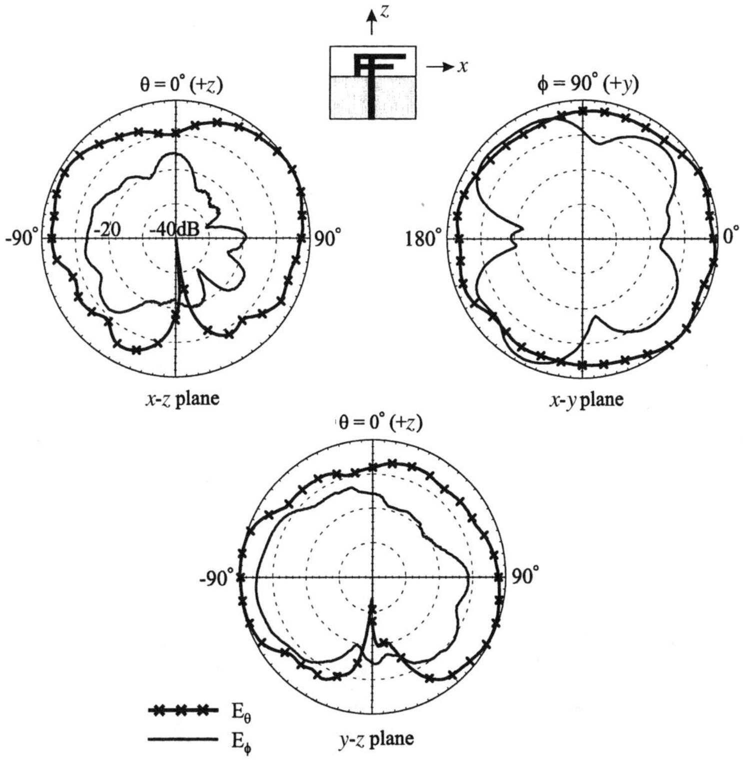

FIGURE 5.31 Measured radiation patterns at 2450 MHz for the antenna studied in Figure 5.30. (From Ref. 10, © 2001 John Wiley & Sons, Inc.)

note that there is a ground plane printed on the back of the substrate, except in the region (size $15 \times 15\,\text{mm}^2$) below the proposed monopole.

The proposed monopole consists of a vertical metal line and two (one upper and one lower) horizontal lines. To make the design simple, these three metal lines were chosen to be of the same width (2 mm in this study). Also note that, in the present design, the lower horizontal metal line is positioned with a distance of 8 mm from the upper horizontal metal line. However, other distances such as 6 or 7 mm are also promising and can be used in the design. That is, the distance between the upper and lower horizontal metal lines is not critical in the design considerations, as long as the lower horizontal metal line is not too close to the ground plane (should be at least 3 mm in this design) and not too close to the upper horizontal metal line to avoid the possible coupling effect to degrade antenna performance. The proposed monopole can also be easily fed by a 50-Ω microstrip line printed on the same substrate.

There are two resonant paths in this F-shaped monopole: The longer path, starting from the feed point to the open end of the upper horizontal metal line, generates the first or lower operating band; the shorter path, starting from the feed point to the

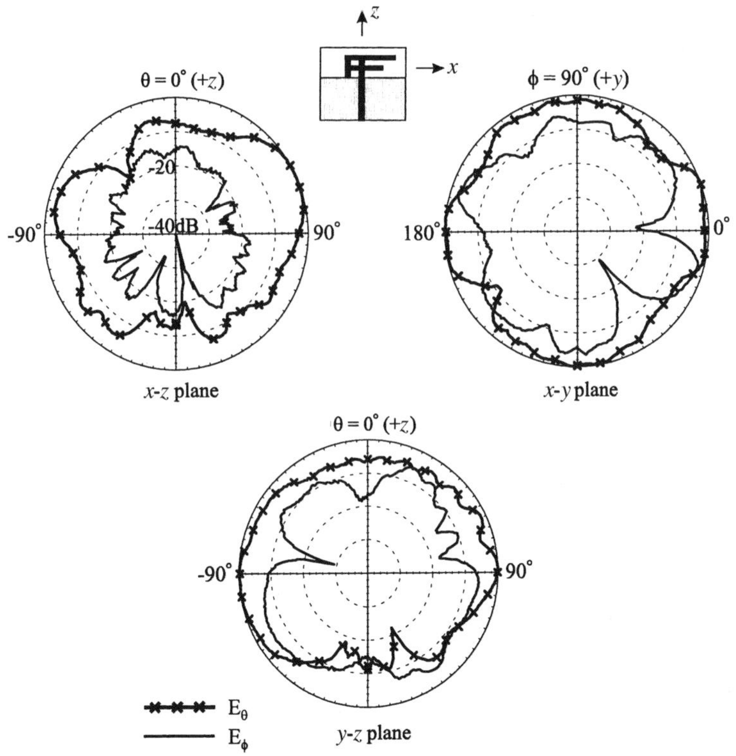

FIGURE 5.32 Measured radiation patterns at 5250 MHz for the antenna studied in Figure 5.30. (From Ref. 10, © 2001 John Wiley & Sons, Inc.)

open end of the lower horizontal metal line, controls the second or upper operating mode. In the proposed design, the longer resonant path has a length of about 23 mm, which corresponds to about 19% of the wavelength at 2.45 GHz. On the other hand, the shorter path has a length of about 12 mm, which is about 21% of the wavelength at 5.25 GHz. The lengths of both the longer and shorter resonant paths are smaller than one-quarter wavelength of a conventional straight monopole in free space. This behavior is largely due to the effect of the microwave substrate supporting the proposed monopole, which leads to a decreased resonant length for the proposed monopole. This effect is also helpful for achieving a smaller antenna size for a fixed operating frequency. It should also be noted that, by tuning the length of the upper and the lower horizontal metal lines, the antenna's longer and shorter resonant lengths can be effectively varied, thereby making it easy to achieve a desired dual-band operation.

Figure 5.35 shows the measured and simulated return loss of the constructed prototype. Two separate resonant modes at about 2.4 and 5.2 GHz are excited with good impedance matching. The lower band has an impedance bandwidth

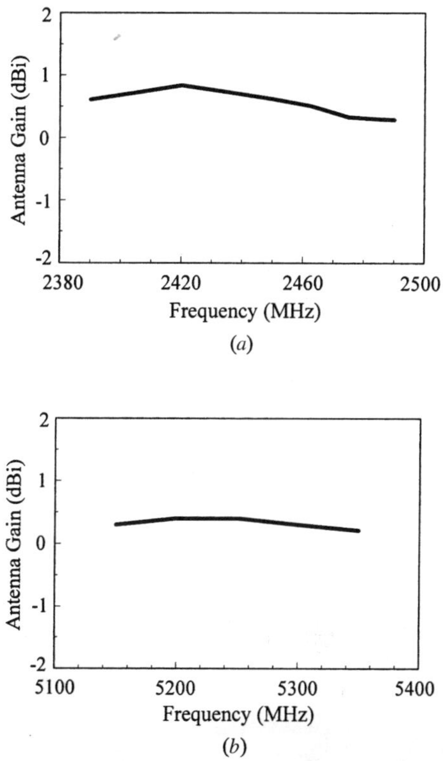

FIGURE 5.33 Measured antenna gain in (*a*) the 2.4 GHz band and (*b*) the 5.2 GHz band for the antenna studied in Figure 5.30. (From Ref. 10, © 2001 John Wiley & Sons, Inc.)

(2.5:1 VSWR) of 570 MHz, which covers the required bandwidth of the 2.4 GHz band for WLAN operation (2.4–2.484 GHz). The upper band shows an impedance bandwidth of 280 MHz, which meets the bandwidth requirement of the 5.2 GHz band for WLAN operation (5.15–5.35 GHz).

Radiation characteristics have also been studied. Figure 5.36 plots the measured radiation patterns at 2.45 GHz, and the results at 5.25 GHz are shown in Figure 5.37. Measured results at other frequencies in the 2.4 and 5.2 GHz bands also show radiation patterns similar to those shown in Figures 5.41 and 5.42, respectively. Relatively more smooth radiation patterns are observed at 2.45 GHz than at 5.25 GHz. This behavior occurs largely because the ground plane effects on radiation patterns are relatively much smaller for operating at lower frequencies. Figure 5.38 presents the measured antenna gain for frequencies across the WLAN bands at 2.4 and 5.2 GHz. For the 2.4 GHz band, the peak antenna gain is about 2 dBi, with gain variations less than about 0.6 dBi. For the 5.2 GHz band, the antenna gain is about 2.3–2.7 dBi.

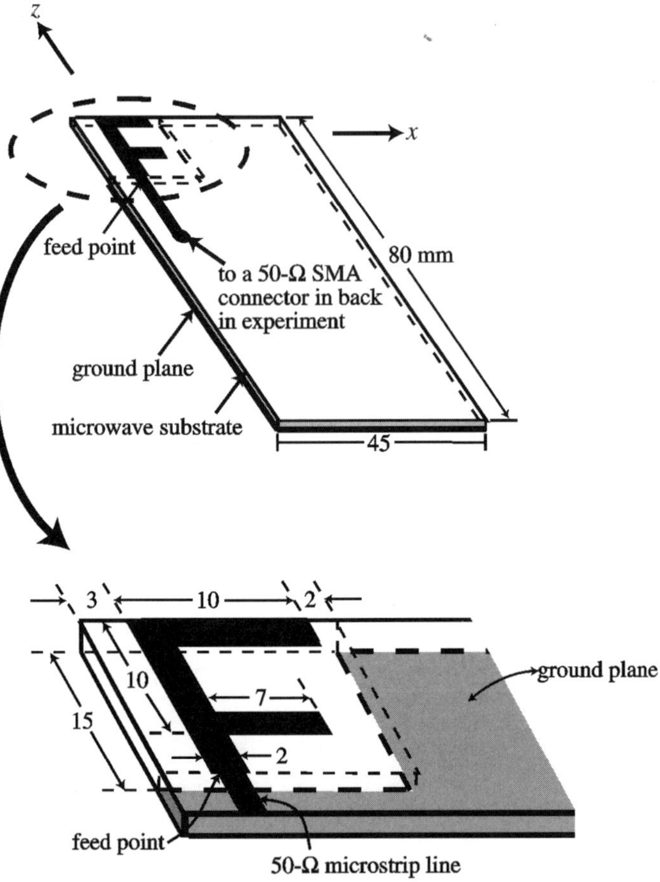

FIGURE 5.34 Geometry of the dual-band F-shaped monopole antenna integrated on a microwave substrate for 2.4/5.2 GHz operations. (From Ref. 11, © 2002 John Wiley & Sons, Inc.)

5.4.3 Integrated Spiral Monopole

In Sections 5.4.1 and 5.4.2, two different resonant paths are utilized to obtain the desired dual-frequency operation. In this section, we demonstrate another promising design using a rectangular spiral printed monopole to achieve the 2.4 and 5.2 GHz dual-band operation [12]. In this design, the monopole's first two resonant frequencies are tuned to be around the desired 2.4 and 5.2 GHz; that is, a single resonant path is used to obtain the dual-band operation, which gives the proposed design a more compact required realty space.

Figure 5.39 shows the proposed dual-band monopole printed and integrated on a grounded FR4 substrate (thickness 0.8 mm and relative permittivity 4.4) of dimensions $45 \times 80\,\text{mm}^2$. The proposed rectangular spiral monopole occupies an area of

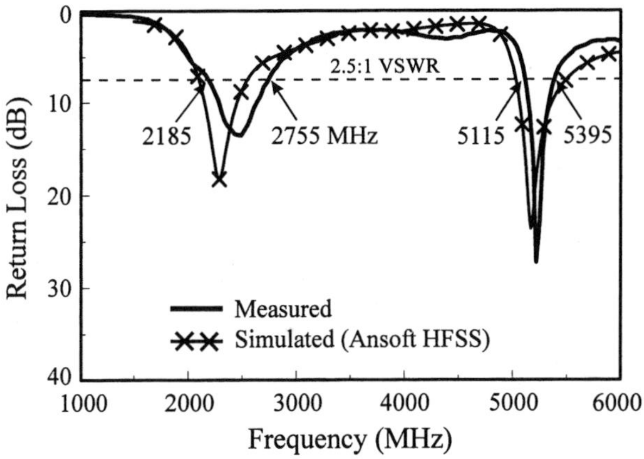

FIGURE 5.35 Measured and simulated return loss of the antenna shown in Figure 5.34. (From Ref. 11, © 2002 John Wiley & Sons, Inc.)

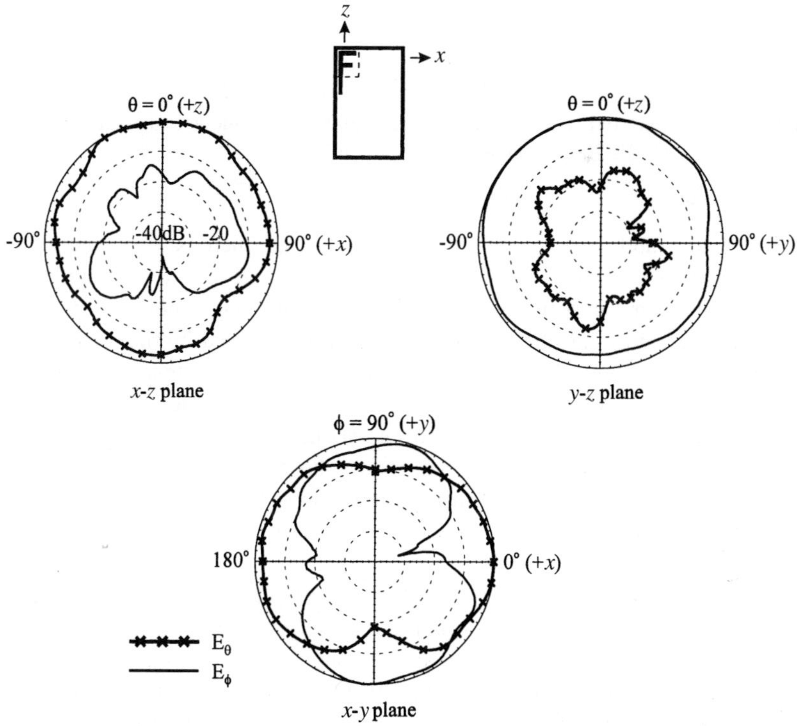

FIGURE 5.36 Measured radiation patterns at 2450 MHz for the antenna shown in Figure 5.34. (From Ref. 11, © 2002 John Wiley & Sons, Inc.)

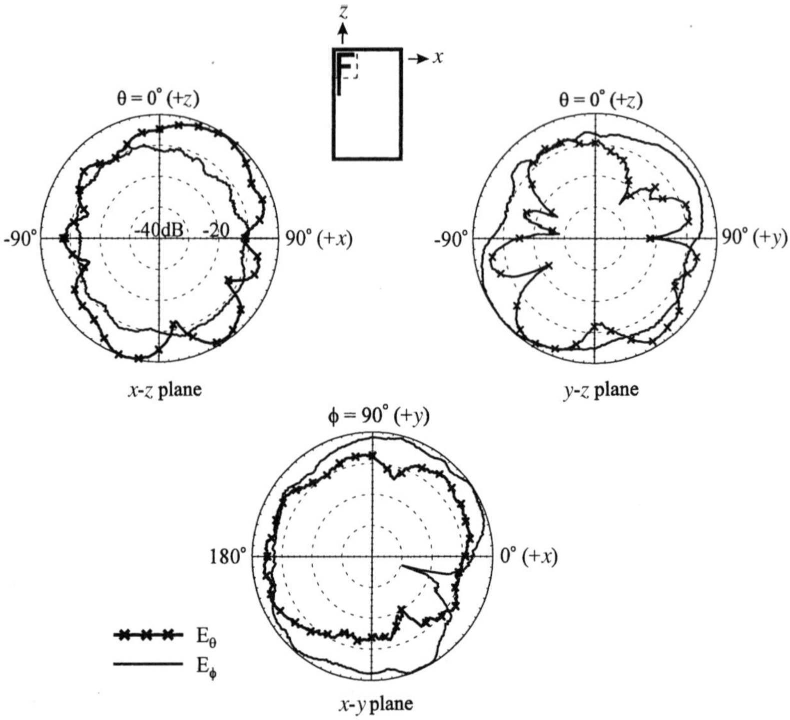

FIGURE 5.37 Measured radiation patterns at 5250 MHz for the antenna shown in Figure 5.34. (From Ref. 11, © 2002 John Wiley & Sons, Inc.)

only $5 \times 10\,\text{mm}^2$, and on the other side of the substrate, there is no ground plane in the region (size $8 \times 10\,\text{mm}^2$) below the printed monopole. In the rectangular spiral monopole, there are five linear sections with a right-angle connection between two adjacent sections, and the innermost section is encircled by the other sections. Their total length (about 30 mm in this design) can be determined approximately from one-quarter wavelength of the lower operating frequency (2.4 GHz in this design). Next, to fine-tune the monopole's second resonant frequency to around 5.2 GHz, the widths of these linear sections are chosen to increase from the section having the feed point to the innermost section having the open end; in this design, 0.5 mm is chosen for the widths of the first two linear sections and 1 mm is for the widths of the other linear sections.

Note that there is a small tuning stub at the feed point of the proposed rectangular spiral monopole. This tuning stub is found to be very helpful in obtaining improved impedance matching in both the lower and the upper resonant modes. The proposed monopole can also be easily fed by using a 50-Ω microstrip line printed on the same substrate, and no external impedance circuitry is required. Also note that, in the experiment, the microstrip feed line is connected, through a via-hole, to a 50-Ω SMA connector placed on the ground-plane side of the substrate, and a coax feed

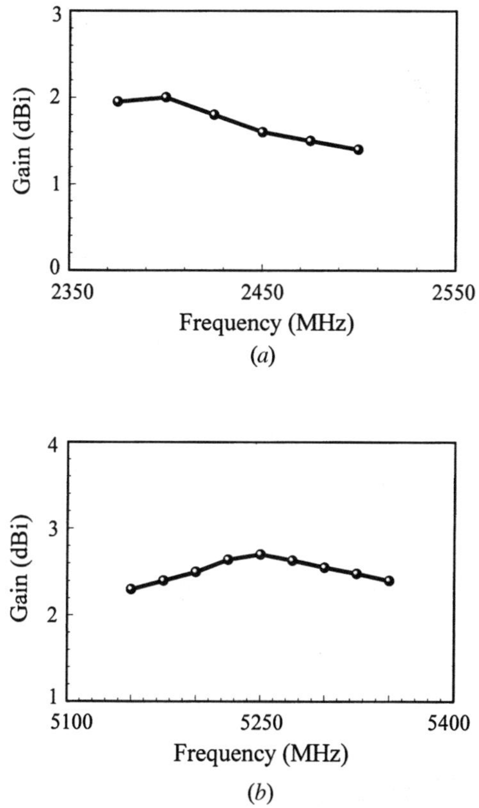

FIGURE 5.38 Measured antenna gain across (*a*) the 2.4 GHz band and (*b*) the 5.2 GHz band for the antenna shown in Figure 5.34. (From Ref. 11, © 2002 John Wiley & Sons, Inc.)

is used to excite the proposed monopole. In practical applications, the microstrip feed line can be directly connected to the transmitter/receiver of the system.

Figure 5.40 shows the measured and simulated return loss of the constructed prototype. Satisfactory agreement between the measurement and the Ansoft HFSS simulation is obtained. Two separate wide bandwidths at about 2.4 and 5.2 GHz are obtained. The lower band has an impedance bandwidth (1.5:1 VSWR) of 156 MHz (2354–2510 MHz), and covers the required bandwidth of the 2.4 GHz band. The upper band, on the other hand, has an even larger bandwidth of 540 MHz (4845–5385 MHz), also covering the required bandwidth of the 5.2 GHz band.

Figures 5.41 and 5.42 plot the measured radiation patterns at 2450 and 5250 MHz, respectively. Note that, because there are ground planes in two sides of the proposed monopole, the obtained patterns show some asymmetry. The measured antenna gain for frequencies across the 2.4 and 5.2 GHz bands is shown in Figure 5.43. In the 2.4 GHz band, stable antenna gain is seen and the antenna

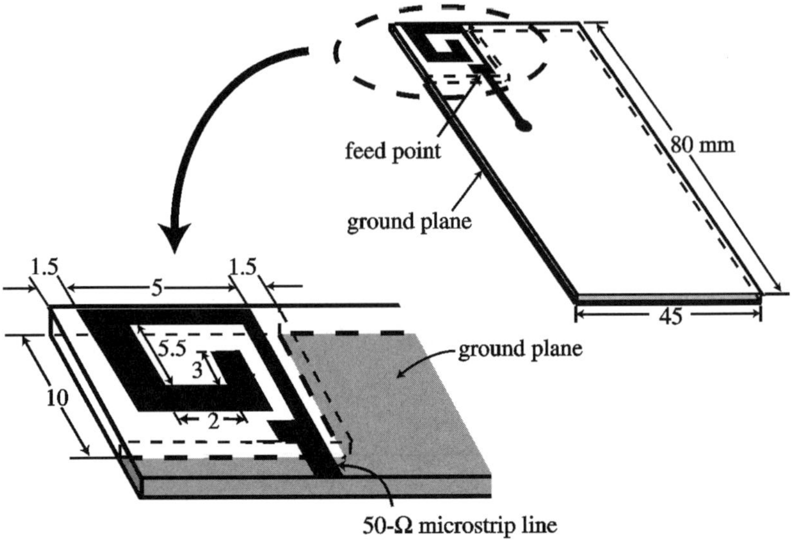

FIGURE 5.39 Geometry of the dual-band rectangular spiral monopole antenna for 2.4/5.2 GHz operations. (From Ref. 12, © 2002 IEEE, reprinted with permission)

gain level is about 1.5–1.7 dBi. For the 5.2 GHz band, the antenna gain is about 1.2–1.7 dBi.

5.4.4 Printed Spiral Monopole

The dual-band rectangular spiral monopole antenna studied in Section 5.4.3 can also be applied to a notebook computer for WLAN operations in the 2.4/5.2 GHz bands.

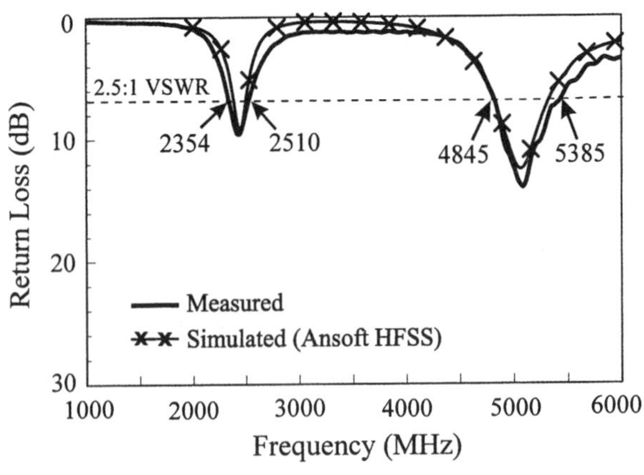

FIGURE 5.40 Measured and simulated return loss for the antenna shown in Figure 5.39. (From Ref. 12, © 2002 IEEE, reprinted with permission)

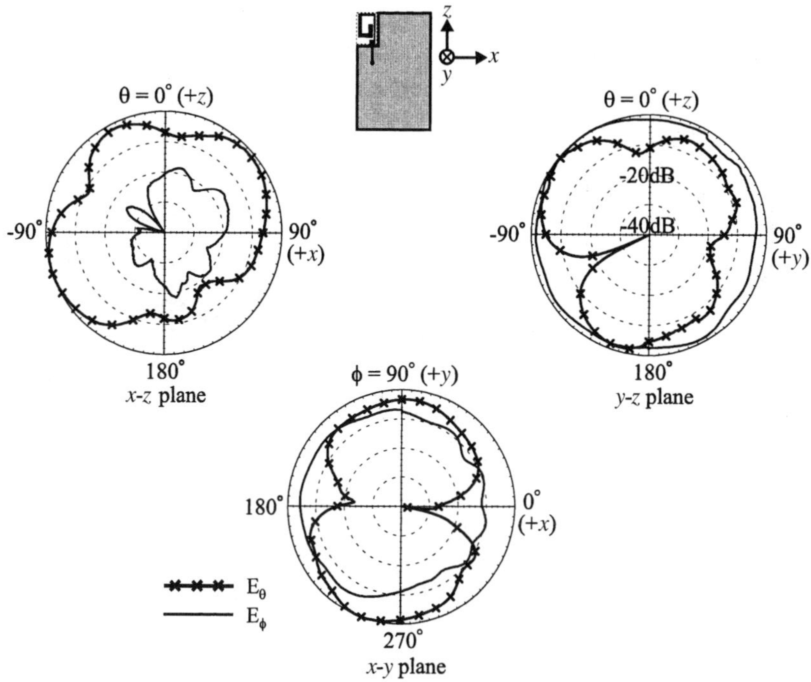

FIGURE 5.41 Measured radiation patterns at 2450 MHz for the antenna shown in Figure 5.39. (From Ref. 12, © 2002 IEEE, reprinted with permission)

For this application, the monopole antenna should be narrow in width so that it can fit into the corners [15] or the hinge [16] of the notebook computer as a concealed antenna. Figure 5.44 shows the geometry of a promising design. The monopole antenna is printed on a rectangular FR4 substrate of dimensions $8.5 \times 19\,\text{mm}^2$ and is connected to a 50-Ω microstrip line printed on the same substrate. The microstrip line is further connected to a 50-Ω coaxial line, whose one end is connected to the receiver/transmitter of the system. Also note that the rectangular spiral monopole occupies a compact area of $5 \times 11\,\text{mm}^2$, and a ground plane of dimensions $8.5 \times 8\,\text{mm}^2$ is printed on the back of the FR4 substrate (see the figure). Figure 5.45 shows the measured return loss of the constructed prototype. The obtained bandwidths (2.5:1 VSWR) are 328 MHz (2200–2528 MHz) in the 2.4 GHz band (lower band) and 915 MHz (4635–5550 MHz) in the 5.2 GHz band (upper band). These two bandwidths satisfy the required WLAN operations in the 2.4 and 5.2 GHz bands. The present monopole antenna has been applied in a notebook computer for WLAN operations, and good system performance has been obtained.

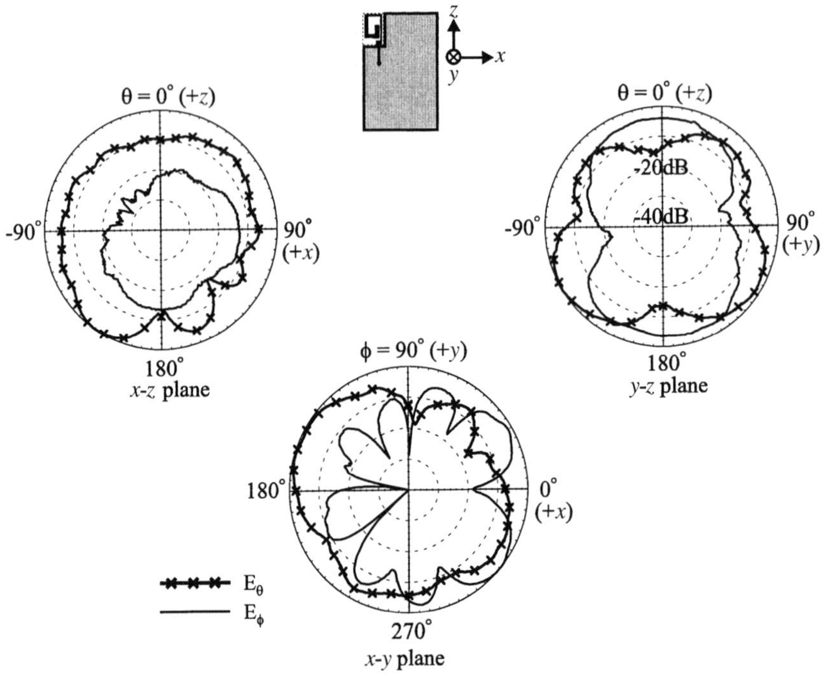

FIGURE 5.42 Measured radiation patterns at 5250 MHz for the antenna shown in Figure 5.39. (From Ref. 12, © 2002 IEEE, reprinted with permission)

5.4.5 Printed Double-T Monopole

By using a double-T monopole (two stacked T-shaped monopoles), good dual-band operations in the 2.4 and 5.2 GHz bands can also be obtained. Figure 5.46 shows the geometry of a double-T monopole printed on a dielectric substrate. Both the upper and lower T-shaped monopoles are operated as quarter-wavelength structures. By varying the size of the upper T-shaped monopole, the resonant mode for the lower operating band can be controlled. On the other hand, one can control the resonant mode for the upper operating band by varying the size of the lower T-shaped monopole.

In comparison to the stacked inverted-F monopole antenna studied in Section 5.4.1, the double-T monopole antenna has a simpler structure, because no shorting pins are required to short-circuit the two horizontal metal lines to the ground plane printed on the back of the dielectric substrate. The double-T monopole antenna can also be excited by a 50-Ω microstrip line printed on the same substrate, and no external circuitry is required to achieve good impedance matching for operating frequencies across the 2.4 and 5.2 GHz bands. In addition, it has been observed that, by increasing the strip width of the double-T monopole, impedance bandwidths of the antenna's two operating bands can be greatly enhanced, and two separate wide operating bands covering the 2.4 and 5.2 GHz bands can be obtained.

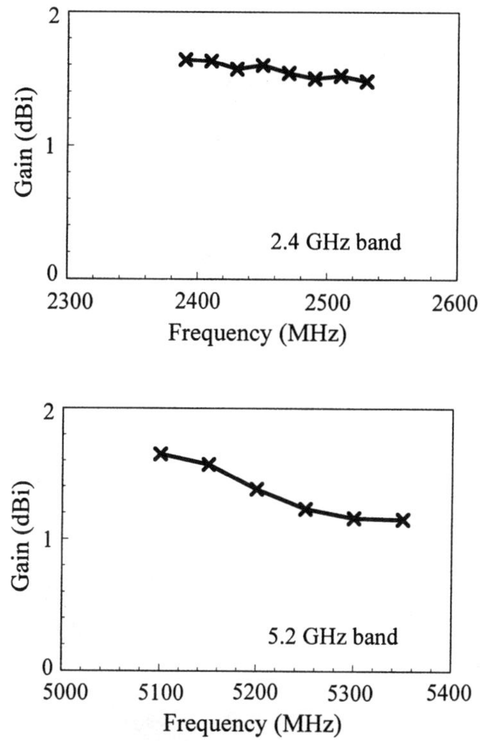

FIGURE 5.43 Measured antenna gain across the 2.4 and 5.2 GHz bands for the antenna shown in Figure 5.39. (From Ref. 12, © 2002 IEEE, reprinted with permission).

5.5 PRINTED DIVERSITY MONOPOLE ANTENNA

The printed diversity monopole antenna for operating in the 2.4 GHz band has been implemented [13], and the antenna geometry is shown in Figure 5.47. In this design, two identical planar monopoles of length ℓ are printed on an FR4 substrate of thickness 0.8 mm, relative permittivity 4.4, and size $100 \times 80\,\mathrm{mm}^2$, which can be considered to be the circuit board of the system. Note that, on the other side of the substrate, there is a ground plane consisting of a bottom ground portion (length L and width W) and a top protruded rectangular ground portion (length a and width b). On the bottom ground portion, the two upper corners (size $d \times d$) are removed such that the two monopoles slanted at $\pm 45°$ to the centerline of the ground plane are perpendicular to the ground-plane edge, similar to the configuration of the conventional printed monopole antenna. To specify the monopole position, the distance from the monopole to the top edge ($\overline{AB}$) of the bottom ground portion is denoted as c. Also notice that, with the proposed arrangement of two monopoles spaced 90° apart, good polarization diversity with high port decoupling can be obtained. In addition, because of the presence of the protruded ground plane, which makes the obtained radiation patterns of the two ortho-

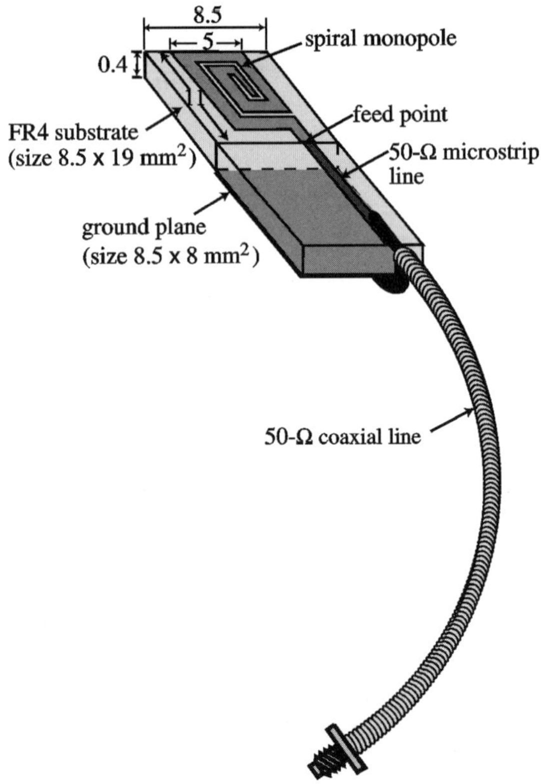

FIGURE 5.44 Geometry of the printed spiral monopole antenna fed by a coaxial cable.

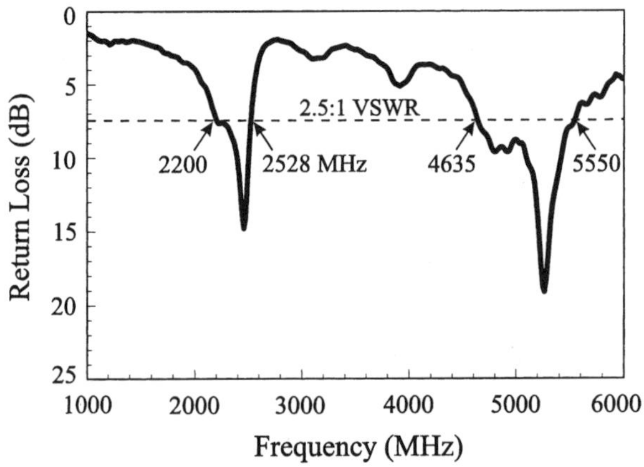

FIGURE 5.45 Measured return loss of the antenna shown in Figure 5.44.

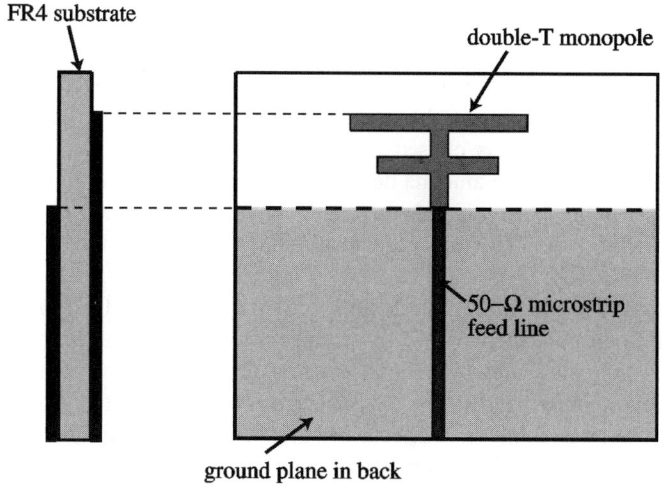

FIGURE 5.46 Geometry of the printed double-T monopole antenna for the 2.4 and 5.2 GHz bands.

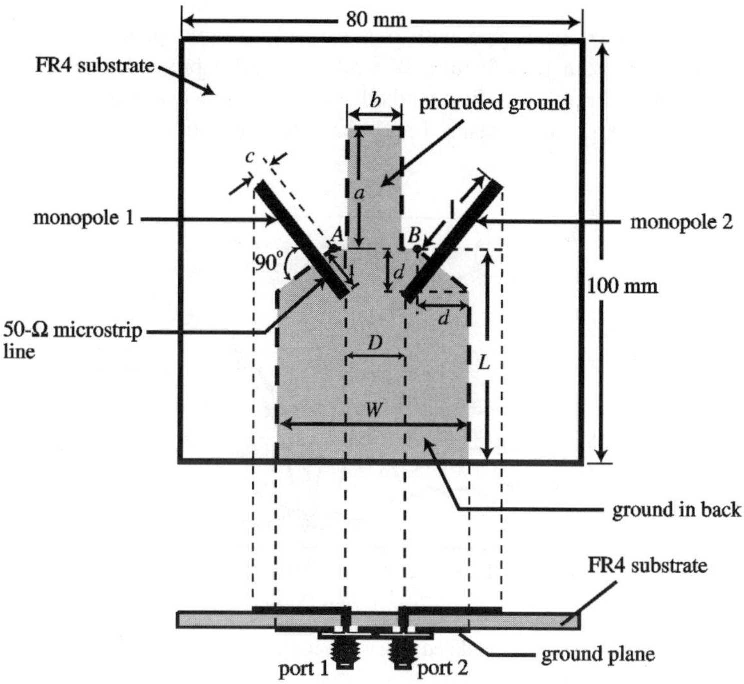

FIGURE 5.47 Geometry of the printed diversity monopole antenna for the 2.4 GHz band. (From Ref. 13, © 2002 IEEE, reprinted with permission)

gonal monopoles tend to be complementary to each other, the antenna studied here can also provide spatial diversity.

The two monopoles are also oriented symmetrically with respect to the centerline of the ground plane and are fed by using a 50-Ω microstrip line, respectively. The width of the monopoles is chosen to be the same as that of the microstrip feed lines, which simplifies the antenna design. Through a via-hole in the FR4 substrate, each microstrip feed line, having a length of ℓ_1, is connected to a 50-Ω SMA connector (port 1 or 2 in the present study) on the other side of the dielectric substrate.

It is noted that, by varying the length ℓ_1 of the microstrip feed line (that is, the distance D between the two SMA connectors is varied) with other antenna parameters fixed, the measured S parameters for ports 1 and 2 are almost not affected. On the other hand, when the dimensions (a and b) of the protruded ground portion are varied, significant effects on the measured S parameters have been observed. There exist optimal dimensions of a and b for measured reflection coefficient S_{11} and isolation S_{21}. For the position (c) of the monopole to the top edge of the bottom ground portion, large effects on the measured S parameters have also been seen and should be considered in the practical design.

For WLAN operations in the 2.4 GHz band, the two printed monopoles were both designed to operate as quarter-wavelength structures. For this purpose and to consider the substrate effect on reducing the antenna's resonant length, the monopole length was chosen to be 21 mm (about 17% of the wavelength at 2.4 GHz) in this study. A constructed prototype with a reasonable ground plane consisting of a bottom ground portion ($L = 56$ mm, $W = 50$ mm) and a protruded ground portion ($a = 32$ mm, $b = 17$ mm) was first studied. Other antenna parameters are given in Figure 5.48, in which the measured and simulated reflection coefficient (S_{11}) and

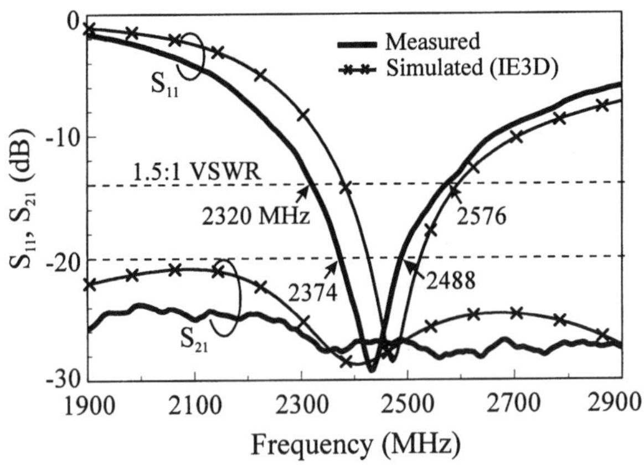

FIGURE 5.48 Measured and simulated reflection coefficient (S_{11}) and isolation (S_{21}) for the antenna shown in Figure 5.47 with $\ell = 21$ mm, $L = 56$ mm, $W = 50$ mm, $a = 32$ mm, $b = 17$ mm, $c = 5$ mm, $\ell_1 = 8$ mm, $d = 16$ mm, $D = 13$ mm. (From Ref. 13, © 2002 IEEE, reprinted with permission)

isolation (S_{21}) for the constructed prototype are shown. [Note that, because of the symmetric structures of monopoles 1 and 2, the measured S_{22} is almost the same as S_{11} and is thus not shown in the figure for brevity.] The simulated results are obtained from IE3D simulation software, and the results in general agree with the measured data. The observed minor differences between the simulated and measured results are probably because the substrate size used in the IE3D simulation is of infinite dimensions. From the measured results, the obtained impedance bandwidth (1.5:1 VSWR) reaches 256 MHz (2320–2576 MHz), which covers the WLAN band at 2.4 GHz. In addition, for operating frequencies within the WLAN band, the impedance matching is always better than −20 dB. As for the measured isolation, the operating frequencies across the WLAN band are all less than −27 dB.

Figure 5.49 shows the effect of various lengths of the protruded ground portion on the measured S_{11} and S_{21} for the proposed antenna with other parameters fixed. It is first seen that, for the case of $a = 0$ (without the protruded ground portion), the impedance matching is degraded and the isolation between ports 1 and 2 is no better than −12 dB within the desired WLAN band. On the other hand, with the presence of the protruded ground portion between the two monopoles, the impedance matching improves and the port isolation is also enhanced. This behavior is probably because the protruded ground portion reduces the mutual coupling between the two monopoles, which enhances the isolation between ports 1 and 2 and is also helpful in achieving good impedance matching for the proposed antenna. However, it is noted that there exists an optimal value for achieving optimal port isolation. In this study, it is found that, when $a = 32$ mm (about 1.5 times the monopole length; i.e., $a \cong 1.5\ell$), which is the case studied in Figure 5.48, optimal port isolation (S_{21} less than −27 dB) for the operating frequencies across the WLAN band is obtained. Also notice that, by comparing to the case with $a = 0$, the improvement in S_{21} across

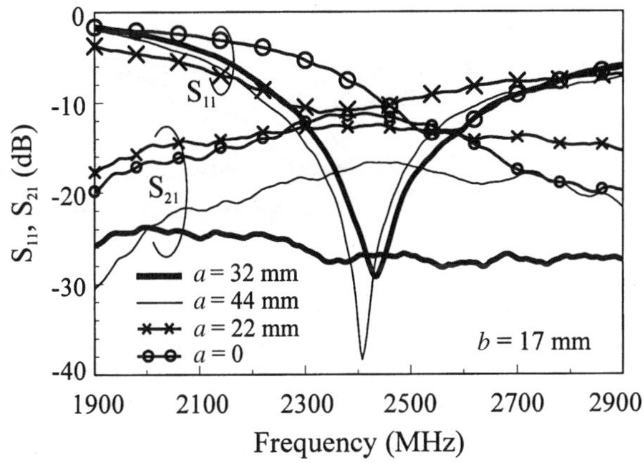

FIGURE 5.49 Measured S_{11} and S_{21} versus the protruded ground length for the antenna shown in Figure 5.47. Other antenna dimensions are given in Figure 5.48. (From Ref. 13, © 2002 IEEE, reprinted with permission)

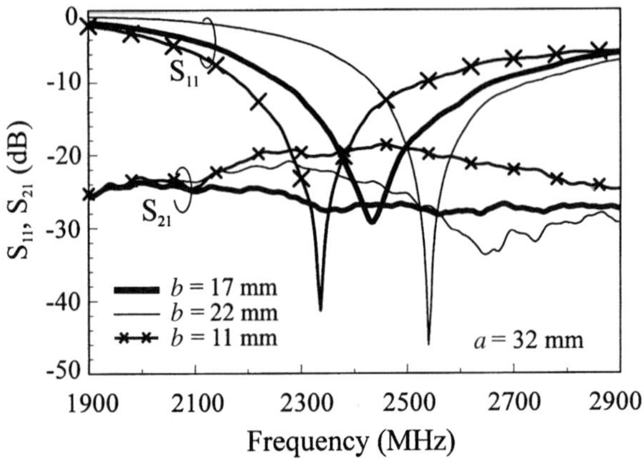

FIGURE 5.50 Measured S_{11} and S_{21} versus the protruded ground width for the antenna shown in Figure 5.47. Other antenna dimensions are given in Figure 5.48. (From Ref. 13, © 2002 IEEE, reprinted with permission)

the WLAN band for the case with $a = 32$ mm is as large as about 15 dB (−27 vs. −12 dB).

The effect of various protruded ground widths on the measured S_{11} and S_{21} is presented in Figure 5.50. In this case, the protruded ground length is fixed at 32 mm. Large variations of the measured S parameters due to various values of b are also seen. There also exists an optimal value of b for optimal port isolation. In this study, when $b = 17$ mm (about 0.8 times the monopole length; i.e., $b \cong 0.8\ell$), also the case studied in Figure 5.48, optimal port isolation is obtained.

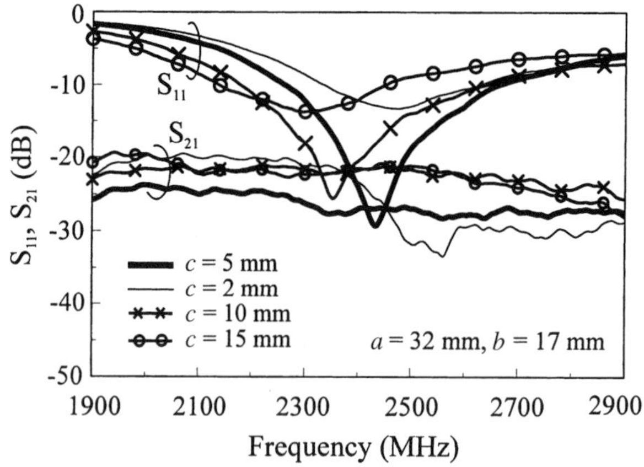

FIGURE 5.51 Measured S_{11} and S_{21} versus the monopole position for the antenna shown in Figure 5.47. Other antenna dimensions are given in Figure 5.48. (From Ref. 13, © 2002 IEEE, reprinted with permission)

Figure 5.51 shows the measured S_{11} and S_{21} for the cases with various monopole positions. In this case, the dimensions of the protruded ground portion are fixed at $a = 32$ mm and $b = 17$ mm, the optimal cases studied in Figures 5.49 and 5.50. It is seen that, for the case with $c = 5$ mm (about 0.25 times the monopole length; i.e., $c \cong 0.25\ell$), also the case studied in Figure 2, optimal port isolation with good impedance matching for the operating frequencies across the WLAN band is obtained. When c is smaller, the two monopoles are closer to the protruded ground portion and good impedance matching becomes difficult to achieve, although the port isolation is better for higher operating frequencies starting from about 2.7 GHz. On the other hand, when c is larger, the port isolation becomes worse. The difference in S_{21} between the cases of $c = 5$ and 10 mm is about 7 dB.

Figures 5.52 and 5.53 show the measured radiation patterns at 2450 MHz for port 1 and port 2 excitation, respectively, of the antenna with optimal dimensions in this

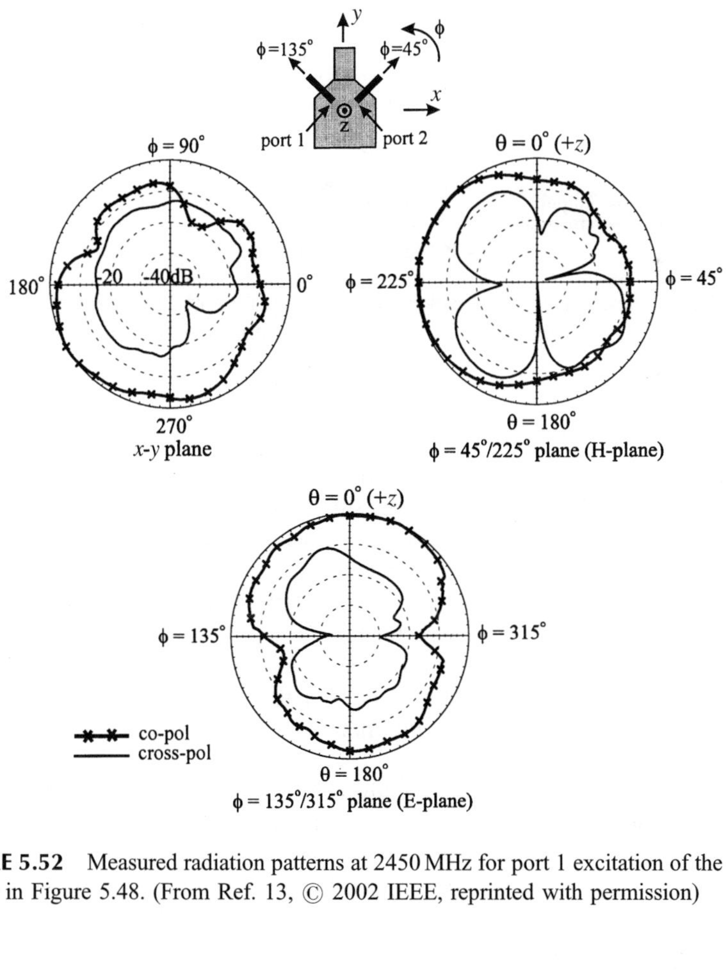

FIGURE 5.52 Measured radiation patterns at 2450 MHz for port 1 excitation of the antenna studied in Figure 5.48. (From Ref. 13, © 2002 IEEE, reprinted with permission)

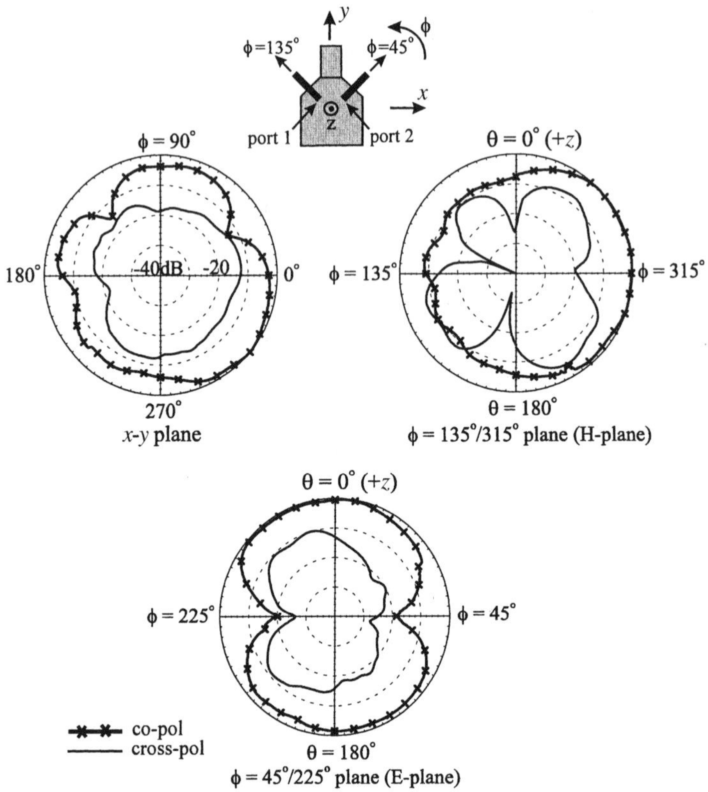

FIGURE 5.53 Measured radiation patterns at 2450 MHz for port 2 excitation of the antenna studied in Figure 5.48. (From Ref. 13, © 2002 IEEE, reprinted with permission)

study (the case shown in Fig. 5.48). For other operating frequencies within the WLAN band, measured radiation patterns similar to those shown here are observed. Also note that, in the measurements, the antenna was placed vertically in the anechoic chamber, and the probe (a waveguide antenna) is aligned to be in parallel to monopole 1 or 2 under test. That is, the copolarized component shown in Figure 5.52 for port 1 excitation is in the direction of $\varphi = 135°$; on the other hand, for port 2 excitation shown in Figure 5.53, the copolarized component is in the direction of $\varphi = 45°$. From the obtained results, it is seen that the copolarized radiation for monopole 1 is orthogonal to that for monopole 2, which provides polarization diversity for the proposed antenna.

In addition, it is observed that, mainly because of the presence of the ground plane (the bottom ground portion and the protruded ground portion), the radiation patterns in the *x–y* plane and the H-plane ($\varphi = 45°/225°$ plane for port 1 and $\varphi = 135°/315°$ for port 2) are asymmetric. However, the two monopoles together

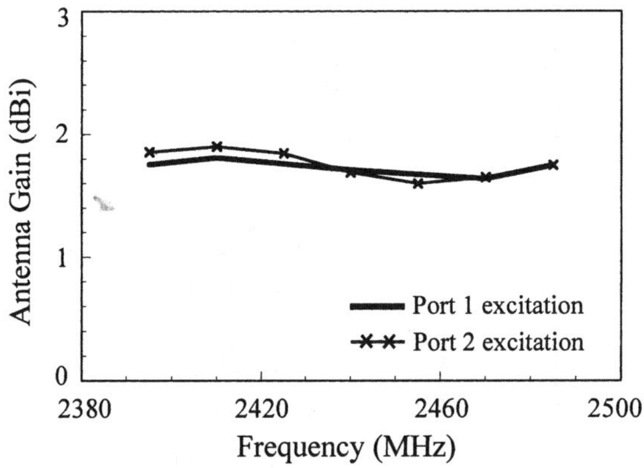

FIGURE 5.54 Measured antenna gain for port 1 and port 2 excitation of the antenna studied in Figure 5.48. (From Ref. 13, © 2002 IEEE, reprinted with permission)

make the proposed antenna with wide radiation coverage. Also note that, in the E-plane patterns ($\varphi = 135°/315°$ plane for port 1 and $\varphi = 45°/225°$ for port 2), slight asymmetry with respect to the grounded dielectric substrate is seen. This behavior is largely due to the effect of the grounded dielectric substrate, which makes the radiation on the substrate side ($\theta = 0°$) slightly larger than that on the ground plane side ($\theta = 180°$). The measured antenna gain for port 1 and port 2 excitation is shown in Figure 5.54. Because of the symmetric structures of monopoles 1 and 2, the measured antenna gain for frequencies within the WLAN band is about the same and shows a gain level of about 1.7 dBi, with very small gain variations (within 0.3 dBi).

Dual-band operations of this kind of diversity monopole antenna are also possible. Figure 5.55 shows a promising design of the diversity dual-band monopole antenna for WLAN operations in the 2.4 and 5.2 GHz bands. In this design, instead of the straight monopoles used in Figure 5.47, two F-shaped monopoles as discussed in Section 5.4.2 are used. With the design dimensions shown in Figure 5.55, a prototype has been constructed and studied. Figure 5.56 shows the measured reflection coefficient (S_{11}) and isolation (S_{21}) of the constructed prototype. It is seen that two resonant modes at about 2.4 and 5.2 GHz are excited with good impedance matching (reflection coefficients all less than -10 dB). For operating frequencies across the 2.4 and 5.2 GHz bands, very good isolation ($S_{21} < -28$ dB) is also obtained.

Figures 5.57 and 5.58 plot the measured radiation patterns at 2450 MHz for port 1 and port 2 excitation, respectively. The corresponding results at 5250 MHz for port 1 and port 2 excitation are shown in Figures 5.59 and 5.60, respectively. Results similar to those shown in Figures 5.52 and 5.53 are obtained. The measured antenna gain is presented in Figure 5.61. Figure 5.61*a* shows the measured antenna gain in the 2.4 GHz band, and the antenna gain levels for ports 1 and 2 are about 1.6–1.9 dBi. For the measured antenna gain in the 5.2 GHz band (see Fig. 5.61*b*), the antenna gain levels are about 1.2–1.9 dBi for ports 1 and 2.

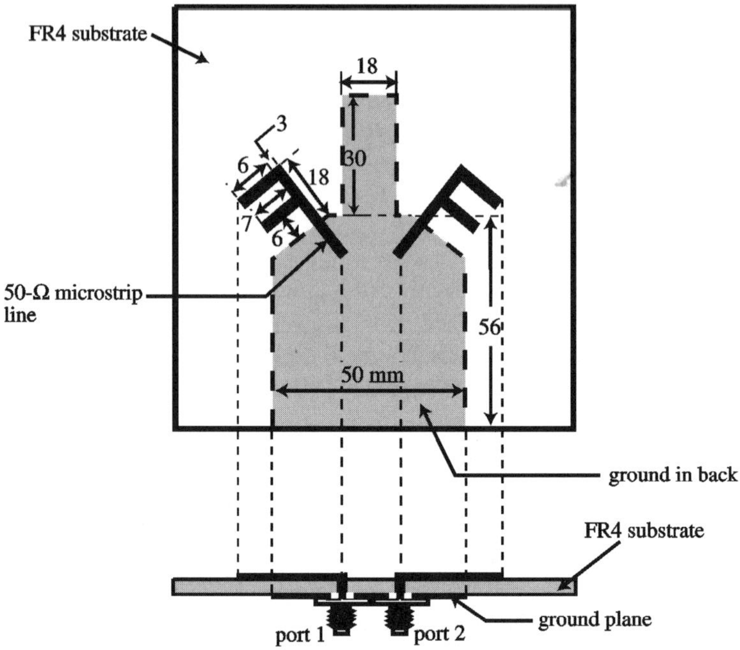

FIGURE 5.55 Geometry of a printed diversity dual-band monopole antenna for the 2.4 and 5.2 GHz bands.

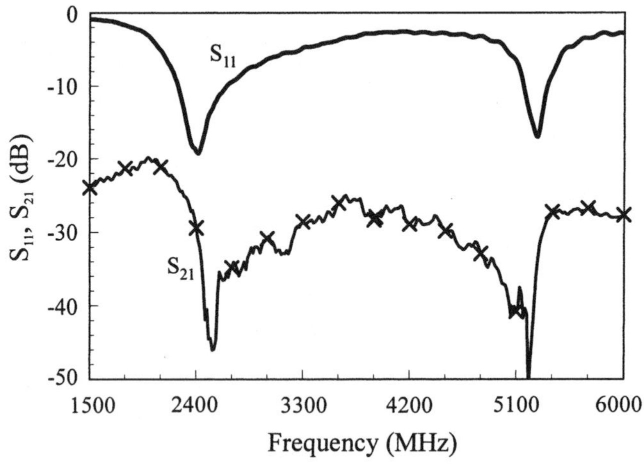

FIGURE 5.56 Measured reflection coefficient (S_{11}) and isolation (S_{21}) for the antenna shown in Figure 5.55.

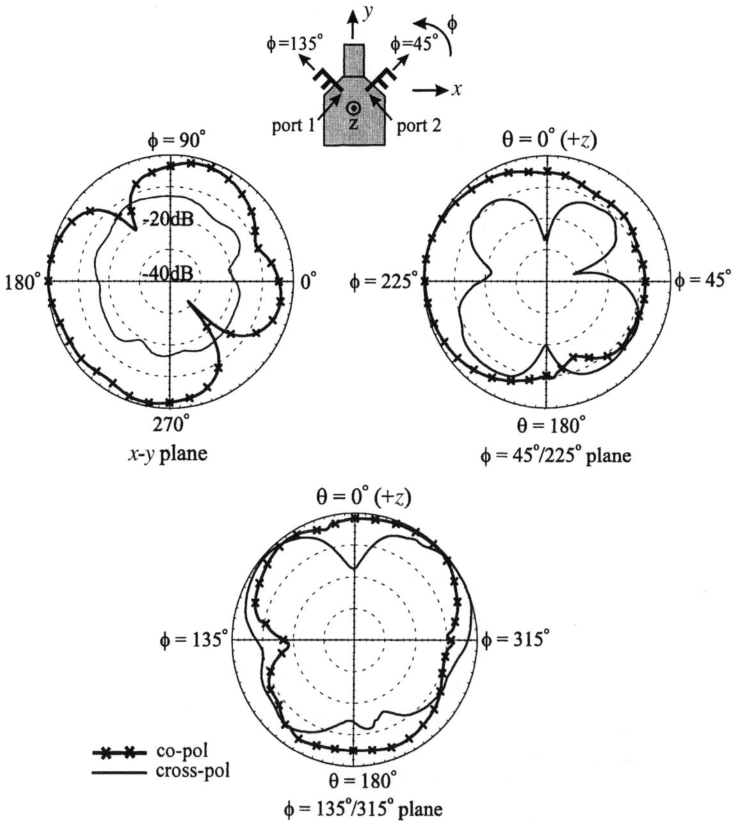

FIGURE 5.57 Measured radiation patterns at 2450 MHz for port 1 excitation of the antenna shown in Figure 5.55.

5.6 PRINTED DIPOLE ANTENNA FOR DUAL-BAND OPERATION

The printed dipole antenna is also very suitable for applications in a notebook computer or a PDA for WLAN operations [15, 16]. Figures 5.62 and 5.63 show two promising dual-band dipole antenna designs, and their possible applications in a notebook computer are shown in Figure 5.64. In these two designs, two dipole arms are printed on a dielectric substrate of a narrow width (usually about 5 mm or less), which makes them very easy to install within the housing of the notebook computer or the PDA as a concealed antenna for WLAN operations. In Figure 5.62, the two long dipole arms are for generating the lower operating mode at about 2.4 GHz, and the four short dipole arms are for obtaining the upper operating mode at about 5.2 GHz. No external matching circuitry is required for obtaining good impedance matching for the two desired operating bands.

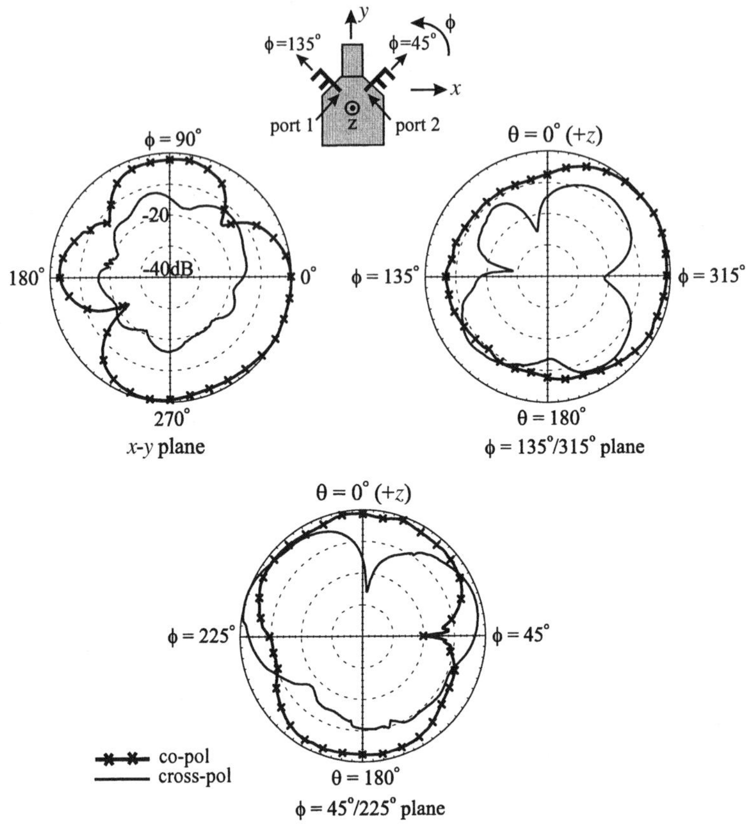

FIGURE 5.58 Measured radiation patterns at 2450 MHz for port 2 excitation of the antenna shown in Figure 5.55.

In Figure 5.63, two open rectangular ring strips are used as the two arms for the dual-band dipole antenna. This dual-band dipole antenna can have a narrower width than that shown in Figure 5.62. It has been found that, because of the presence of the opening in the rectangular ring strip, two different resonant paths are formed. The longer path starts from the feed point and experiences two right-angle turns to the opening, and the shorter path has a straight path beginning from the feed point to the opening. The longer and shorter paths determine the lower (2.4 GHz) and upper (5.2 GHz) resonant modes, respectively. In this design, external matching circuitry is also not required. From the experimental study, very good antenna performance for operating frequencies across the 2.4 and 5.2 GHz bands has been observed for the proposed antenna.

The printed dipole antenna can also be printed and integrated on the system circuit board for WLAN operations. Figure 5.65 shows the geometry of a promising on-board dual-band printed dipole antenna. In this design, a U-shaped slot is embedded in each dipole arm to achieve dual-band operation. Because of the

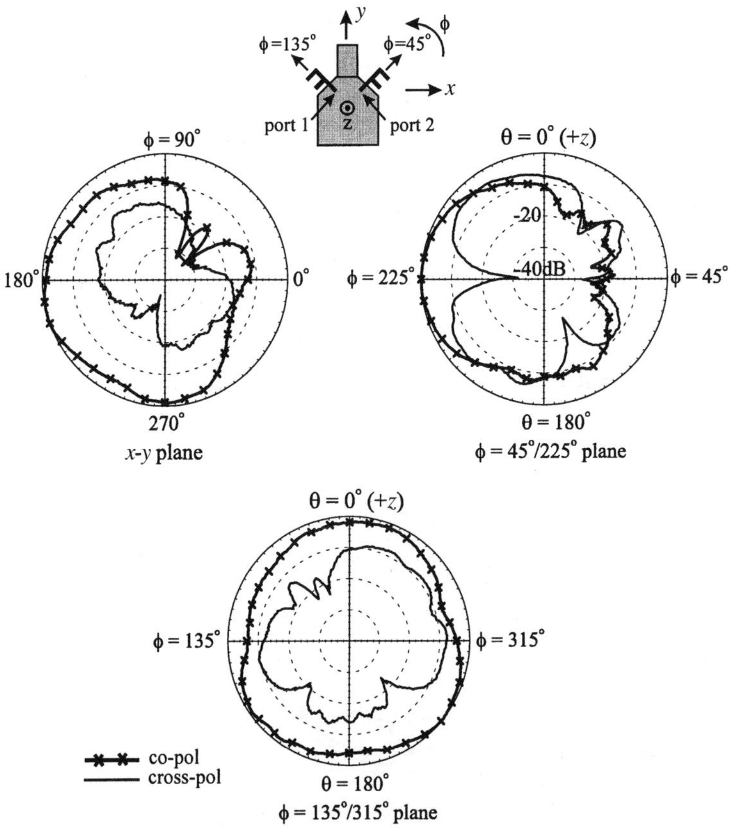

FIGURE 5.59 Measured radiation patterns at 5250 MHz for port 1 excitation of the antenna shown in Figure 5.55.

embedded U-shaped slots, two additional dipole arms (the patch regions encircled by the slots) are obtained, which form a second dipole antenna with a smaller size and can generate a higher resonant mode for the 5.2 GHz band operation. By incorporating the original dipole arms designed for the 2.4 GHz band operation, this on-board dipole antenna can perform WLAN operations in the 2.4 and 5.2 GHz bands.

5.7 PIFA FOR DUAL WLAN OR ISM BANDS

5.7.1 PIFA with Two Shorting Strips

Many 900/1800 MHz dual-frequency PIFAs discussed in Chapter 2 are also very suitable for applications in WLAN operations in the 2.4/5.2 GHz bands. In this section, another promising design [18] is presented. The antenna geometry is shown in Figure 5.66. The top patch occupies an area of $18 \times 27\ \text{mm}^2$ and is printed

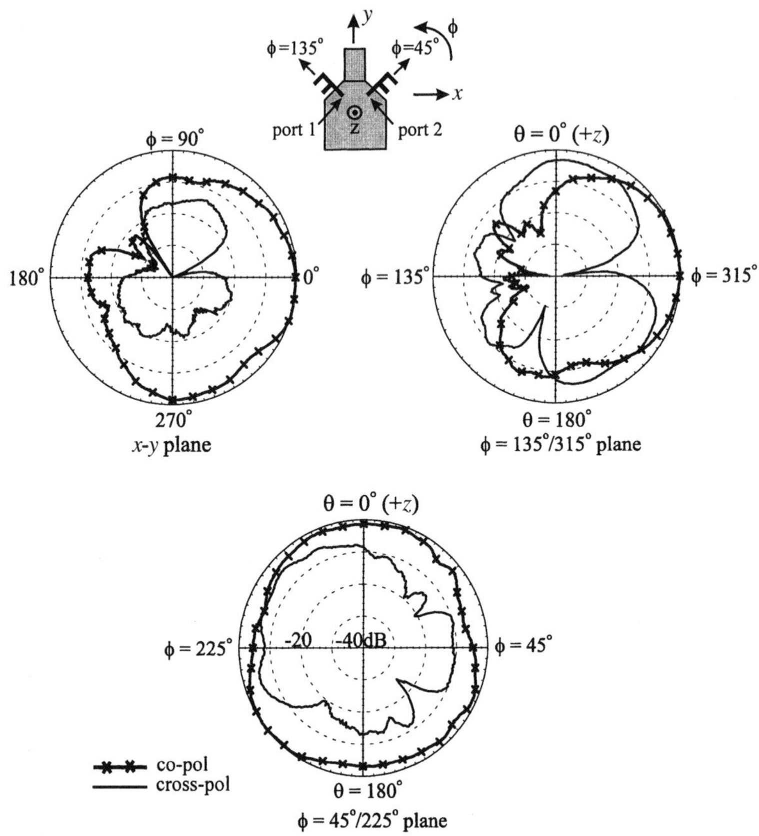

FIGURE 5.60 Measured radiation patterns at 5250 MHz for port 2 excitation of the antenna shown in Figure 5.55.

on a thin FR4 substrate of thickness 0.8 mm and relative permittivity 4.4. This FR4 substrate serves as a support for the antenna's top patch. Between the FR4 substrate and the ground plane is an air-layer substrate of thickness 3.2 mm. The two operating frequencies, 2.4 and 5.2 GHz, for the proposed design are the antenna's first two resonant frequencies, which are mainly controlled by the shorting pin position in the patch's centerline and the four inserted slits at the patch's nonradiating edges. Their relative positions are shown in the figure. Also note that the length of the patch's radiating edge close to the inserted slits is adjusted to be slightly different from that of the opposite radiating edge, which is also used for fine-tuning the antenna's first two resonant frequencies.

In this design, the feed position is required to be placed close to the shorting pin. In this case, although excitation of the antenna's first two resonant frequencies at 2.4 and 5.2 GHz can be obtained, it is still not easy to achieve good impedance matching for both of the two frequencies (when the impedance matching for one frequency

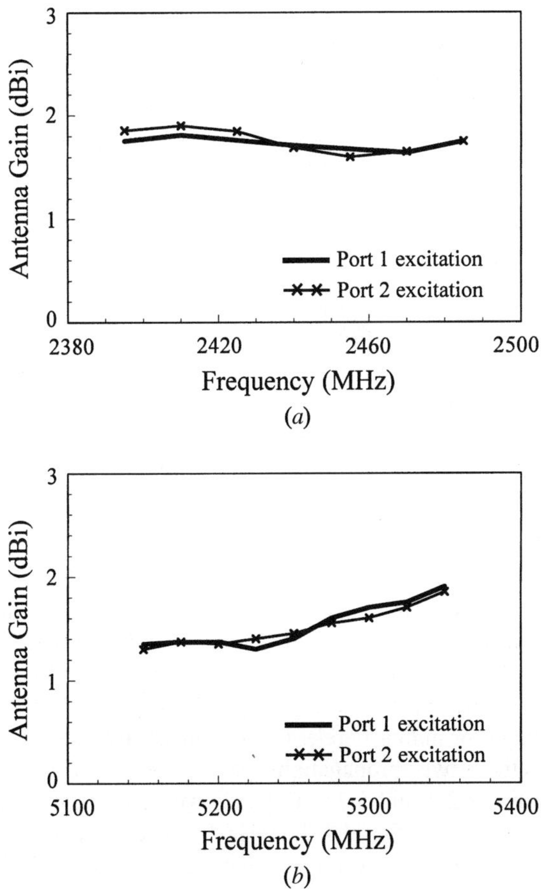

FIGURE 5.61 Measured antenna gain for port 1 and port 2 excitation of the antenna shown in Figure 5.55. (*a*) The 2.4 GHz WLAN band; (*b*) the 5.2 GHz WLAN band.

improves, that for the other frequency usually degrades). However, it is found that, by adding a shorting plate centered at the patch's radiating edge that is close to the feed position and the shorting pin, good impedance matching for both the two operating frequencies is obtained.

With the design dimensions given in Figure 5.66, the proposed antenna was constructed and studied. A ground plane with a size of $35 \times 35\,\text{mm}^2$ was used in the study. Figure 5.67 shows the measured return loss of the proposed antenna. It is clearly seen that two resonant modes at about 2.4 and 5.2 GHz are excited with good impedance matching. The impedance bandwidth (2.5:1 VSWR) for the lower band covers the operating bandwidth (2400–2484 MHz) of the WLAN system and reaches 90 MHz, or 3.7% with respect to 2.45 GHz. On the other hand, the upper

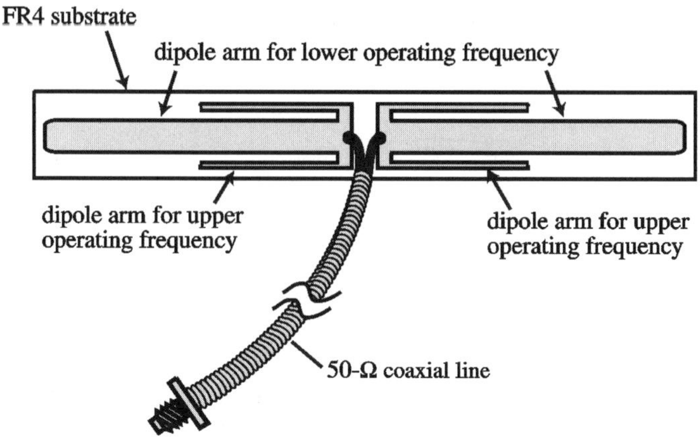

FIGURE 5.62 Geometry of the dual-band printed dipole antenna with multiple arms.

band meets the bandwidth requirement (5150–5350 MHz) of the 5.2 GHz band, and has a bandwidth of 277 MHz, or 5.3% referenced to 5.25 GHz.

Radiation characteristics have also been studied. Figures 5.68 and 5.69 show the measured radiation patterns of the proposed antenna at 2.45 and 5.25 GHz, respectively. For both operating frequencies, the cross-polarization radiation is seen to be large, especially for the E-plane (x–z plane) radiation. However, this is not necessarily a disadvantage for indoor wireless communications where multiple reflections for electromagnetic wave propagation usually occur. The antenna gain for frequencies within the 2.4 and 5.2 GHz bands was also measured, and the results are shown in Figure 5.70. For the 2.4 GHz band, the antenna gain has a peak value of 4.1 dBi and the gain variations are less than 1.5 dBi. For the 5.2 GHz

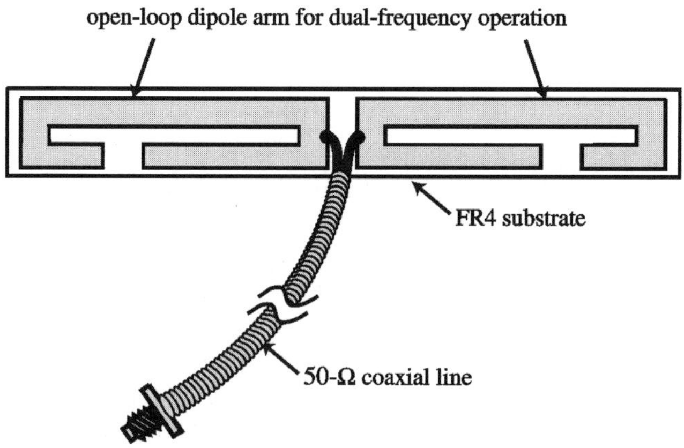

FIGURE 5.63 Geometry of the dual-band printed dipole antenna with two open rectangular ring strips.

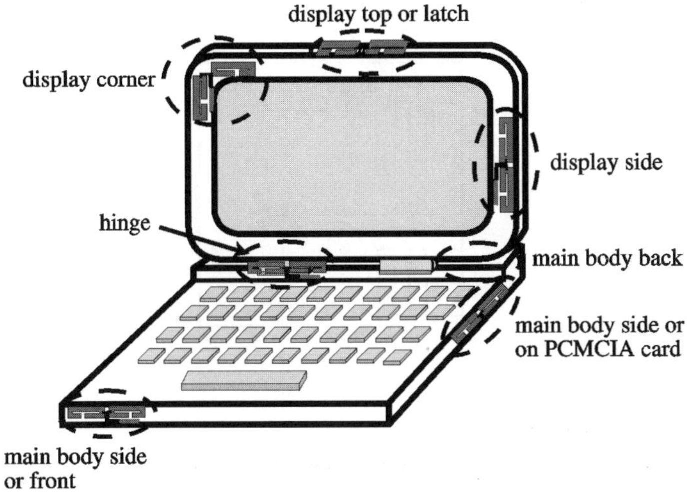

FIGURE 5.64 Printed dipole antennas applied to a notebook computer for WLAN operations.

band, the peak antenna gain is about 1.4 dBi and the gain variations are less than 2 dBi. The smaller antenna gain observed for the 5.2 GHz band is largely because the FR4 substrate, whose dielectric loss increases rapidly with increasing operating frequencies, was used in the design as a support for the radiating patch. If a low-loss dielectric substrate is used to support the radiating patch in the proposed antenna, it is expected that the antenna gain for frequencies in the 5.2 GHz band can be improved.

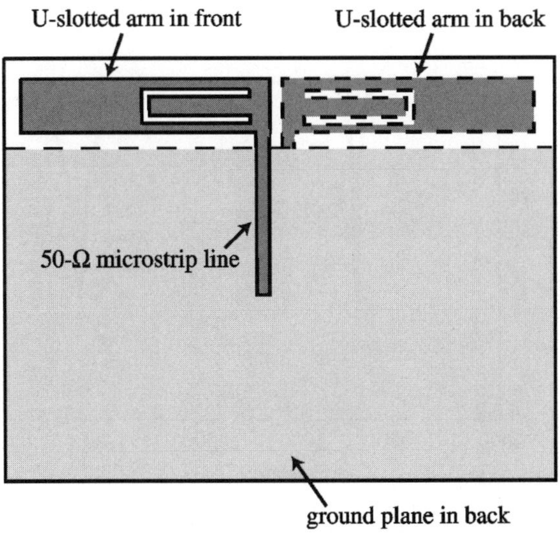

FIGURE 5.65 Geometry of an on-board dual-band printed dipole antenna.

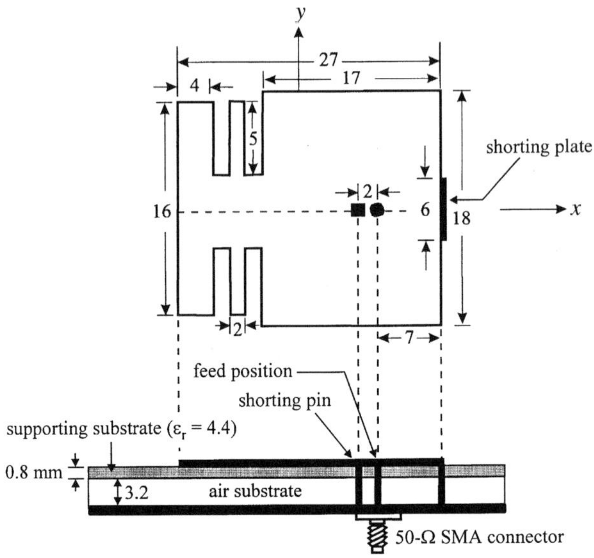

FIGURE 5.66 Geometry of the PIFA with two shorting strips for the 2.4 and 5.2 GHz bands. Dimensions given in the figure are in millimeters. (From Ref. 18, © 2001 John Wiley & Sons, Inc.)

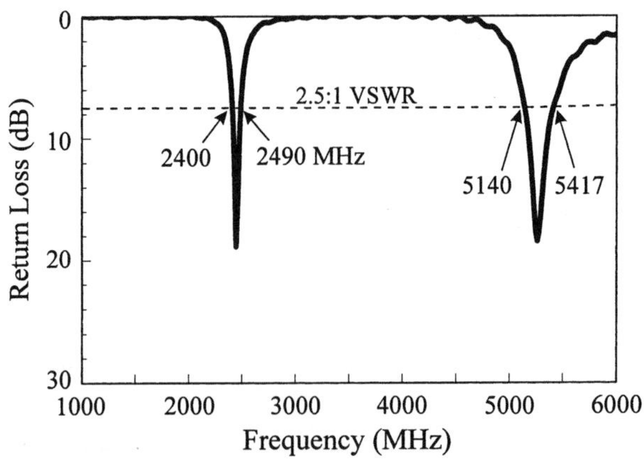

FIGURE 5.67 Measured return loss for the antenna shown in Figure 5.66. (From Ref. 18, © 2001 John Wiley & Sons, Inc.)

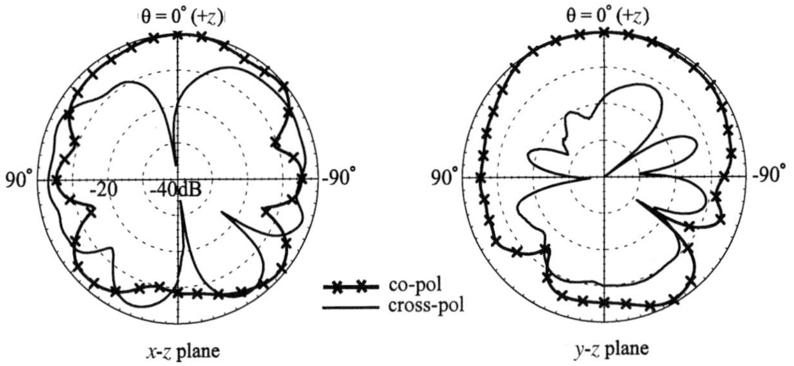

FIGURE 5.68 Measured x–z plane (E-plane) and y–z plane (H-plane) radiation patterns at 2.45 GHz for the antenna shown in Figure 5.66. (From Ref. 18, © 2001 John Wiley & Sons, Inc.)

5.7.2 PIFA with a Parasitic Shorted Patch

Figure 5.71 shows another promising dual-frequency PIFA design [19]. Note that the dimensions shown in the figure are for achieving dual ISM (industrial, scientific, and medical) band operations at 2.45 GHz (2400–2500 MHz) and 5.8 GHz (5725–5875 MHz). The driven and parasitic patches are printed on an inexpensive FR4 substrate of thickness 0.4 mm and relative permittivity 4.4. This supporting FR4 substrate is then placed above a ground plane with an air-layer thickness of 4.6 mm, which makes the total height of the constructed prototype 5.0 mm. Also note that both driven and parasitic patches are short-circuited to the ground plane, which makes the two patches operate as quarter-wavelength structures and reduces

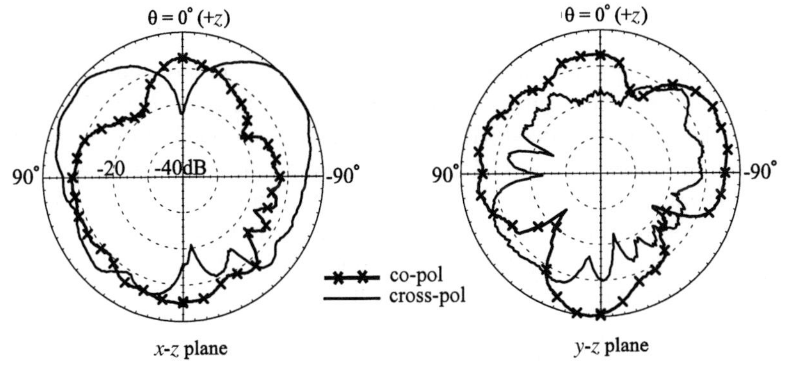

FIGURE 5.69 Measured x–z plane (E-plane) and y–z plane (H-plane) radiation patterns at 5.25 GHz for the antenna shown in Figure 5.66. (From Ref. 18, © 2001 John Wiley & Sons, Inc.)

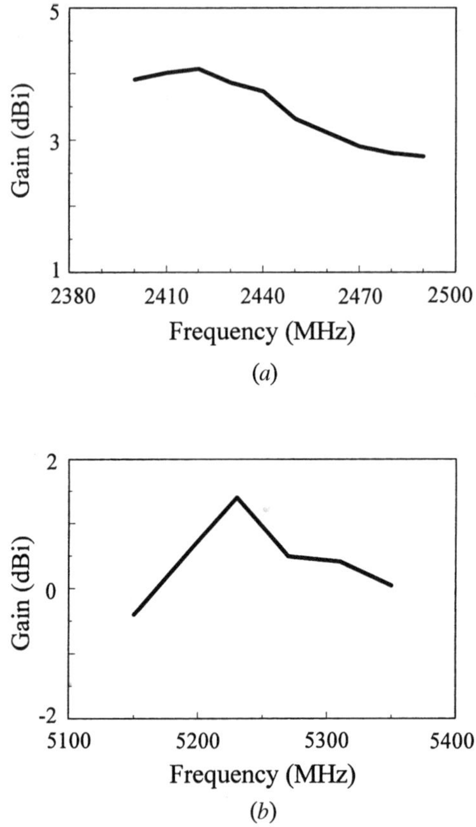

FIGURE 5.70 Measured antenna gain across (*a*) the 2.4 GHz band and (*b*) the 5.2 GHz band for the antenna shown in Figure 5.66. (From Ref. 18, © 2001 John Wiley & Sons, Inc.)

the required patch sizes for operating at 2.45 and 5.8 GHz in this study. In addition, to achieve a compact configuration, the driven patch, which generates the 5.8 GHz band in the design, is embedded within a large rectangular notch cut in the parasitic patch, which generates the 2.45 GHz band. The two patches together occupy a small area of $9\times20\,mm^2$ in this design. Also note that the dimensions of the driven and parasitic patches are first roughly determined from Equations 2.1 and 2.2 in Chapter 2. The design dimensions (see the dimensions given in Fig. 5.71*b*) are then optimized with the simulation software Ansoft HFSS.

A prototype of the proposed design for achieving 2.4/5.8 GHz dual ISM-band operations has been constructed. Figure 5.72 shows the measured and simulated return loss of the constructed prototype. Good agreement between the measured and simulated results is seen. The lower band has an impedance bandwidth (2.5:1 VSWR) of 132 MHz (2383–2515 MHz), covering the ISM band at 2.4 GHz. For

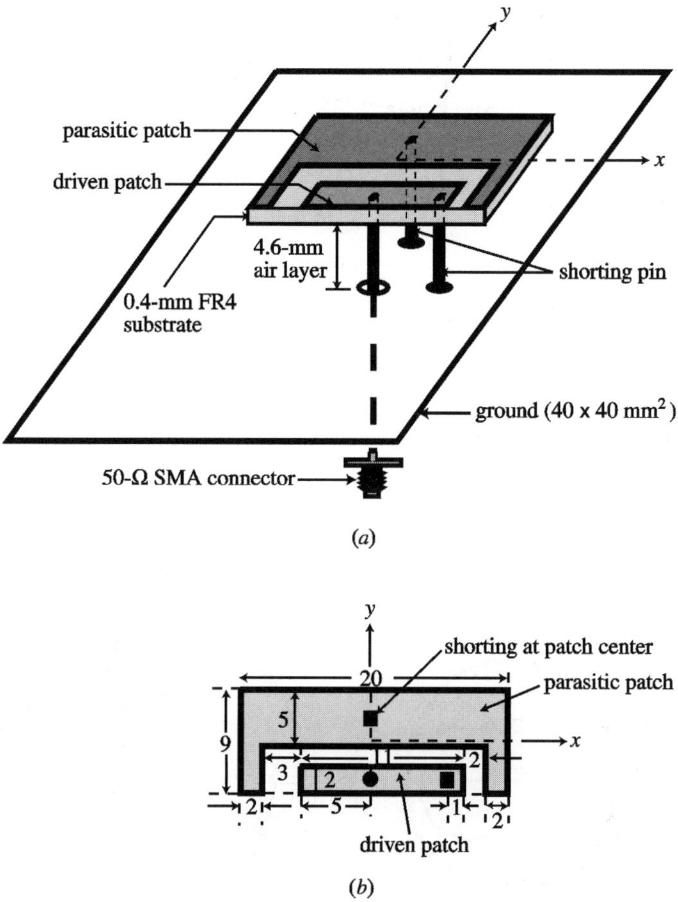

FIGURE 5.71 (*a*) Geometry of the PIFA with a parasitic shorted patch for the 2.4 and 5.8 GHz bands; (*b*) layout of the top patch of the PIFA. (From Ref. 19, © 2002 John Wiley & Sons, Inc.)

the upper band at 5.8 GHz, a much larger bandwidth is obtained. For the frequencies within the 5.8 GHz ISM band, the return loss is better than 10 dB (VSWR < 2). Also note that the resonant mode occurring at about 4.6 GHz is the higher-order mode of the parasitic patch, which is not used in the proposed design.

Measured radiation patterns at 2.45 and 5.8 GHz are plotted in Figure 5.73, and the measured antenna gain for frequencies across the 2.4 and 5.8 GHz ISM bands is presented in Figure 5.74. It is observed that the measured antenna gain is stable for the frequencies across the lower and upper ISM bands. The antenna gain level is about 2 and 3 dBi for the 2.4 and 5.8 GHz ISM bands, respectively.

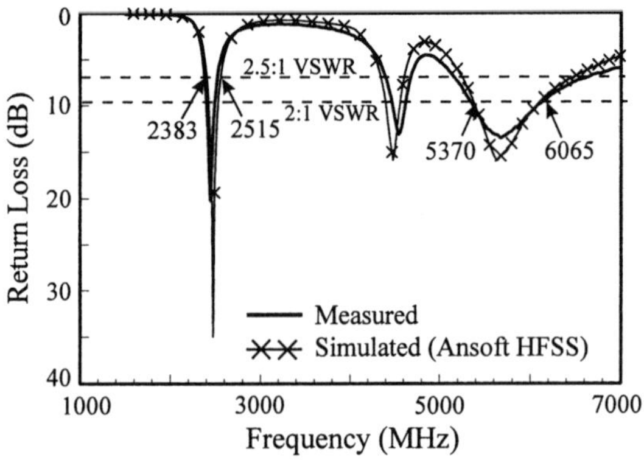

FIGURE 5.72 Measured and simulated return loss for the antenna shown in Figure 5.71. (From Ref. 19, © 2002 John Wiley & Sons, Inc.)

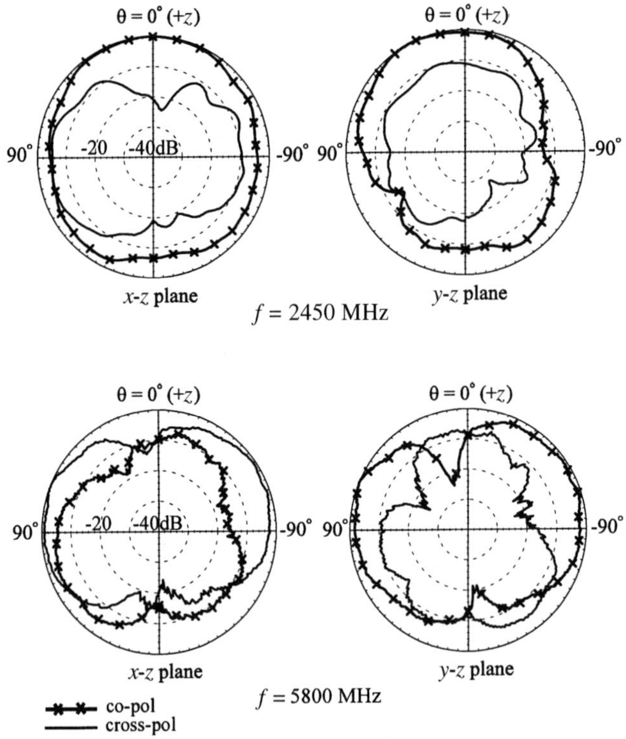

FIGURE 5.73 Measured radiation patterns at 2450 and 5850 MHz for the antenna shown in Figure 5.71. (From Ref. 19, © 2002 John Wiley & Sons, Inc.)

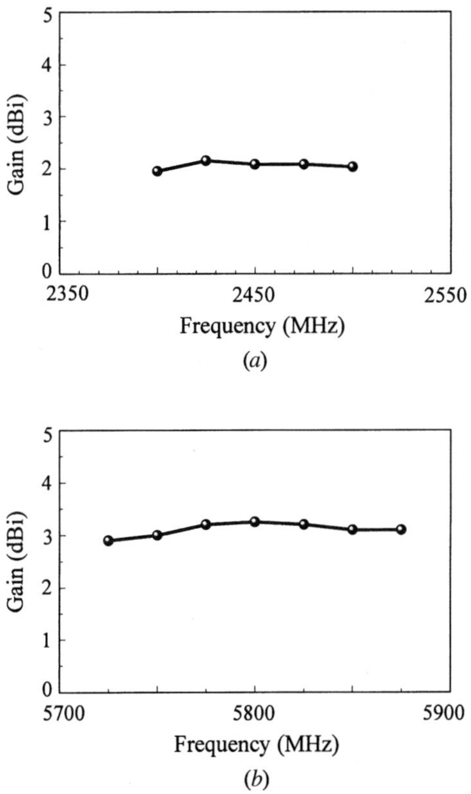

FIGURE 5.74 Measured antenna gain across (*a*) the 2.4 GHz band and (*b*) the 5.8 GHz band for the antenna shown in Figure 5.71. (From Ref. 19, © 2002 John Wiley & Sons, Inc.)

5.8 LOW-COST COMPACT DUAL-POLARIZED MICROSTRIP ANTENNA FOR WLAN OPERATION

With the use of the inexpensive FR4 substrate, a low-cost compact dual-polarized microstrip antenna [20, 21] suitable for WLAN operations in the 2.4 GHz band has been implemented. The antenna design is shown in Figure 5.75. The antenna has a simple square patch and a slotted ground plane with four narrow linear slots spaced orthogonally and is printed on a very common and inexpensive FR4 substrate (thickness 1.6 mm and relative permittivity 4.4). Two probe feeds are fed at two orthogonal positions to provide $\pm 45^\circ$ slanted polarizations. The side lengths of the square patch and the ground plane are 23 and 35 mm, respectively. The size of the proposed antenna, compared with the case without embedded slots in the ground plane, can be significantly reduced by about 40%, which makes it suitable to be integrated into a restricted area of the system. In addition, good dual-polarized performance of the proposed antenna is obtained. The two linear polarizations have

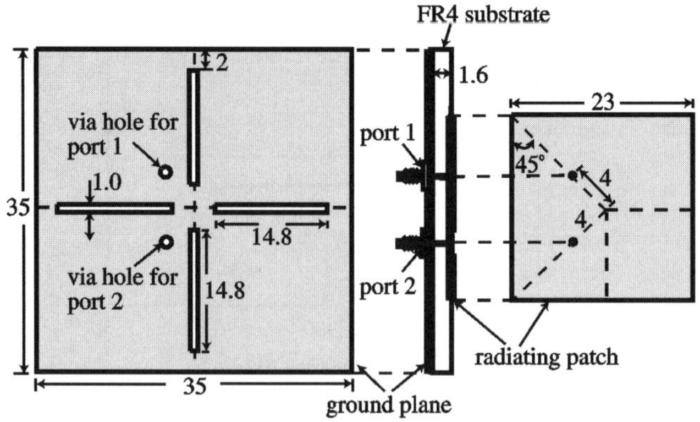

FIGURE 5.75 Geometry of the compact dual-polarized microstrip antenna for WLAN operation in the 2.4 GHz band.

broadside radiation patterns, and the proposed antenna has an impedance bandwidth (10 dB return loss) of about 3.5% covering the 2.4 GHz ISM band. A peak antenna gain of about 2.5 dBi for both linear polarizations is obtained, and the isolation between the two feeding ports is less than −20 dB.

REFERENCES

1. K. L. Wong, F. S. Chang, and T. W. Chiou, "Low-cost broadband circularly polarized probe-fed patch antenna for WLAN base station," in 2002 IEEE Antennas Propagat. Soc. Int. Symp. Dig., vol. 2, pp. 526–529.
2. Y. L. Kuo and K. L. Wong, "A planar inverted-L patch antenna for 2.4/5.2-GHz dual-band operation," Microwave Opt. Technol. Lett., vol. 31, pp. 394–396, Dec. 5, 2001.
3. K. Kawahata, K. Okada, A. Yuasa, and S. Nagumo, "Surface-mountable type antenna, antenna device, and communication device including the antenna device," U.S. Patent No. 6177908, Jan. 23, 2001.
4. K. Kawahata, S. Itoh, Y. Kushihi, M. Shikata, and H. Matsumoto, "Surface mount type circularly polarized wave antenna and communication apparatus using the same," U.S. Patent No. 6140968, Oct. 31, 2000.
5. H. Mandai, K. Asakura, T. Tsuru, S. Kanba, and T. Suesada, "Chip antenna," U.S. Patent No. 5977927, Nov. 2, 1999.
6. Y. Dakeya, T. Suesada, K. Asakura, N. Nakajima, and H. Mandai, "Chip multilayer antenna for 2.45 GHz-band application using LTCC technology," in 2000 IEEE MTT-S Int. Microwave Symp. Dig., pp. 1693–1696.
7. W. Choi, S. Kwon, and B. Lee, "Ceramic chip antenna using meander conductor line," Electron. Lett., vol. 37, pp. 933–934, July 19, 2001.
8. F. S. Chang and K. L. Wong, "Planar monopole folded into a rectangular-disk-like structure as surface-mountable antenna for 2.4/5.2 GHz dual-band operation," Microwave Opt. Technol. Lett., vol. 34, pp. 166–169, Aug. 5, 2002.

9. M. Ali and G. J. Hayes, "Analysis of integrated inverted-F antennas for Bluetooth applications," in Proc. 2000 IEEE-APS Conf. On Antennas Propagat. Wireless Communications, pp. 21–24.
10. Y. L. Kuo, T. W. Chiou, and K. L. Wong, "A novel dual-band printed inverted-F antenna," Microwave Opt. Technol. Lett., vol. 31, pp. 353–355, Dec. 5, 2001.
11. S. H. Yeh and K. L. Wong, "Integrated F-shaped monopole antenna for 2.4/5.2 GHz dual-band operation," Microwave Opt. Technol. Lett., vol. 34, pp. 24–26, July 5, 2002.
12. H. C. Tung, W. S. Chen, and K. L. Wong, "Integrated rectangular spiral monopole antenna for 2.4/5.2 GHz dual-band operation," in 2002 IEEE Antennas Propagat. Soc. Int. Symp. Dig., vol. 3, pp. 446–449.
13. Y. L. Kuo and K. L. Wong, "Dual-polarized monopole antenna for WLAN application," in 2002 IEEE Antennas Propagat. Soc. Int. Symp. Dig., vol. 4, pp. 80–83.
14. I. F. Chen and M. L. Lee, "Miniature microstrip antenna," U.S. Patent No. 6295029 B1, Sept. 25, 2001.
15. S. N. Tsai, "Antenna module for portable computer," U.S. Patent No. 6297779, Oct. 2, 2001.
16. W. L. Cheng and C. H. Fang, "Antenna for portable device," U.S. Patent Application Publication No. US 2001/0040529 A1.
17. Y. H. Suh and K. Chang, "Low cost microstrip-fed dual frequency printed dipole antenna for wireless communications," Electron. Lett., vol. 36, pp. 1177–1179, July 6, 2000.
18. H. C. Tung and K. L. Wong, "A shorted microstrip antenna for 2.4/5.2-GHz dual-band operation," Microwave Opt. Technol. Lett., vol. 30, pp. 401–402, Sept. 20, 2001.
19. S. H. Yeh, S. T. Fang, and K. L. Wong, "Dual-band shorted patch antenna for dual ISM-band application," Microwave Opt. Technol. Lett., vol. 32, pp. 79–80, Jan. 5, 2001.
20. T. W. Chiou and K. L. Wong, "Designs of compact microstrip antennas with a slotted ground plane," in 2001 IEEE Antennas Propagat. Soc. Int. Symp. Dig., pp. 732–735.
21. K. L. Wong, Compact and Broadband Microstrip Antennas, Chap. 4, John Wiley & Sons, New York, NY, 2002.

CHAPTER SIX

Dielectric Resonator Antennas for Wireless Communications

6.1 INTRODUCTION

Because of their advantages of low loss, simple structure, compact size, and good antenna performance, dielectric resonator (DR) antennas are very promising for application in wireless communications. Advanced designs in achieving circular polarization (CP), broadband, and dual-polarized operations for DR antennas have also been reported in recent years. Several interesting designs have been demonstrated to achieve CP operation. They include the use of DR elements with special configurations [1, 2] and the use of the two-feed design incorporating a cylindrical ring DR [3] or a cylindrical DR [4–6]. For the designs in References 1 and 2, the requirement for special configurations of the DR elements, such as the nearly cubic DR and the rectangular DR with a special length-to-width ratio, increases the complexity in antenna fabrication. As for the two-feed designs [3–6], an external phase shifter to obtain a 90° phase difference between the two feeds is required, which also increases the antenna fabrication complexity. Recently, several promising CP designs using a single-feed method have been demonstrated [7–10]. These designs make CP operations of the DR antenna easier and simpler to achieve. Details of these single-feed CP designs are presented in Section 6.2. In Section 6.3, the technique of using two linearly polarized DR elements to obtain CP radiation [11] is described.

In Section 6.4, a broadband design of a very-high-permittivity DR antenna for WLAN operation in the 5.2 GHz band [12] is introduced. The broadband operation is achieved by loading the DR with two narrow metal strips of suitable lengths, which results in dual-resonance excitation and leads to an enhanced operating bandwidth to cover the 5.2 GHz band. Finally, in Section 6.5, two feeding designs of placing the microstrip feed line directly under the DR and conformal to the surface of the DR to obtain dual-polarized operations of the DR antenna [13] are demonstrated.

6.2 SINGLE-FEED SINGLE-ELEMENT CIRCULARLY POLARIZED DR ANTENNA

6.2.1 With Conformal Loading Strips

By loading conformal strips of suitable dimensions to the surface of a cylindrical DR antenna, CP radiation has been obtained with a microstrip-line feed [7] and a probe feed [8]. The antenna geometries are shown in Figure 6.1. In Figure 6.1a, two conformal strips are printed on the surface of the cylindrical DR and a microstrip feed line incorporating a vertical strip is used to excite the DR. The two conformal strips are spaced 180° apart, and one of them is 45° away from the microstrip feed

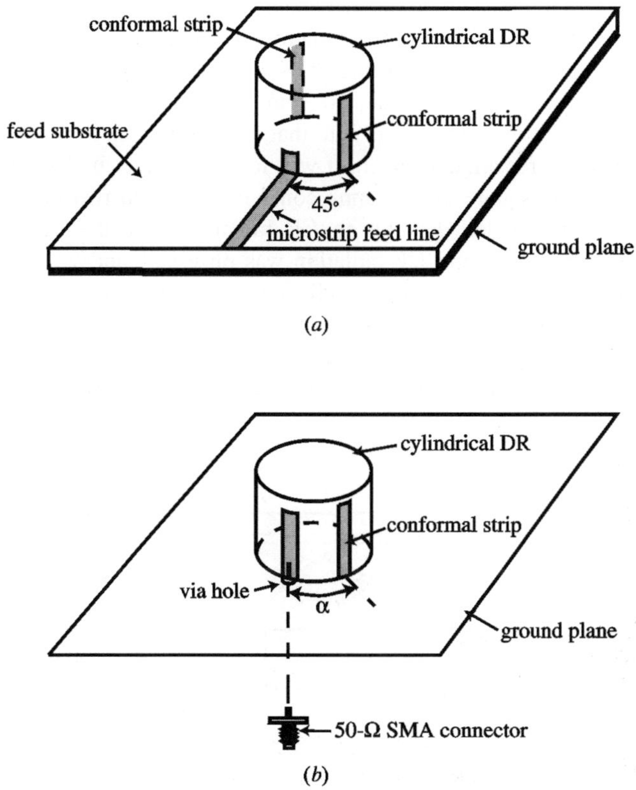

FIGURE 6.1 Geometries of the circularly polarized DR antenna with conformal loading strips. (a) The design with a microstrip-line feed; (b) the design with a probe feed.

line. With the proposed design, two orthogonal modes with a 90° phase difference can be excited, which leads to CP radiation of the DR antenna. In Figure 6.1b, a single probe feed is used to achieve CP radiation of the DR antenna. In this case, the probe pin is connected to a conformal strip, and by varying the angular position (α) of the other conformal strip, which is short-circuited to the ground plane, two near-degenerate orthogonal resonant modes can be excited to obtain CP radiation [8].

6.2.2 With a Loading Patch

Figure 6.2 shows the geometry of the circularly polarized low-profile square DR antenna with a loading patch. The square DR has a side length of L (=28.2 mm) and a height of h (=4.9 mm). The relative permittivity of the DR was chosen to be a high value of 79 in this study, which makes it possible for the DR to have a lower height at a fixed operating frequency (h is only about $0.17L$ in this study), compared with the case with a lower-permittivity DR. A rectangular patch, having a length of ℓ and a width of w, is centered on top of the square DR, and the patch's centerlines are also oriented along the diagonals of the DR. This patch-loaded square DR can then be easily excited by a 50-Ω microstrip line printed on an FR4 substrate of thickness 1.6 mm and relative permittivity 4.4. By selecting a suitable tuning stub length (t) of the microstrip feed line, good impedance matching of the proposed antenna can easily be obtained. Also note that, simply by selecting suitable aspect ratios of the loading rectangular patch, it was found that the fundamental resonant mode (TE_{111}) of the square DR antenna could be split into two orthogonal near-degenerate resonant modes, leading to CP radiation. For the case with $\ell > w$ shown in Figure 6.2, left-hand CP radiation was observed, and when ℓ and w are interchanged in Figure 6.2, right-hand radiation can be obtained.

Prototypes (antennas 1 to 5) of the proposed CP design have been constructed and studied. The experimental results of antennas 1 to 5 are given in Table 6.1, and the reference antenna (the case without the loading patch) is also shown for

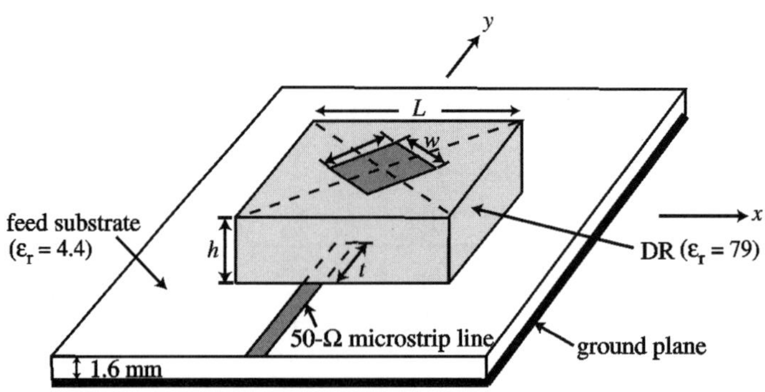

FIGURE 6.2 Geometry of the circularly polarized low-profile square DR antenna with a loading patch. (From Ref. 9, © 2001 John Wiley & Sons, Inc.)

TABLE 6.1 Performance of the Antenna Shown in Figure 6.2 [9]; $L = 28.2$ m, $h = 4.9$ mm, $\ell = 7$ mm, $w = 5$ mm, $t = 11$ mm, Ground-Plane Size $= 50 \times 50$ mm^2. Aspect Ratio is Determined From ℓ/w, and f_c is the Center Frequency of the Obtained CP Bandwidth Determined by 3 dB Axial Ratio.

	ℓ (mm)	w (mm)	Aspect Ratio	t (mm)	f_c (MHz)	CP BW (%)
Reference	0	0	—	11	2284*	—
Antenna 1	6.5	4.0	1.63	11	2245*	—
Antenna 2	7.0	5.0	1.40	11	2220	1.1
Antenna 3	9.0	8.0	1.13	11	2126	1.1
Antenna 4	9.5	8.8	1.08	11	2087	1.2
Antenna 5	10.0	9.5	1.05	11	2040	1.1

*Center frequency of the impedance bandwidth (2:1 VSWR).

comparison. The measured input impedance on the Smith chart for antenna 2 and the reference antenna is also shown in Figure 6.3. The loading patch for antenna 2 has dimensions of 7×5 mm^2, with an aspect ratio of 1.4 (ℓ/w), and it is noted that there is a small dip in the impedance locus near the center of the Smith chart for antenna 2. In contrast, the impedance locus for the reference antenna shows a smooth curve near the center of the Smith chart. This behavior indicates that two near-degenerate resonant modes have been generated for antenna 2, and from the measured axial ratio shown in Figure 6.4, it also indicates that the two near-degenerate modes are orthogonal, which makes possible the circularly polarized radiation for antenna 2. The obtained CP bandwidth, determined from 3 dB axial ratio, is about 25 MHz, or about 1.1% referenced to the center frequency at 2220 MHz. Figure 6.5 also presents the measured antenna gain for the frequencies within the CP bandwidth. The peak antenna gain obtained is about 3.3 dBi, with small gain variations less than 0.5 dBi. The measured radiation patterns for antenna 2 are plotted in Figure 6.6, and good left-hand CP radiation is observed.

It was also found that, in addition to the dimensions of the loading rectangular patch used for antenna 2, there are still other possible patch dimensions capable of splitting the antenna's fundamental resonant mode into two near-degenerate modes for CP radiation. Table 6.1 shows the experimental results of three other possible designs (antennas 3, 4, and 5), in which the lengths of the loading patch increase from $\ell = 7.0$ mm for antenna 2 to $\ell = 9.0$ mm for antenna 3, 9.5 mm for antenna 4, and 10.0 mm for antenna 5. For these designs, the obtained CP characteristics are similar to those obtained for antenna 2. However, it should be noted that, by increasing the length of the loading patch, the required patch's aspect ratio for CP radiation quickly decreases and becomes close to unity, which will give the proposed CP design strict manufacturing tolerances in practical fabrication. On the other hand, when the length ℓ is decreased from 7.0 to 6.5 mm (see antenna 1 in Table 6.1) or smaller and the width w is also decreased to achieve various aspect ratios for the loading patch, no splitting of the antenna's fundamental modes can be obtained in the present study. Thus, to achieve CP radiation for the proposed antenna studied here, it is

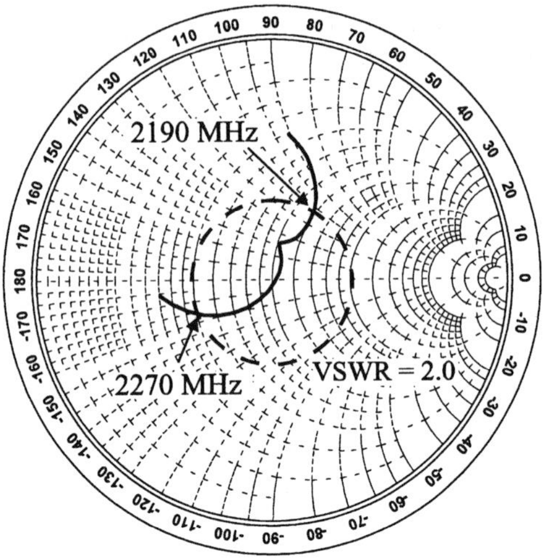

(*a*)

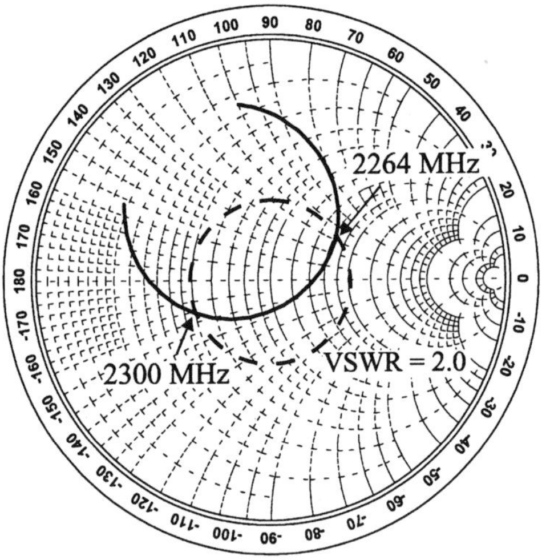

(*b*)

FIGURE 6.3 Measured input impedance on the Smith chart for (a) antenna 2 shown in Table 6.1 and (b) the reference antenna shown in Table 6.1. (From Ref. 9, © 2001 John Wiley & Sons, Inc.)

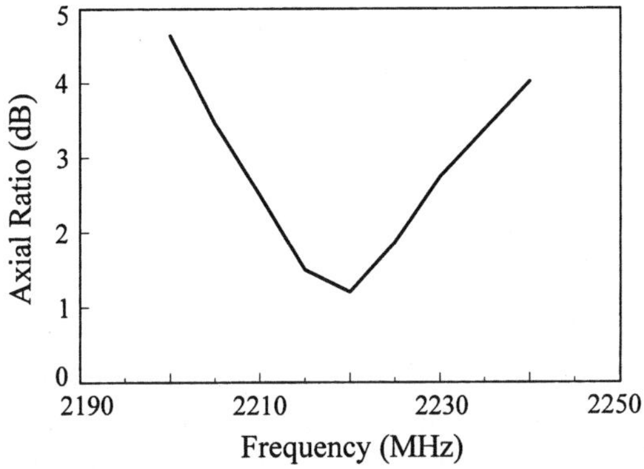

FIGURE 6.4 Measured axial ratio in the broadside direction for antenna 2 shown in Table 6.1. (From Ref. 9, © 2001 John Wiley & Sons, Inc.)

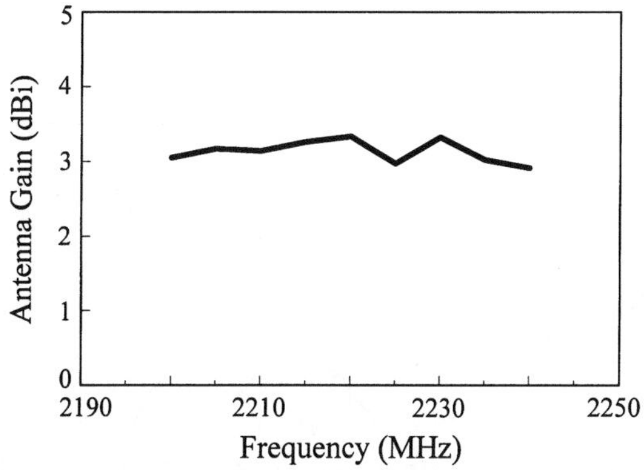

FIGURE 6.5 Measured antenna gain for antenna 2 shown in Table 6.1. (From Ref. 9, © 2001 John Wiley & Sons, Inc.)

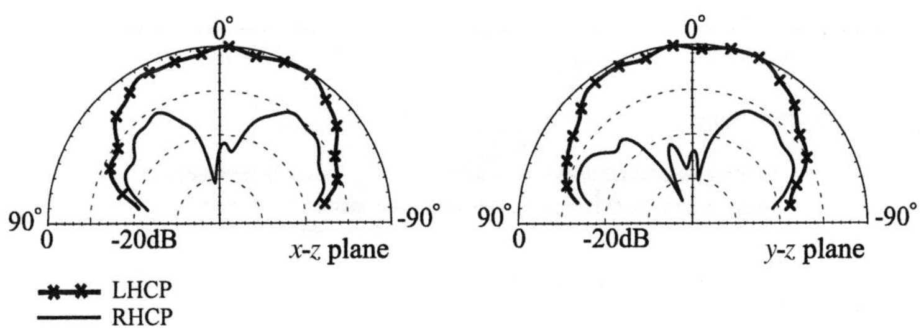

FIGURE 6.6 Measured radiation patterns at 2220 MHz for antenna 2 shown in Table 6.1. (From Ref. 9, © 2001 John Wiley & Sons, Inc.)

suggested that the length of the loading rectangular patch be chosen from about 7 to 10 mm, or about 0.25–0.35 times the side length of the square DR element.

In addition, it should be noted that, independent of various sizes of the loading rectangular patch chosen, the tuning stub length for achieving good impedance matching for the proposed antenna remained the same, which also makes impedance matching of the proposed antenna very easy to achieve.

6.2.3 With a Cross-Slot-Coupled Feed

Figure 6.7 shows the CP design for a low-profile circular disk DR antenna excited by a microstrip line through a coupling cross-slot of unequal arm lengths [10]. The DR has a radius R and a height h_2. The DR also has a very high relative permittivity of $\varepsilon_{ra} = 79$, and its diameter-to-height ratio is 5.77 ($h_2 = 5.1$ mm and $R = 14.72$ mm). In this design, by carefully adjusting the arm lengths (L_{ap1} and L_{ap2}) of the cross-slot, CP operation of the proposed antenna can be obtained. For the condition of $L_{ap1} > L_{ap2}$, right-hand CP operation is obatined, and $L_{ap1} < L_{ap2}$ is for left-hand CP operation.

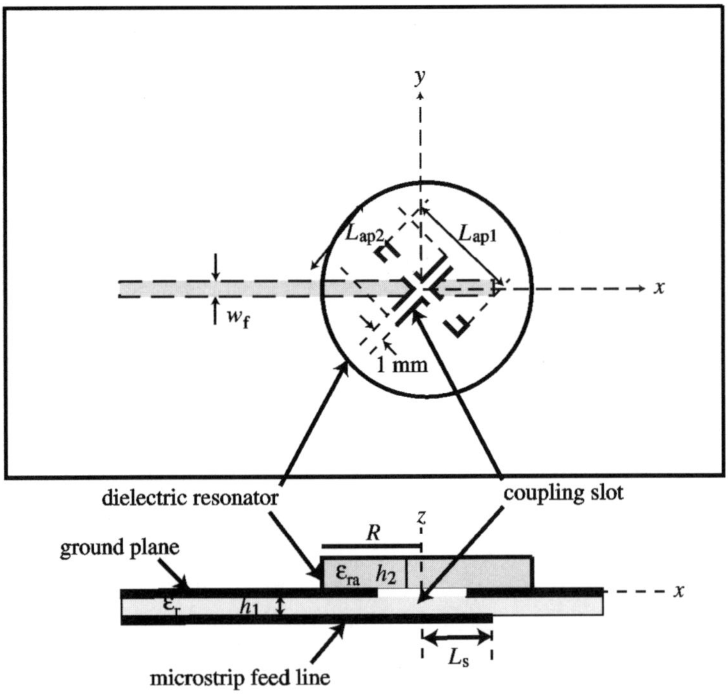

FIGURE 6.7 Geometry of the circularly polarized cross-slot-coupled circular disk DR antenna; $L_{ap1} > L_{ap2}$ is for right-hand CP operation, and $L_{ap1} < L_{ap2}$ is for left-hand CP operation. (From Ref. 10, © 1999 IEEE, reprinted with permission)

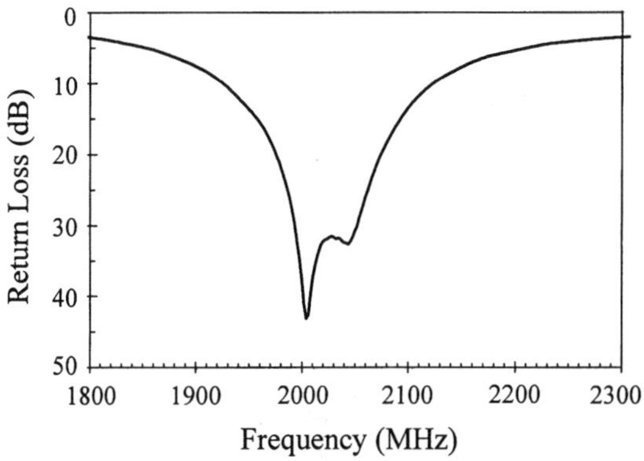

FIGURE 6.8 Measured return loss for the antenna shown in Figure 6.7 with $h_1 = 0.8$ mm, $\varepsilon_r = 4.4$, $h_2 = 5.1$ mm, $\varepsilon_{ra} = 79$, $W_f = 1.5$ mm, $L_s = 10$ mm, $L_{ap1} = 13$ mm, $L_{ap2} = 12$ mm, $R = 14.72$ mm. (From Ref. 10, © 1999 IEEE, reprinted with permission)

A prototype of the proposed design with right-hand CP operation has been implemented. The measured return loss and input impedance on the Smith chart are shown in Figures 6.8 and 6.9, respectively. A wide impedance bandwidth (2:1 VSWR) of 207 MHz (or about 10.1% with respect to the center frequency at 2044 MHz) is obtained. In this design, the two arm lengths of the cross-slot are $L_{ap1} = 13$ mm and $L_{ap2} = 12$ mm, and the arm length ratio (L_{ap1}/L_{ap2}) is 1.083, slightly larger than unity. This cross-slot of unequal arm lengths makes the fundamental resonant mode ($\mathrm{HEM}_{11\delta}$) of the circular disk DR split into two near-degenerate orthogonal modes for CP operation, and the resonant frequency of the $\mathrm{HEM}_{11\delta}$ mode can be roughly estimated from Reference 14

$$f = \frac{c}{2\pi R\sqrt{\varepsilon_{ra}}}\sqrt{(1.841)^2 + (\pi R/2h_2)^2}, \qquad (6.1)$$

where c is the speed of light in air.

Figure 6.10 shows the measured axial ratio of the CP radiation. A large 3 dB CP bandwidth of about 80 MHz or 3.9% is obtained. This large CP bandwidth gives the present design relaxed manufacturing tolerances. Measured radiation patterns are plotted in Figure 6.11, and Figure 6.12 shows the antenna gain in the broadside direction. Good right-hand CP radiation is obtained, and at resonance, the measured antenna gain is about 6.4 dBi. The antenna gain variation across the 3 dB CP bandwidth is seen to be small, less than 1.0 dBi. It can also be expected that this CP design is applicable for the DR antenna with a simple square cross section. This advantage makes the present design much easier to be implemented, compared with conventional designs that require the use of the DR with special configurations [1, 2]. Also, the present design allows postadjustments, by fine-tuning the coupling

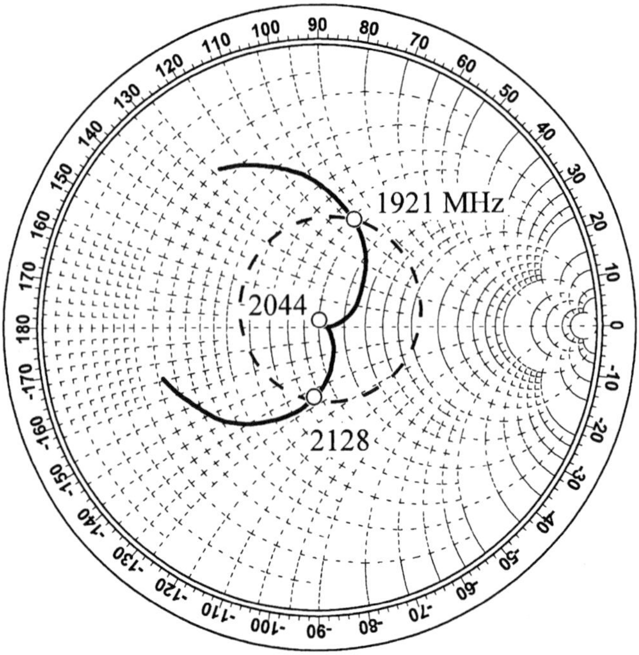

FIGURE 6.9 Measured input impedance on the Smith chart for the antenna studied in Figure 6.8. (From Ref. 10, © 1999 IEEE, reprinted with permission)

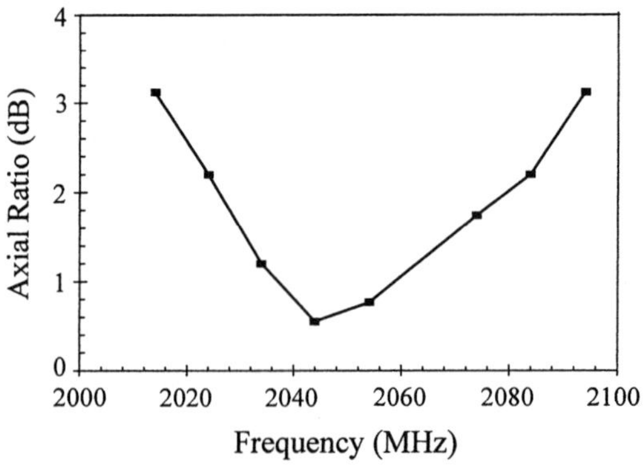

FIGURE 6.10 Measured axial ratio in the broadside direction for the antenna studied in Figure 6.8. (From Ref. 10, © 1999 IEEE, reprinted with permission)

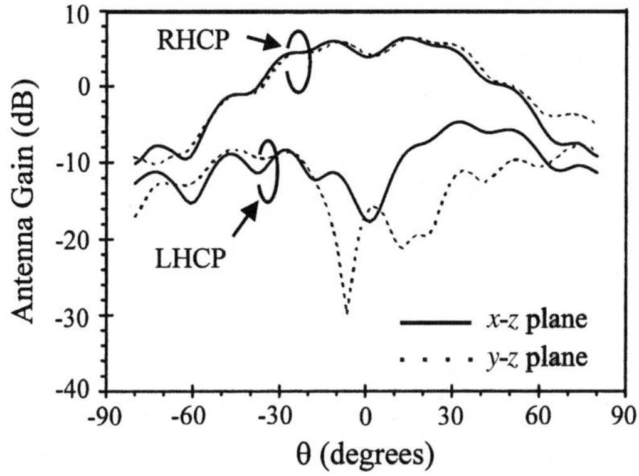

FIGURE 6.11 Measured radiation patterns at 2044 MHz for the antenna studied in Figure 6.8. (From Ref. 10, © 1999 IEEE, reprinted with permission)

cross-slot size, to compensate possible manufacturing errors to meet precise frequency specifications.

6.2.4 With a Slotted Ground Plane

By placing a DR element directly on a 50-Ω microstrip line and embedding two pairs of narrow slots of unequal lengths in the ground plane as shown in Figure 6.13, a circularly polarized DR antenna can also be obtained. By adjusting the slot lengths of the two pairs of linear slots, the fundamental resonant mode of the DR antenna can also be split into two near-degenerate orthogonal modes for CP operation, similar to the design technique discussed in Section 6.2.3. A prototype based on the design dimensions shown in Figure 6.13 has been successfully implemented,

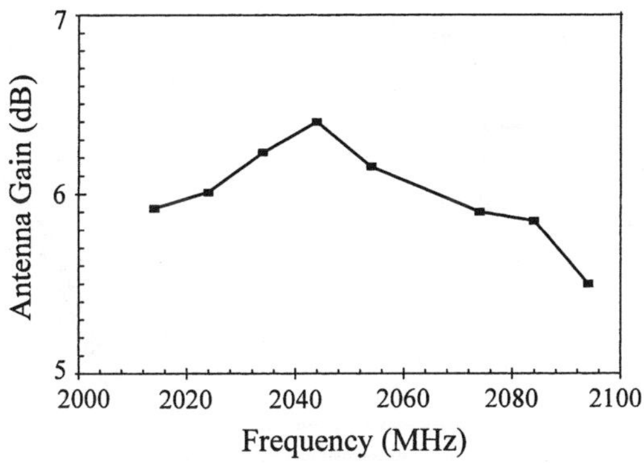

FIGURE 6.12 Measured antenna gain for the antenna studied in Figure 6.8. (From Ref. 10, © 1999 IEEE, reprinted with permission)

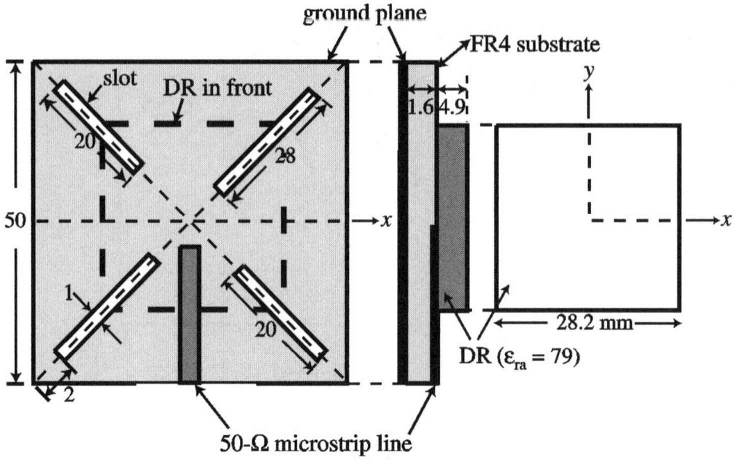

FIGURE 6.13 Geometry of the circularly polarized DR antenna with a slotted ground plane.

and good CP radiation characteristics have been obtained. The obtained CP operation has a center operating frequency at about 2 GHz.

6.3 TWO-ELEMENT CIRCULARLY POLARIZED DR ANTENNA

By using two linearly polarized DR elements, CP operation has also been obtained [11]. Figure 6.14 shows the CP antenna design. The two cylindrical DRs have the

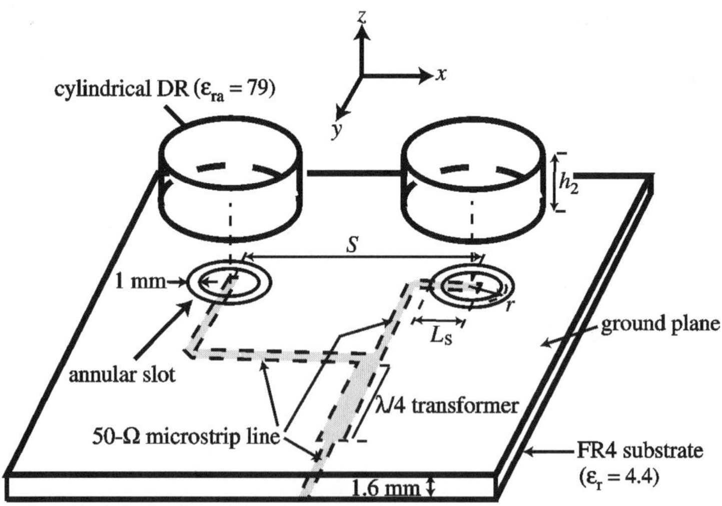

FIGURE 6.14 Geometry of a two-element DR antenna for CP radiation. (From Ref. 11, © 1999 John Wiley & Sons, Inc.)

same relative permittivity ε_{ra} (79 in this study) and the same dimensions of radius R and height h. The distance between the centers of the two DRs is S. The feed network, consisting of three 50-Ω microstrip lines and a quarter-wavelength impedance transformer, is printed on the substrate of thickness 1.6 mm and relative permittivity 4.4. For the feed network shown in the figure, right-hand CP radiation can be excited. The electromagnetic energy from the microstrip feed line is coupled from an annular slot in the ground plane of the microstrip line to each DR antenna. By adjusting the tuning stub length, L_S, extending from the slot edge into the slot center, the input impedance of the DR antenna can easily be tuned.

In the experiment, the microstrip line was designed to have a 50-Ω characteristic impedance. Low-profile cylindrical DRs (h/R about 0.3) were used in the present design. To obtain good impedance matching, the dimensions of the coupling annular slot and tuning stub length are first determined. The annular slot is centered below the DR antenna. Figure 6.15 shows measured return loss for a single linearly polarized DR antenna with various tuning stub lengths. In this case, the slot radius r is 6.5 mm and the slot width is 1 mm. From the results, it can be seen that varying the tuning stub length is very effective in achieving good impedance matching. A return loss of about 40 dB can be obtained with $L_S = 9$ mm, and the impedance bandwidth, determined from 10 dB return loss, is 92 MHz (or about 5.9% with respect to the center frequency at 1563 MHz).

With the parameters of a single DR antenna determined, the proposed two-element circularly polarized DR antenna has been constructed and tested. The distance S between the two DR antennas is chosen to be 83 mm (about 43% of the wavelength at the center frequency). With the feed-network layout shown in Figure 6.14, the two DR antennas can be excited with equal amplitudes and a 90° phase difference. The measured input impedance on the Smith chart is shown in Figure 6.16, and the axial ratio of the CP radiation is presented in Figure 6.17, in which the CP bandwidth, determined from 3 dB axial ratio, is found to be about

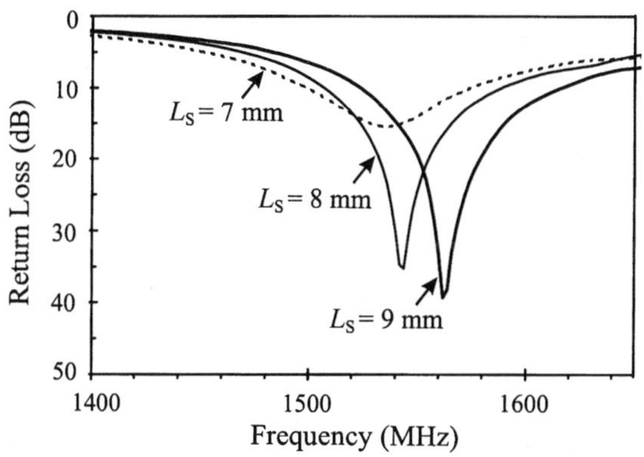

FIGURE 6.15 Measured return loss for the antenna shown in Figure 6.14 with a single DR element for linearly polarized radiation; $h = 5$ mm, $R = 15.95$ mm, $\varepsilon_{ra} = 79$, $r = 6.5$ mm. (From Ref. 11, © 1999 John Wiley & Sons, Inc.)

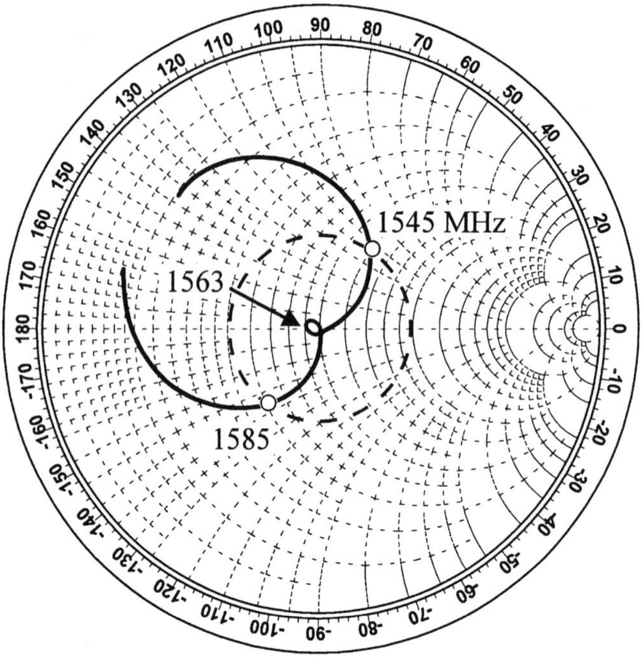

FIGURE 6.16 Measured input impedance on the Smith chart for the antenna shown in Figure 6.14; $L_S = 9$ mm. Other parameters are the same as given in Figure 6.15. (From Ref. 11, © 1999 John Wiley & Sons, Inc.)

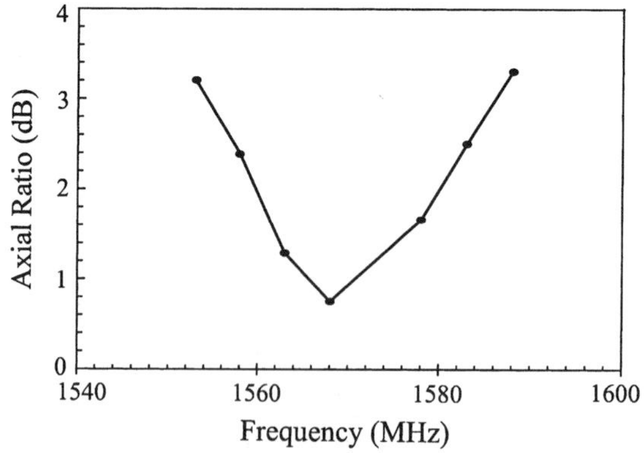

FIGURE 6.17 Measured axial ratio in the broadside direction for the antenna studied in Figure 6.16. (From Ref. 11, © 1999 John Wiley & Sons, Inc.)

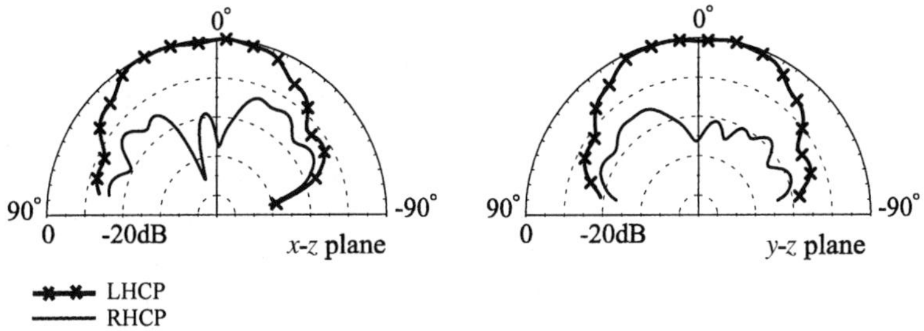

FIGURE 6.18 Measured radiation patterns at 1563 MHz for the antenna studied in Figure 6.16. (From Ref. 11, © 1999 John Wiley & Sons, Inc.)

31 MHz, or 2.0%. Measured radiation patterns in two orthogonal planes at 1563 MHz are plotted in Figure 6.18, and good right-hand CP radiation is observed.

6.4 BROADBAND DR ANTENNA

To meet the WLAN operation in the 5.2 GHz band, a broadband DR antenna using the strip-loading technique has been reported [12]. Figure 6.19 shows the proposed broadband design for the DR antenna with a very high relative permittivity of 90.5. The dimensions shown in the figure are for WLAN operations in the 5.2 GHz band. The DR antenna is of a square disk shape and has a cross section of $10 \times 10\ \text{mm}^2$ and a very low profile of 1.9 mm. The proposed DR antenna can be easily fed by using a 50-Ω microstrip line, which is printed on a grounded FR4 substrate in the present study. The microstrip feed line is placed below the centerline (*y*-axis in the figure) of the DR, and good impedance matching can be easily obtained by moving the DR

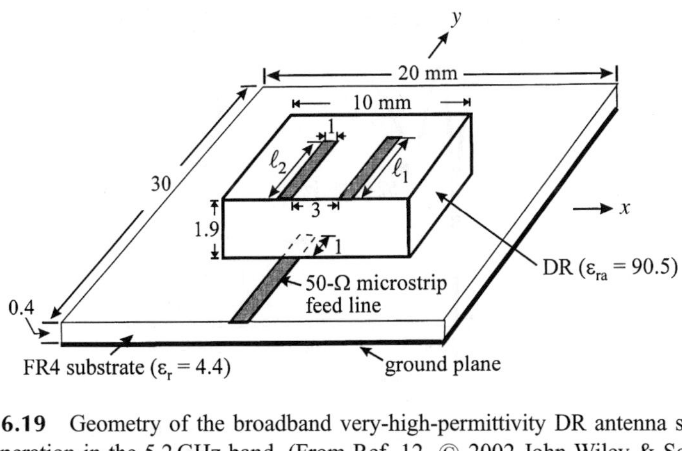

FIGURE 6.19 Geometry of the broadband very-high-permittivity DR antenna suitable for WLAN operation in the 5.2 GHz band. (From Ref. 12, © 2002 John Wiley & Sons, Inc.)

along the y-axis, that is, by selecting a suitable length of the microstrip feed line under the DR (in this study, the length is 1 mm).

Two narrow conducting strips (width chosen to be 1 mm in the study) of different lengths ℓ_1 and ℓ_2 are printed on the DR. The two strips are also oriented in parallel with the microstrip feed line, with one end of the strips coinciding with the edge of the DR. The two strips have the same distance to the centerline of the DR, and the spacing between them is selected to be 3 mm in this study. It is found that, by using the proposed strip-loading technique, the antenna's fundamental resonant mode (TE_{111} mode), whose resonant frequency can be roughly determined from

$$f_{111} \approx \frac{c}{2\pi\sqrt{\varepsilon_{ra}}}\sqrt{2\left(\frac{\pi}{L}\right)^2+\left(\frac{\pi}{2t}\right)^2}, \tag{6.2}$$

becomes a dual-resonance mode, making possible a wider impedance bandwidth. (In Equation 6.2, c is the speed of light in air, and ε_{ra}, L, and t are the DR's relative permittivity, side length, and height, respectively.) The obtained dual-resonance excitation is probably due to the introduced inductance, contributing from the loading strips (self-inductance), and the introduced capacitance, contributing from between the loading strips and the antenna's ground plane, which makes it possible to induce an additional resonance to the antenna's fundamental resonant mode. It should also be noted that the resonant frequencies of the two resonances are greatly dependent on the lengths (ℓ_1 and ℓ_2) of the two loading strips, and there exist optimal values of ℓ_1 and ℓ_2 for forming the two resonances into a wider impedance bandwidth for the proposed antenna.

Figure 6.20 shows the measured return loss for the proposed antenna. The results for both the cases with and without the strip loading are presented. It is first observed

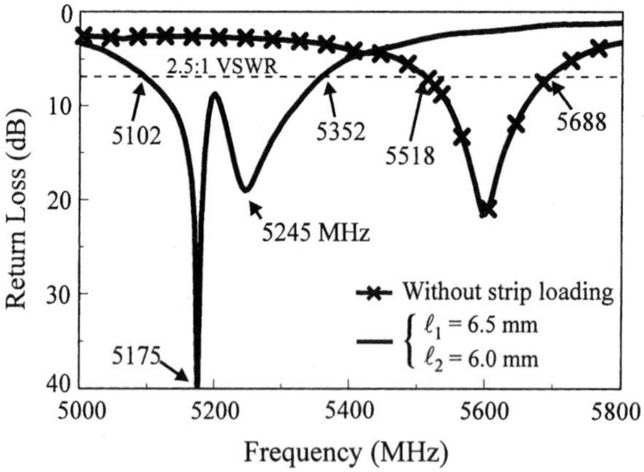

FIGURE 6.20 Measured return loss for the antenna shown in Figure 6.19. (From Ref. 12, © 2002 John Wiley & Sons, Inc.)

that a dual-resonance resonant mode is obtained for the proposed antenna, whereas for the case without the strip loading, the resonant mode shows only a single resonance. It is also noted that the impedance bandwidth (2.5:1 VSWR) of the proposed antenna reaches 250 MHz (5102–5352 MHz), or about 4.8%, which covers the required bandwidth (5150–5350 MHz) for WLAN operation in the 5.2 GHz band. The obtained impedance bandwidth is also much larger than that of the case without strip loading (170 MHz or about 3%). In addition, it is seen that the proposed antenna has resonant frequencies lower than that of the case without strip loading. From Equation 6.2, this indicates that a larger DR height ($t > 1.9$ mm) is required for the case without strip loading to achieve an operating bandwidth in the 5.2 GHz band.

Radiation characteristics of the proposed antenna have also been studied. Figure 6.21 shows the measured radiation patterns in two principal planes at two resonant frequencies of the proposed antenna. Good broadside radiation patterns are obtained. For other frequencies within the WLAN band, measured radiation patterns similar to those shown here are observed. Also note that some asymmetry in the radiation patterns is observed, which is largely due to the effect of the loading strips. Large

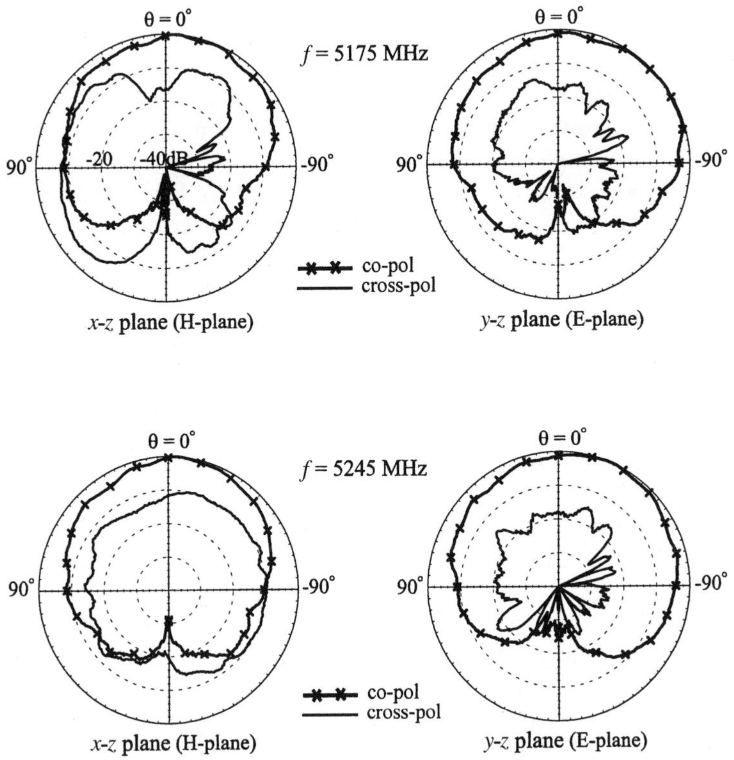

FIGURE 6.21 Measured radiation patterns at 5175 and 5245 MHz for the antenna shown in Figure 6.19. (From Ref. 12, © 2002 John Wiley & Sons, Inc.)

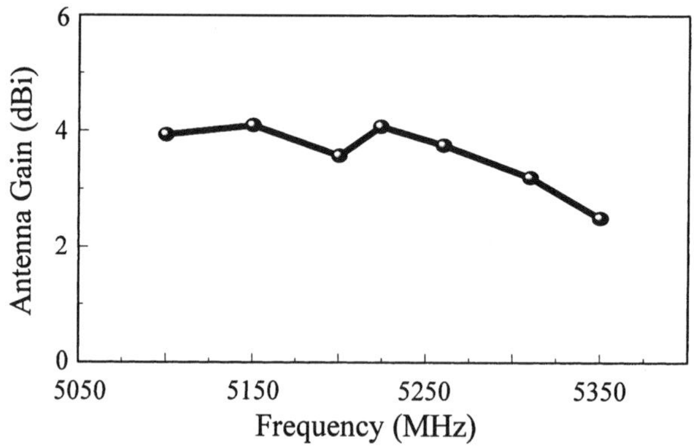

FIGURE 6.22 Measured antenna gain across the 2.4 GHz WLAN band for the antenna shown in Figure 6.19. (From Ref. 12, © 2002 John Wiley & Sons, Inc.)

cross-polarization radiation is also seen, especially in the H-plane patterns. However, for practical WLAN applications where the wave propagation usually experiences multiple reflections between the transmitter and the receiver, especially in indoors applications, large cross-polarization radiation can become an advantage for the proposed antenna. The measured antenna gain against frequency for the proposed antenna is also shown in Figure 6.22. For the frequencies across the WLAN band, the peak antenna gain is about 4.1 dBi, where the radiation efficiency is estimated to be about 57%, and the gain variations are less than 1.6 dBi.

Figure 6.23 shows other promising broadband designs for the very-high-permittivity DR antenna loaded with conducting strips. In Figure 6.23a, a single conducting strip printed on top of the DR antenna and oriented in perpendicular to the microstrip feed line, a broadband operation can also be obtained. By using two conducting strips as seen in Figure 6.23b, a similar broadband operation is also possible. The loading strip geometry shown in Figure 6.23c is also promising for obtaining a broadband operation of the DR antenna.

6.5 DUAL-POLARIZED DR ANTENNA

Dual-polarized operation of the DR antenna has also been demonstrated [13]. The dual-polarized DR antennas are shown in Figure 6.24a and b and are denoted as designs A and B, respectively. In design A, the DR is placed directly on two 50-Ω microstrip feed lines, which are arranged in orthogonal directions for the excitation of two orthogonal linearly polarized waves. The microstrip lines are printed on a microwave substrate of thickness 1.6 mm and relative permittivity of ε_r. The DR has a height of h and a square cross section of dimensions $L\times L$ and was constructed from ceramic material of relative permittivity $\varepsilon_{ra} = 79$. The portion of the microstrip lines under the DR has a length of d_1. Simply by selecting a proper value of d_1, good impedance matching of the dual-polarization excitation can be

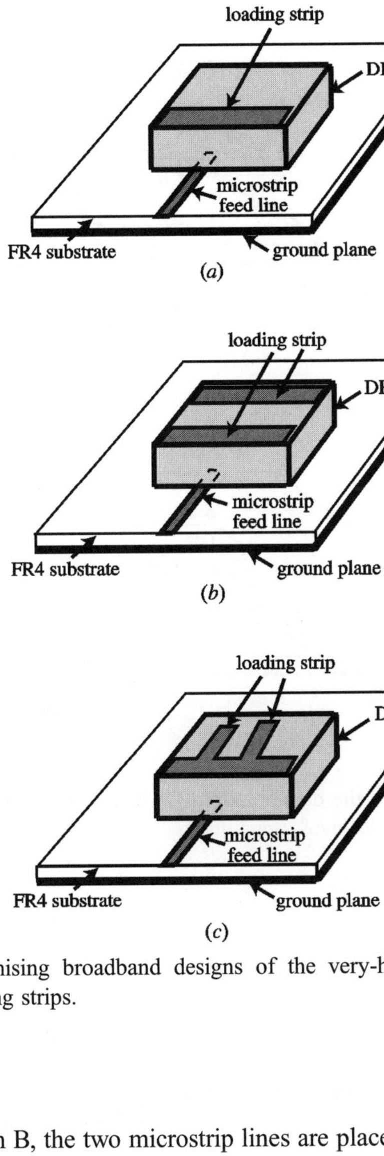

FIGURE 6.23 Promising broadband designs of the very-high-permittivity DR antenna loaded with conducting strips.

obtained. For design B, the two microstrip lines are placed conformal to the surface of the DR, and the length of the conformal microstrip-line section is denoted as d_2 in this study.

Prototypes of designs A and B have been constructed and studied. Measured reflection coefficient (S_{11}) and isolation (S_{21}) for designs A and B are shown in Figures 6.25 and 6.26, respectively. The dimensions of the DR studied were $4.9 \times 28.2 \times 28.2$ mm ($h \times L \times L$), and an FR4 substrate ($\varepsilon_r = 4.4$) was used. Optimal d_1 and d_2 for designs A and B with good impedance matching were found to be 6.5 and 3 mm, respectively. First note that, for design A, the obtained impedance bandwidth is 58 MHz, or about 2.6% referenced to the resonant frequency

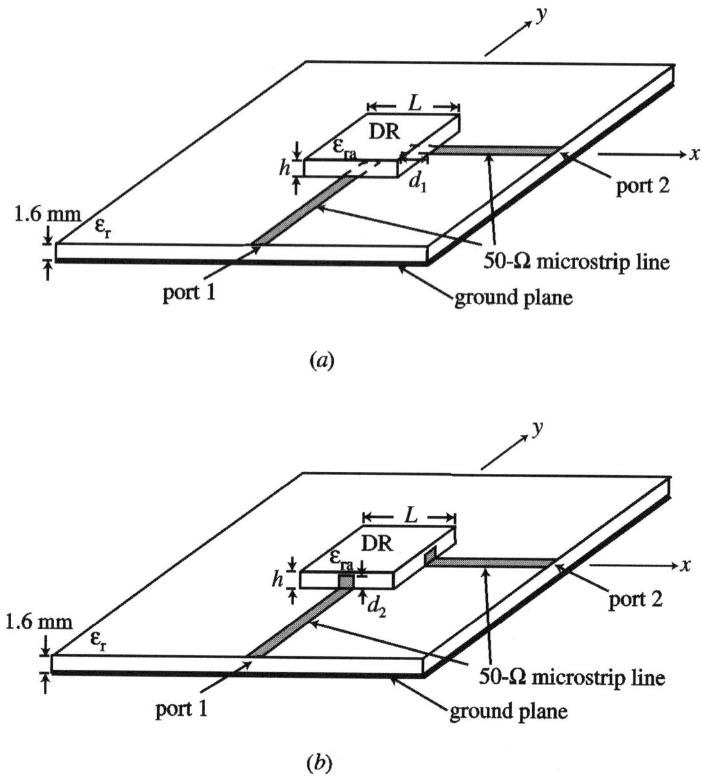

FIGURE 6.24 Geometries of the dual-polarized DR antenna. (a) Design A; (b) design B. (From Ref. 13, © 2001 John Wiley & Sons, Inc.)

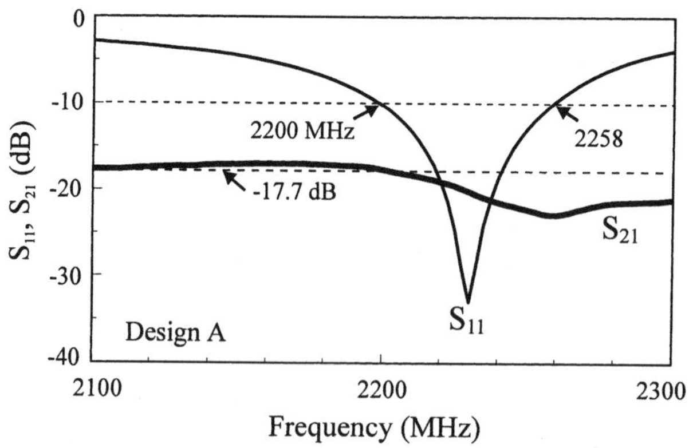

FIGURE 6.25 Measured reflection coefficient (S_{11}) and isolation (S_{21}) for design A shown in Figure 6.24a; $L = 28.2$ mm, $d_1 = 6.5$ mm, $\varepsilon_r = 4.4$, $\varepsilon_{ra} = 79$, $h = 4.9$ mm, ground plane size $= 100 \times 100$ mm^2. (From Ref. 13, © 2001 John Wiley & Sons, Inc.)

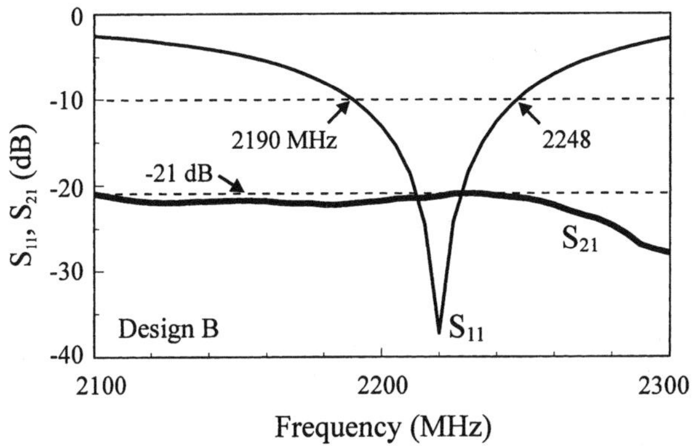

FIGURE 6.26 Measured reflection coefficient (S_{11}) and isolation (S_{21}) for design B shown in Figure 6.24b with $d_2 = 3$ mm. Other parameters are the same as in Figure 6.25. (From Ref. 13, © 2001 John Wiley & Sons, Inc.)

at 2230 MHz. For the isolation between the two feeding ports, the measured isolation across the entire impedance bandwidth is less than −17.7 dB. For design B, the obtained impedance bandwidth is also 58 MHz, or about 2.6% referenced to the resonant frequency at 2220 MHz. The measured isolation between the two feeding ports is less than −21 dB within the impedance bandwidth, which is better than design A. This behavior probably occurs because the separation between the two feeding ports is larger for design B, which helps improve the port decoupling. Measured radiation patterns of port 1 excitation for designs A and B are presneted in Figures 6.27 and 6.28, respectively. It is observed that design B has much lower cross-polarization radiation than design A (about 10 dB and 8 dB lower in the E- and H-plane patterns, respectively). From the obtained results, it can be concluded that, in view of the measured isolation and cross-polarization results, DR antennas with two conformal microstrip line feeds (design B) have better dual-polarized performance than those with two planar microstrip line feeds (design A).

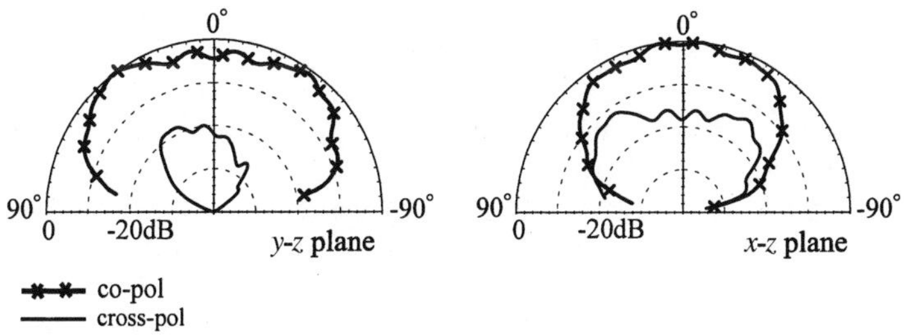

FIGURE 6.27 Measured radiation patterns at 2230 MHz for port 1 excitation of the antenna studied in Figure 6.25. (From Ref. 13, © 2001 John Wiley & Sons, Inc.)

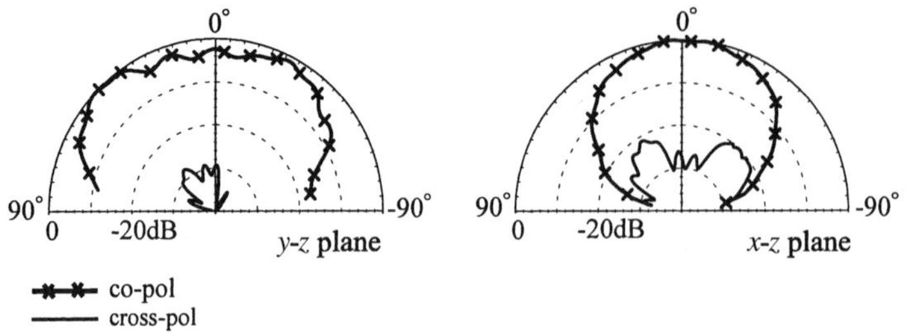

FIGURE 6.28 Measured radiation patterns at 2220 MHz for port 1 excitation of the antenna studied in Figure 6.26. (From Ref. 13, © 2001 John Wiley & Sons, Inc.)

REFERENCES

1. M. B. Oliver, Y. M. M. Antar, R. K. Mongia, and A. Ittipiboon, "Circularly polarized rectangular dielectric resonator antenna," Electron. Lett., vol. 31, pp. 418–419, March 16, 1995.
2. K. P. Esselle, "Circularly polarized higher-order rectangular dielectric-resonator antenna," Electron. Lett., vol. 32, pp. 150–151, Feb. 1, 1996.
3. R. K. Mongia, A. Ittipiboon, M. Cuhaci, and D. Roscoe, "Circularly polarized dielectric resonator antenna," Electron. Lett., vol. 30, pp. 1361–1363, Aug. 18, 1994.
4. G. Drossos, Z. Wu, and L. E. Davis, "Circularly polarized cylindrical dielectric resonator antenna," Electron. Lett., vol. 32, pp. 282–283, Feb. 15, 1996.
5. K. W. Leung, W. C. Wong, K. M. Luk, and E. K. N. Yung, "Circular-polarized dielectric resonator antenna excited by dual conformal strips," Electron. Lett., vol. 37, pp. 484–486, March 16, 2000.
6. W. C. Wong, and K. W. Leung, "Circularly polarized dielectric resonator antenna excited by dual conformal strips of unequal lengths," Microwave Opt. Technol. Lett., vol. 29, pp. 348–350, June 5, 2001.
7. M. T. Lee, K. M. Luk, E. K. N. Yung, and K.W. Leung, "Circularly polarized dielectric resonator antenna with a microstrip feed," in Proc. 1999 Asia-Pacific Microwave Conf., pp. 722–723.
8. R. T. Long, R. J. Dorris, S. A. Long, M. A. Khayat, and J. T. Williams, "Use of parasitic strip to produce circular polarization and increased bandwidth for cylindrical dielectric resonator antenna," Electron. Lett., vol. 37, pp. 406–408, March 29, 2001.
9. F. R. Hsiao, T. W. Chiou, and K. L. Wong, "Circularly polarized low-profile square dielectric resonator antenna with a loading patch," Microwave Opt. Technol. Lett., vol. 31, pp. 157–159, Nov. 5, 2001.
10. C. Y. Huang, J. Y. Wu, and K. L. Wong, "Cross-slot-coupled microstrip antenna and dielectric resonator antenna for circular polarization," IEEE Trans. Antennas Propagat., vol. 47, 605–609, April 1999.
11. C. Y. Huang, K. L. Wong , C. F. Yang, and J. Y. Wu, "Planar array composed of two linearly polarized dielectric resonator antennas for circular polarization," Microwave Opt. Technol. Lett., vol. 21, pp. 323–324, June 5, 1999.

12. F. S. Hsiao, J. S. Kuo, T. W. Chiou, and K. L. Wong, "A very-high-permittivity broadband dielectric resonator antenna for WLAN application in the 5.2 GHz Band," Microwave Opt. Technol. Lett., vol. 32, pp. 426–427, March 20, 2002.
13. C. Y. Huang, T. W. Chiou, and K. L. Wong, "Dual-polarized dielectric resonator antennas," Microwave Opt. Technol. Lett., vol. 31, pp. 222–223, Nov. 5, 2001.
14. S. A. Long, M. W. McAllister, and L. C. Shen, "The resonant cylindrical dielectric cavity antenna," IEEE Trans. Antennas Propagat., vol. 31, pp. 406–412, May 1983.

CHAPTER SEVEN

Integration of Antennas for Different Operating Bands

7.1 INTRODUCTION

For operating in different systems, for example, the cellular, the WLAN, and the GPS (global positioning system) systems, and so on, different antennas with separate feeds are usually used. These different antennas together may occupy a large realty space in the system. Thus the compact integration of these antennas to occupy a minimum volume in the system becomes an important subject. In this chapter, several possible integrations of a DCS antenna with a WLAN antenna or a GPS antenna are demonstrated.

In Section 7.2, an integration example of a DCS antenna and a WLAN antenna [1] is described. In Section 7.3, the integration between a conical pattern DCS antenna and a GPS antenna is discussed. The conical pattern DCS antenna is capable of generating a conical pattern linearly polarized wave to provide low-angle coverage and is very suitable for mounting on a vehicle for cellular communications [2]. The GPS antenna constructed with a ceramic substrate of a very high relative permittivity of 90.5 is also described, which has a compact size of $12 \times 12 \times 4\ \mathrm{mm}^3$ and is very suitable for integration with other antennas.

7.2 INTEGRATION OF TWO SEPARATE DCS AND WLAN ANTENNAS

Figure 7.1 shows the integration of two stacked PIFAs [3, 4] for the DCS and WLAN operations. The side views of the DCS and WLAN antennas are shown in

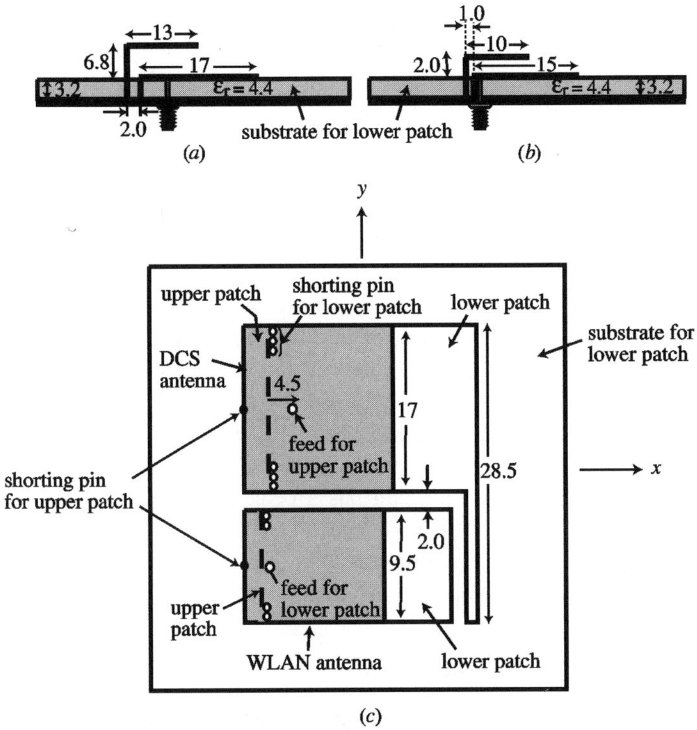

FIGURE 7.1 (a) Side view of the stacked PIFA for DCS operation; (b) side view of the stacked PIFA for WLAN operation; (c) top view of the DCS and WLAN antennas integrated on the same microwave substrate. Dimensions given in the figures are in millimeters, and the ground plane size is $50 \times 50\,mm^2$. (From Ref. 1, © 2001 John Wiley & Sons, Inc.)

Figure 7.1a and b, respectively. Figure 7.1c depicts the top view of the integrated configuration of the two antennas, whose lower or driven patches are printed on the same microwave substrate (thickness 3.2 mm and relative permittivity 4.4 in this design) and short-circuited to the same ground plane through the microwave substrate. Six and four shorting pins, symmetrically arranged close to the two ends of the shorted edge, are used for short-circuiting the lower patches of the DCS and WLAN antennas to the ground plane, respectively. On the other hand, the upper or parasitic patches of the two antennas are short-circuited to the same ground plane with a single shorting pin at the center of the shorted edge. The region between the shorted lower and upper patches is an air layer, and plastic posts (not shown in the figure) are used to support the upper patch above the lower patch. Also note that separate probe feeds are used for the two antennas and the feed position is selected along the centerline of the lower patch.

To achieve a compact integrated configuration for the two antennas, the lower patch of the DCS antenna is transformed to be of an L shape, which effectively reduces the required lower patch length to 19 mm (without the patch shape transformation, the required lower-patch is about 23 mm), and placed close to that of the WLAN antenna with a small gap of 2 mm. This arrangement makes the two antennas together occupy a small area of $19 \times 28.5\,\text{mm}^2$. As for the total antenna height, 10 and 5.2 mm for the DCS and WLAN antennas are chosen, respectively, to provide sufficient operating bandwidths (about 9.5 and 3.5% for the DCS and WLAN operations, respectively).

The two integrated DCS and WLAN antennas shown in Figure 7.1 have been constructed and studied. Figure 7.2 shows the measured return loss and antenna gain for the DCS antenna, and those for the WLAN antenna are shown in Figure 7.3. From the measured return loss, it is clearly seen that two resonant modes are excited at close frequencies for both of the two antennas. The two excited modes are associated to the lower and upper shorted patches, and this characteristic leads to enhanced impedance bandwidths obtained for the DCS and WLAN operations. The obtained bandwidths, determined from 10 dB return loss, cover the DCS band (1710–1880 MHz) and the 2.4 GHz WLAN band (2400–2484 MHz). For the DCS band, the measured antenna gain has a peak value of about 3.8 dBi and the gain variations are less than 1.8 dBi. For the 2.4 GHz WLAN band, the peak antenna gain is 2.5 dBi and the gain variations are less than 1.2 dBi.

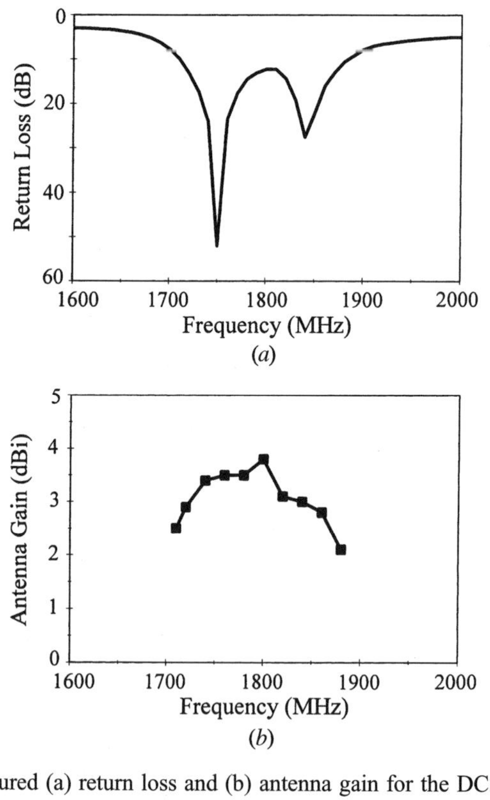

FIGURE 7.2 Measured (a) return loss and (b) antenna gain for the DCS antenna shown in Figure 7.1. (From Ref. 1, © 2001 John Wiley & Sons, Inc.)

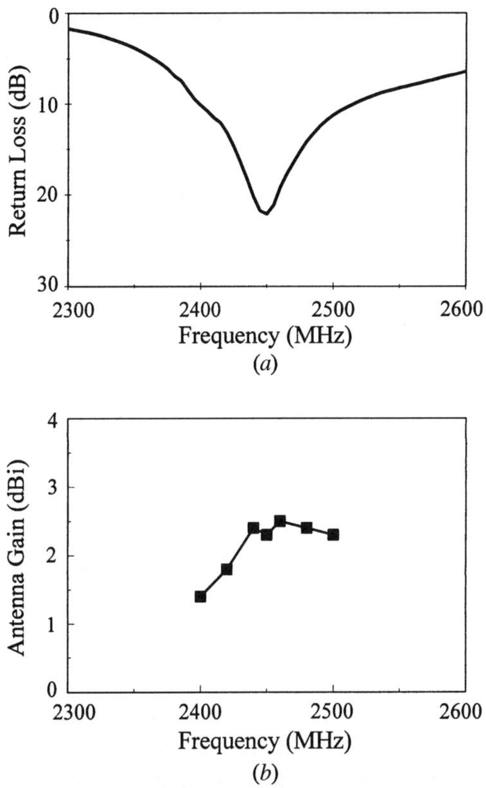

FIGURE 7.3 Measured (a) return loss and (b) antenna gain for the WLAN antenna shown in Figure 7.1. (From Ref. 1, © 2001 John Wiley & Sons, Inc.)

The radiation patterns have also been measured. Figure 7.4a and b shows the measured results of the DCS antenna at 1800 MHz and the WLAN antenna at 2450 MHz, respectively. Broadside radiation patterns are in general obtained, with slight asymmetry seen in the E-plane (*x*–*y* plane) patterns. Large cross-polarization radiation in both E- and H-plane patterns is also observed, which is, however, not a disadvantage for indoor or urban mobile communications with a multipath propagation environment. The isolation between the two antennas has also been studied. Figure 7.5 presents the measured S_{21} against frequency. The measured results indicate that the two antennas have an isolation better than −16 and −21 dB, respectively, for the proposed antenna operated in the DCS and WLAN bands.

7.3 INTEGRATION OF TWO SEPARATE DCS AND GPS ANTENNAS

7.3.1 GPS Ceramic Patch Antenna

The GPS patch antenna constructed with a very-high-permittivity ceramic substrate is very attractive because of its compact size. Figure 7.6 shows several promising

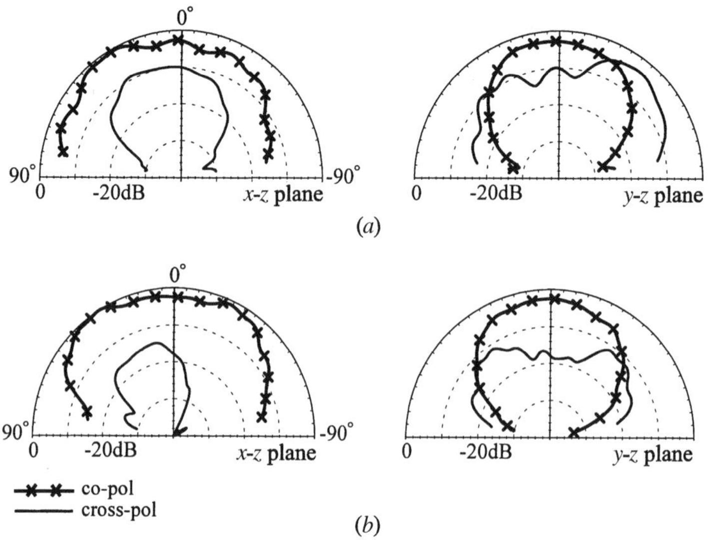

FIGURE 7.4 Measured radiation patterns for the antennas shown in Figure 7.1. (a) The DCS antenna at 1800 MHz; (b) the WLAN antenna at 2450 MHz. (From Ref. 1, © 2001 John Wiley & Sons, Inc.)

geometries of the top patch of such GPS ceramic patch antennas. Note that there is a small annular ring around the feed point on the top patch. This annular ring is very effective in compensating the extra inductance contributed from the long probe pin in the thick ceramic substrate (usually on the order of about 4 mm) to obtain enhanced impedance matching for the GPS operation. For the geometry of the top patch, one can use a corner-truncated square patch (Fig. 7.6a), a nearly square patch (Fig. 7.6b), a square patch with a tuning stub (Fig. 7.6c), or a square patch

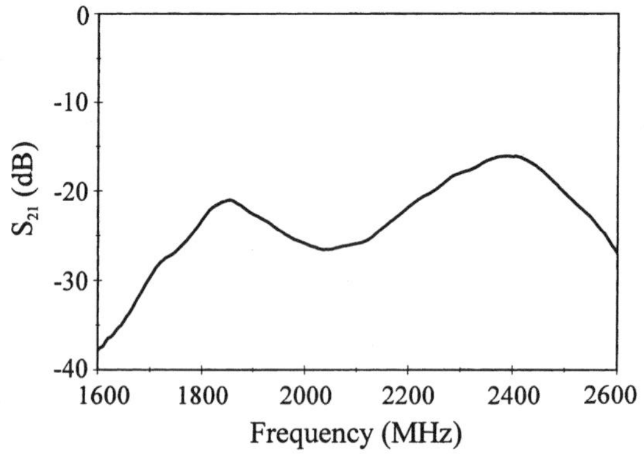

FIGURE 7.5 Measured isolation (S_{21}) between the DCS and the WLAN antennas shown in Figure 7.1. (From Ref. 1, © 2001 John Wiley & Sons, Inc.)

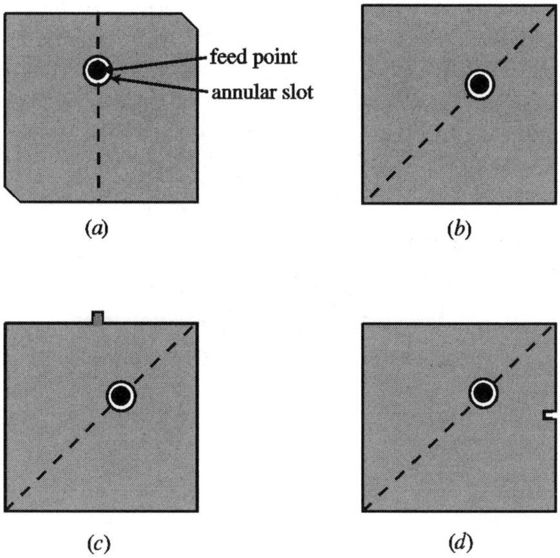

FIGURE 7.6 Promising geometries of the top patch of the GPS patch antenna with a very-high-relative-permittivity ceramic substrate. (a) A corner-truncated square patch; (b) a nearly square patch; (c) a square patch with a tuning stub; (d) a square patch with an inserted slit.

with a slit (Fig. 7.6d). Many other possible top patches capable of CP radiation are shown in Reference 5.

By using a top patch shown in Figure 7.6c and a ceramic substrate of a very high relative permittivity of 90.5, a compact GPS ceramic patch antenna for operating at 1575 MHz has been constructed. The total antenna volume is only $12 \times 12 \times 4\,\text{mm}^3$, and the top patch size is $10 \times 10\,\text{mm}^2$, with a tuning stub of length about 1 mm. The antenna has an impedance bandwidth (2:1 VSWR) of 20 MHz (1565–1585 MHz) and a 3 dB axial-ratio CP bandwidth of about 6 MHz, centered at about 1575 MHz. The measured antenna gain is about 0 dBi.

7.3.2 A GPS Ceramic Patch Antenna and a DCS Conical Pattern Annular Patch Antenna

Figure 7.7 shows a possible integration design of a GPS ceramic patch antenna, discussed in Section 7.3.1, and a DCS conical pattern annular patch antenna. The annular patch antenna is operated in the higher-order mode of TM_{21} for conical pattern radiation, which has null radiation in the broadside direction and strong radiation in the lower elevation angles [6] and is very suitable for application as a vehicle-based antenna. On the other hand, the GPS ceramic patch antenna has a broadside radiation pattern. The two antennas integrated together are very suitable for employing on a vehicle for the DCS and GPS operations.

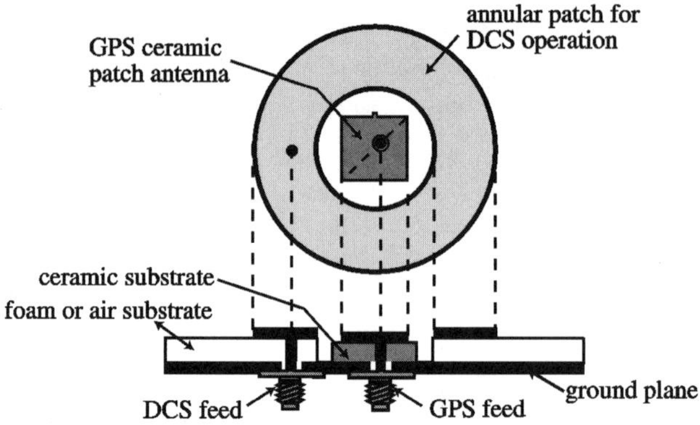

FIGURE 7.7 Geometry of the integrated DCS conical pattern annular patch antenna and GPS ceramic patch antenna.

7.3.3 A GPS Patch Antenna and a DCS Conical Pattern Monopolar Patch Antenna

Another integration design of a GPS antenna and a DCS conical-pattern antenna is shown in Figure 7.8. In this design, both the GPS and DCS antennas have a common ground plane but are fed by separate feeding ports. The GPS antenna, operated at 1575 MHz, is realized by using a corner-truncated square ring microstrip patch [7] placed on a foam substrate of 3 mm thickness (h_1). The inner and outer side lengths of the square ring patch are 72 mm (L_1) and 20 mm (L_2), respectively, and the side length of the truncated corners is 7 mm (ΔL). The feed position of the GPS antenna is placed along the x-axis for receiving right-hand circularly polarized waves, and the distance of the probe feed away from the patch center is 12 mm (d). In the center of the square ring patch is a low-profile cylindrical monopole [8] for the DCS operation, which has a diameter of 6 mm (d_1) and is top-loaded with a shorted square patch. The square patch has a side length of 29 mm (ℓ) and is short-circuited to the antenna ground by using two identical conducting cylinders of diameter 4 mm (d_2) and length 10.8 mm (h_2). Note that the length 10.8 mm corresponds to only about 0.065 wavelength at 1800 MHz, which is much smaller than that of a conventional quarter-wavelength monopole antenna. For the DCS antenna in this design, there are also other possible antenna geometries available, such as the monopolar wire patch antenna [9], the monopolar plate patch antenna [10], the low-profile cylindrical monopole top-loaded with a shorted cross patch [11], and so on.

Also note that, in the design process, the square ring patch for the GPS operation was first designed alone, and then the cylindrical monopole top-loaded with a shorted square patch for the DCS operation was integrated into the center of the square ring patch. In this case, although the shorted square patch of the cylindrical monopole covers some portions of the bottom corner-truncated square-ring patch, the CP performance of the GPS antenna is very slightly affected, and a slight

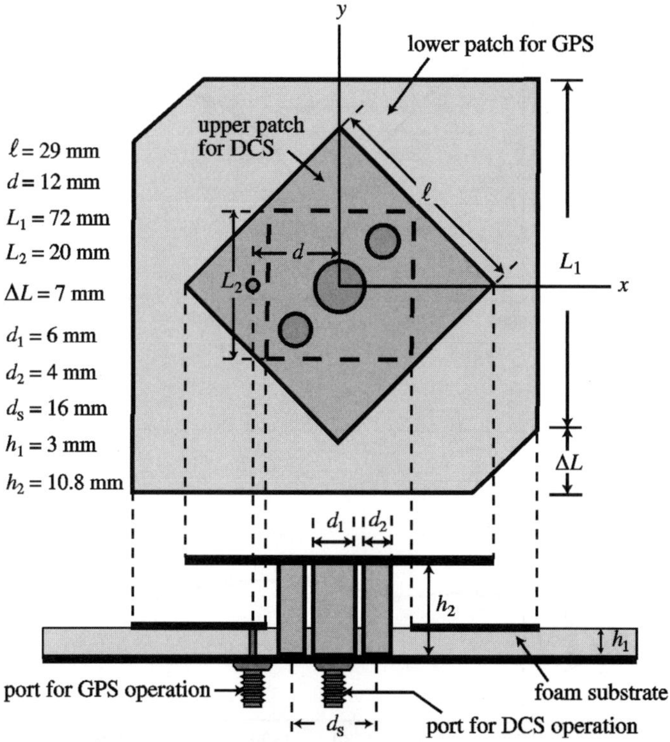

FIGURE 7.8 Geometry of the integrated DCS and GPS patch antennas. (From Ref. 2, © 2001 John Wiley & Sons, Inc.)

fine-tuning of the side length of the truncated corners of the square ring patch is enough to achieve good CP performance at 1575 GHz, after the integration with the DCS antenna.

The integration design shown in Figure 7.8 has been successfully implemented. The measured return loss and antenna gain for the DCS operation are presented in Figure 7.9a and b, respectively. The obtained impedance bandwidth (10 dB return loss) reaches 186 MHz (1702–1888 MHz), or about 10% referenced to 1800 MHz, which covers the bandwidth of the DCS operation. Also note that, because of the use of cylindrical conductors in place of conducting wires, the obtained impedance bandwidth for the DCS antenna studied here is much greater than that (about 3%) of the monopolar wire-patch antenna [9]. For the measured antenna gain within the impedance bandwidth for the DCS antenna, the peak antenna gain is about 3 dBi and the gain variations are within 0.6 dBi.

The performances of the GPS antenna are shown in Figure 7.10. The measured return loss is plotted in Figure 7.10a. From the results obtained, it is clearly seen that two near-degenerate resonant modes are excited. The measured axial ratio

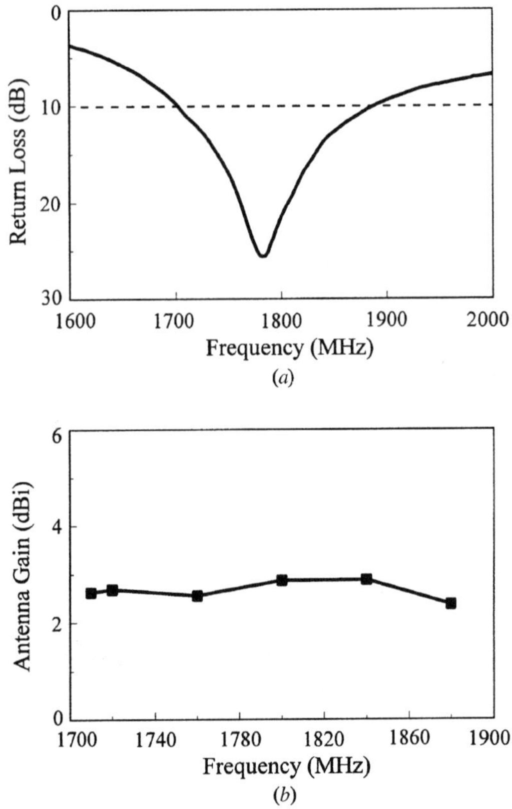

FIGURE 7.9 Measured (a) return loss and (b) antenna gain for the DCS antenna shown in Figure 7.8. (From Ref. 2, © 2001 John Wiley & Sons, Inc.)

against frequency in the broadside direction is shown in Figure 7.10b. The CP bandwidth of 3 dB axial ratio is about 10 MHz, which is much larger than that required for the GPS operation at 1575 MHz (2 MHz bandwidth required). Figure 7.10c presents the measured antenna gain of the GPS antenna for frequencies within the CP bandwidth. The peak antenna gain reaches about 8 dBic, and very small gain variations (less than 0.2 dBic) are observed.

Measured radiation patterns of the proposed integration design at 1800 and 1575 MHz are presented in Figures 7.11 and 7.12, respectively. For the DCS antenna at 1800 MHz, monopole-like radiation is obtained, which is suitable for applications on a vehicle. For the radiation patterns shown in Fig. 5, good broadside radiation patterns with right-hand CP radiation are seen. The isolation between the two feeding ports of the DCS and GPS antennas was also measured, and experimental results show that the isolation between the two ports is less than −20 dB.

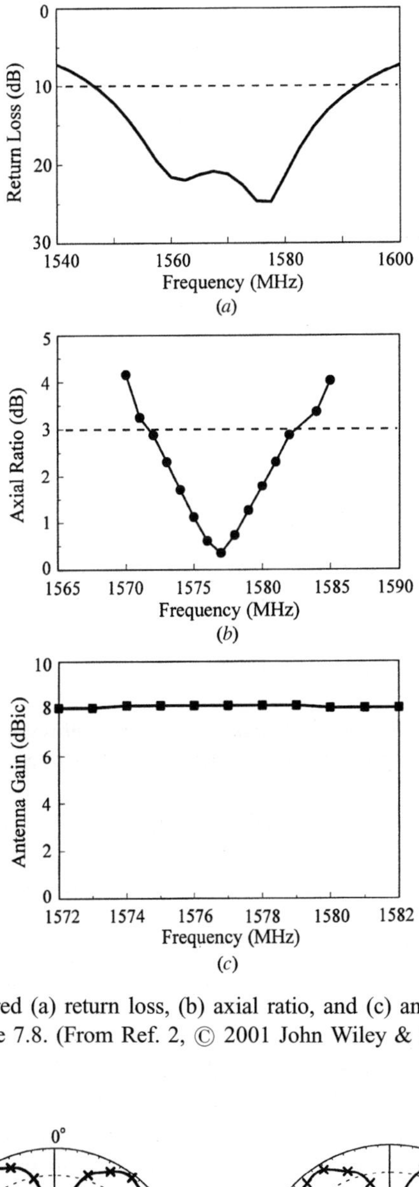

FIGURE 7.10 Measured (a) return loss, (b) axial ratio, and (c) antenna gain for the GPS antenna shown in Figure 7.8. (From Ref. 2, © 2001 John Wiley & Sons, Inc.)

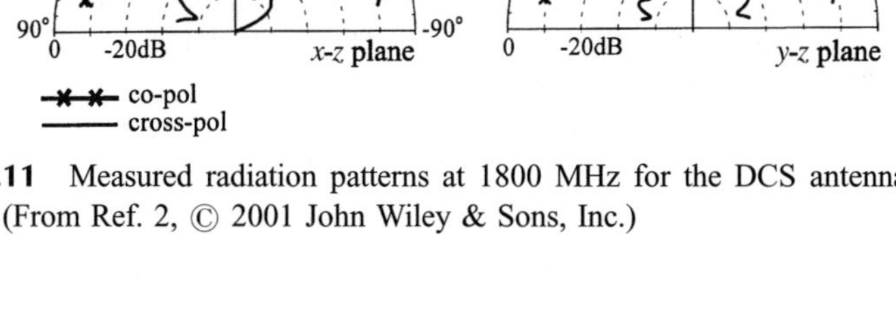

FIGURE 7.11 Measured radiation patterns at 1800 MHz for the DCS antenna shown in Figure 7.8. (From Ref. 2, © 2001 John Wiley & Sons, Inc.)

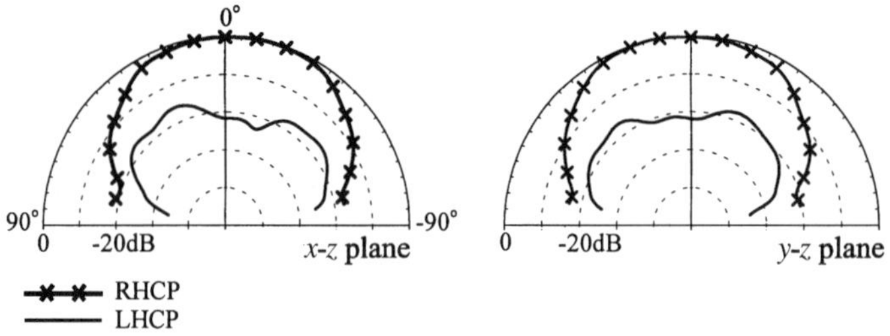

FIGURE 7.12 Measured radiation patterns at 1575 MHz for the GPS antenna shown in Figure 7.8. (From Ref. 2, © 2001 John Wiley & Sons, Inc.)

REFERENCES

1. J. Y. Wu and K. L. Wong, "Two integrated stacked shorted patch antennas for DCS/WLAN dual-band operations," Microwave Opt. Technol. Lett., vol. 30, pp. 134–136, July 20, 2001.
2. J. Y. Wu, J. S. Row, and K. L. Wong, "A compact dual-band microstrip patch antenna suitable for DCS/GPS operations," Microwave Opt. Technol. Lett., vol. 29, pp. 410–412, June 20, 2001.
3. J. Barreiros, P. Cameirao, and C. Peixeiro, "Microstrip patch antenna for GSM 1800 handsets," in 1999 IEEE Antennas Propagat. Soc. Int. Symp. Dig., pp. 2074–2077.
4. J. Ollikainen, M. Fischer, and P. Vainikainen, "Thin dual-resonant stacked shorted patch antenna for mobile communications," Electron. Lett., vol. 35, pp. 437–438, March 18, 1999.
5. K. L. Wong, Compact and Broadband Microstrip Antennas, Chap. 5, John Wiley & Sons, New York, NY, 2002.
6. S. Y. Lin and K. L. Wong, "Improved antenna performances for a conical-pattern annular-ring patch antenna using a slotted ground plane," in Proc. 2001 Asia-Pacific Microwave Conf., pp. 1036–1039.
7. W. S. Chen, C. K. Wu, and K. L. Wong, "Single-feed square-ring microstrip antenna with truncated corners for compact circular polarization operation," Electron. Lett., vol. 34, pp. 1045–1047, May 28, 1998.
8. J. S. Row, S. H. Yeh, and K. L. Wong, "A broadband low-profile cylindrical monopole antenna," Microwave Opt. Technol. Lett., vol. 29, pp. 77–79, April 20, 2001.
9. C. Delaveaud, P. Leveque, and B. Jecko, "New kind of microstrip antenna: the monopolar wire-patch antenna," Electron. Lett., vol. 30, pp. 1–2, Jan. 6, 1994.
10. J. S. Row, S. H. Yeh, and K. L. Wong, "A wideband monopolar plate-patch antenna," in 2001 IEEE Antennas Propagat. Soc. Int. Symp. Dig., vol. 2, pp. 128–131.
11. S. H. Yeh and K. L. Wong, "A broadband low-profile cylindrical monopole antenna top-loaded with a shorted cross patch," Microwave Opt. Technol. Lett., vol. 32, pp. 186–188, Feb. 5, 2002.

Appendix

Summary of Acronyms

CP: circular polarization
DCS: digital communication system
DP: dual polarized
DR: dielectric resonator
GPS: global positioning system
GSM: global system for mobile communication
HFSS: a high-frequency simulation software of Ansoft Corporation
IE3D: a high-frequency simulation software of Zeland Software, Inc.
ISM band: industrial, scientific and medical band
LC: inductor/capacitor
PCMCIA: personal computer memory card international association
PCS: personal communication system
PDA: personal digital assistant
PIFA: planar inverted-F antenna
PILA: planar inverted-L antenna
SAR: specific absorption rate
UMTS: universal mobile telecommunication system
VSWR: voltage standing wave ratio
WLAN: wireless local area network
XPL: cross-polarization level

INDEX

WILEY SERIES IN MICROWAVE AND OPTICAL ENGINEERING

KAI CHANG, Editor
Texas A&M University

FIBER-OPTIC COMMUNICATION SYSTEMS, Third Edition • *Govind P. Agrawal*

COHERENT OPTICAL COMMUNICATIONS SYSTEMS • *Silvello Betti, Giancarlo De Marchis and Eugenio Iannone*

HIGH-FREQUENCY ELECTROMAGNETIC TECHNIQUES: RECENT ADVANCES AND APPLICATIONS • *Asoke K. Bhattacharyya*

COMPUTATIONAL METHODS FOR ELECTROMAGNETICS AND MICROWAVES • *Richard C. Booton, Jr.*

MICROWAVE RING CIRCUITS AND ANTENNAS • *Kai Chang*

MICROWAVE SOLID-STATE CIRCUITS AND APPLICATIONS • *Kai Chang*

RF AND MICROWAVE WIRELESS SYSTEMS • *Kai Chang*

RF AND MICROWAVE CIRCUIT AND COMPONENT DESIGN FOR WIRELESS SYSTEMS • *Kai Chang, Inder Bahl, and Vijay Nair*

DIODE LASERS AND PHOTONIC INTEGRATED CIRCUITS • *Larry Coldren and Scott Corzine*

RADIO FREQUENCY CIRCUIT DESIGN • *W. Alan Davis and Krishna Agarwal*

MULTICONDUCTOR TRANSMISSION-LINE STRUCTURES: MODAL ANALYSIS TECHNIQUES • *J. A. Brandão Faria*

PHASED ARRAY-BASED SYSTEMS AND APPLICATIONS • *Nick Fourikis*

FUNDAMENTALS OF MICROWAVE TRANSMISSION LINES • *Jon C. Freeman*

OPTICAL SEMICONDUCTOR DEVICES • *Mitsuo Fukuda*

MICROSTRIP CIRCUITS • *Fred Gardiol*

HIGH-SPEED VLSI INTERCONNECTIONS: MODELING, ANALYSIS, AND SIMULATION • *A. K. Goel*

FUNDAMENTALS OF WAVELETS: THEORY, ALGORITHMS, AND APPLICATIONS • *Jaideva C. Goswami and Andrew K. Chan*

ANALYSIS AND DESIGN OF INTEGRATED CIRCUIT ANTENNA MODULES • *K. C. Gupta and Peter S. Hall*

PHASED ARRAY ANTENNAS • *R. C. Hansen*

HIGH-FREQUENCY ANALOG INTEGRATED CIRCUIT DESIGN • *Ravender Goyal (ed.)*

MICROSTRIP FILTERS FOR RF/MICROWAVE APPLICATIONS • *Jia-Sheng Hong and M. J. Lancaster*

MICROWAVE APPROACH TO HIGHLY IRREGULAR FIBER OPTICS • *Huang Hung-Chia*

NONLINEAR OPTICAL COMMUNICATION NETWORKS • *Eugenio Iannone, Francesco Matera, Antonio Mecozzi, and Marina Settembre*

FINITE ELEMENT SOFTWARE FOR MICROWAVE ENGINEERING • *Tatsuo Itoh, Giuseppe Pelosi and Peter P. Silvester (eds.)*

INFRARED TECHNOLOGY: APPLICATIONS TO ELECTROOPTICS, PHOTONIC DEVICES, AND SENSORS • *A. R. Jha*

SUPERCONDUCTOR TECHNOLOGY: APPLICATIONS TO MICROWAVE, ELECTRO-OPTICS, ELECTRICAL MACHINES, AND PROPULSION SYSTEMS • *A. R. Jha*

OPTICAL COMPUTING: AN INTRODUCTION • *M. A. Karim and A. S. S. Awwal*

INTRODUCTION TO ELECTROMAGNETIC AND MICROWAVE ENGINEERING • *Paul R. Karmel, Gabriel D. Colef, and Raymond L. Camisa*